Green Energy and Technology

Climate change, environmental impact and the limited natural resources urge scientific research and novel technical solutions. The monograph series Green Energy and Technology serves as a publishing platform for scientific and technological approaches to "green"—i.e. environmentally friendly and sustainable—technologies. While a focus lies on energy and power supply, it also covers "green" solutions in industrial engineering and engineering design. Green Energy and Technology addresses researchers, advanced students, technical consultants as well as decision makers in industries and politics. Hence, the level of presentation spans from instructional to highly technical.

Indexed in Scopus.

Indexed in Ei Compendex.

Alok Kumar Rai · Vivek Mishra ·
Satendra Pal Singh
Editors

Electrochemistry of Next-Generation Solid-State Batteries

Springer

Editors
Alok Kumar Rai
Department of Chemistry
University of Delhi
Delhi, India

Vivek Mishra
Amity Institute of Click-Chemistry
Research and Studies (AICCRS)
Amity University Noida
Noida, Uttar Pradesh, India

Satendra Pal Singh
Department of Physics
University of Lucknow
Lucknow, Uttar Pradesh, India

ISSN 1865-3529 ISSN 1865-3537 (electronic)
Green Energy and Technology
ISBN 978-981-95-7602-9 ISBN 978-981-95-7603-6 (eBook)
https://doi.org/10.1007/978-981-95-7603-6

This Springer imprint is published by the registered company Springer Nature Singapore Pte Ltd.
The registered company address is: 152 Beach Road, #21-01/04 Gateway East, Singapore 189721, Singapore

Contents

About the Editors

Alok Kumar Rai is an Associate Professor in the Department of Chemistry at University of Delhi (North Campus). He joined the Department as an Assistant Professor in September, 2018. Prior to that he was employed as an Associate Professor at Amity University, Noida and Assistant Professor at Amrita Centre for Nanosciences & Molecular Medicine, Kochi. He is also the recipient of prestigious Ramanujan Fellowship from DST-SERB (2016–2021). Dr. Rai received his Ph.D. in Chemistry from Indian Institute of Technology–Banaras Hindu University (IIT–BHU) in 2010. After that he joined Prof. Jaekook Kim lab as a post-doctoral research fellow followed by Research Professor at Chonnam National University, Gwangju, South Korea. Dr. Rai has published more than 80 peer-reviewed research articles including reviews and book chapters. Dr. Rai teaching and research interests involve inorganic chemistry, materials chemistry and electrochemistry. He is currently working on Rechargeable Batteries (Lithium-ion, Sodium-ion, Zinc-ion) and Solid-State Chemistry (metal oxides, novel chemical synthesis, electronic and ionic transport etc).

Vivek Mishra is an Associate Professor at the Amity Institute of Click Chemistry Research and Studies, Amity University, Noida, Uttar Pradesh, India. He received his Ph.D. from the Department of Chemistry, Banaras Hindu University, Varanasi. He pursued post-doctoral research at the Green Process and Material R&D Laboratory, Korea Institute of Industrial Technology, South Korea, and at the Functional Material Research Laboratory, University of Ulsan, South Korea. Dr. Mishra is a recipient of the SERB–Teachers Associateship for Research Excellence (TARE) Fellowship and the SERB–National Postdoctoral Fellowship at the Green Chemistry Network Centre, University of Delhi. His research focuses on forensic sciences, organic chemistry, advanced functional materials, click chemistry, green chemistry, and polymer chemistry. He has authored over 50 articles in international peer-reviewed journals across leading chemistry, forensic, and polymer platforms. He holds four granted international patents and four Indian patents filed in 2022–2023. Dr. Mishra has published several international books and book chapters with reputed publishers. He has contributed more than 15 book chapters of international repute and serves as a reviewer for many prestigious journals. He is also an active member of several international scientific societies.

Satendra Pal Singh received his Ph.D. in Materials Science and Technology from the Indian Institute of Technology–Banaras Hindu University (IIT-BHU), Varanasi. He is currently a Professor in the Department of Physics at the University of Lucknow. His research interests lie in materials science, nanoscience, and nanotechnology, with a focus on crystal structure characterization using powder diffraction techniques and structure–property correlation through X-ray, synchrotron, and neutron diffraction. His work primarily centers on functional ceramics, phosphors, multiferroic materials, dielectric materials, and piezoelectric materials. A significant portion of his research is dedicated to the discovery of novel materials using a combinatorial chemistry approach for energy-saving applications, including phosphors for solid-state lighting (WLEDs), display devices, and cathode/anode materials for energy storage systems. Dr. Singh has published nearly 70 research articles in high-impact international journals and holds several international patents.

Introduction of Next-Generation Solid-State Batteries

Pragati and Alok Kumar Rai

Abstract Solid-state battery has explored a new era of energy storage technology. Compared to traditional liquid electrolytes, solid-state electrolytes offer several advantages, including enhanced safety, improved energy density, and a wider operational temperature range. This chapter provides a comprehensive overview of next-generation solid-state batteries. We delve into the fundamental mechanisms underlying solid-state ion transport, highlighting the key challenges and potential solutions. In addition, the chapter also discovers the diverse applications of solid-state batteries, ranging from electric vehicles and consumer electronics to grid-scale energy storage. By understanding the advantages, challenges, and future prospects of solid-state batteries, we aim to contribute to the advancement of sustainable and high-performance energy storage solutions.

Keywords Solid-state batteries · Solid electrolytes · Interface engineering · Lithium metal anodes

1 Introduction

The development of electrochemical energy storage systems has been a constant quest for increased cycle life, safety, and energy density. Every evolution, from Alessandro Volta's groundbreaking voltaic pile, which illustrated the basic ideas of galvanic cells, to the subsequent development of lead-acid and nickel-based batteries, has addressed the specific technological demands of its era. However, the late-twentieth and early-twenty-first centuries witnessed the transformative impact of lithium-ion batteries (LIBs). LIBs are the most widely used energy storage technology for electric vehicles (EVs), portable devices, and grid-scale storage applications due to their high gravimetric and volumetric energy densities, which are made possible by the intercalation chemistry of lithium ions [1]. The electrochemical mechanism of LIBs relies on the reversible insertion and extraction of lithium ions between a graphitic

Pragati · A. K. Rai (✉)
Department of Chemistry, University of Delhi, Delhi 110007, India
e-mail: akrai@chemistry.du.ac.in

A. K. Rai et al. (eds.), *Electrochemistry of Next-Generation Solid-State Batteries*, Green Energy and Technology, https://doi.org/10.1007/978-981-95-7603-6_1

anode and a transition metal oxide cathode, assisted by a liquid electrolyte [2]. Despite the extraordinary success of this technology, the cost of LIBs is high due to the uneven distribution and limited resources of lithium, resulting it is not enough to satisfy near future energy demands. On the other hand, there are serious safety issues as well, especially with large-format batteries, due to inherent limits with the use of flammable liquid electrolytes like ethylene carbonate and diethyl carbonate. Furthermore, it is becoming more and more evident that the theoretical energy density limitations of traditional LIB chemistries are influenced by the selection of electrode materials and electrolyte stability windows [1, 3]. Since the kinetics of lithium-ion transport in liquid electrolytes is generally adequate, however it can be rate-limiting in high-power applications. The limitations of liquid electrolytes are given below:

- **Safety and Flammability Risks**: The majority of organic liquid electrolytes are extremely flammable and provide a serious risk of fire due to the presence of highly volatile and flammable organic solvents like diethyl carbonate and dimethyl carbonate (DMC). Internal short circuits, overcharging, or external damage can all cause thermal runaway, which led to a sudden and uncontrollable rise in temperature that can result in fires or explosions [3].
- **Limited Electrochemical Stability Window**: The voltage range in which a liquid electrolyte stays stable is known as its electrochemical stability window [3]. If this range is exceeded, electrolyte breakdown may occur, which could cause gas evolution, capacity decline, and elevate internal resistance [4, 5]. The selection of high-voltage electrode materials, which are required to attain greater energy densities, is constrained by this restriction.
- **Electrolyte Leakage and Volatility**: Liquid electrolytes may leak or evaporate during over time, resulting the performance degradation and potential safety hazards [6, 7].
- **Solid–Electrolyte Interphase (SEI) Formation**: A solid–electrolyte interphase (SEI) film formed due to the electrochemical breakdown of liquid electrolytes at the electrode–electrolyte interface [4]. Although the SEI can offer passivation, however the capacity decline and elevated internal resistance are caused by its uncontrolled growth and instability.
- **Dendrite Formation**: During the charging process, lithium ions are reduced and deposited onto the anode. This deposition can be uneven in liquid electrolytes, leading to the formation of needle-like structures called dendrites [4, 8, 9].

1.1 Solid-State Battery

Solid-state battery development has become a crucial and highly active energy storage research area due to the inherent limitations of traditional liquid electrolytes. Solid-state batteries address these issues by replacing flammable organic liquid electrolytes with a solid electrolyte. This simple yet fundamental change eliminates the common problem of electrolyte leakage, which is a frequent failure mode in conventional batteries that raises serious safety concerns. Moreover, the solid electrolyte in SSBs significantly reduces the risk of thermal runaway, an uncontrollable self-heating process that can lead to fires or explosions. Apart of the above, the following characteristics also make solid-state batteries inherently safer, meeting a critical requirement for their widespread adoption across various industries.

- **Wider Electrochemical Stability Window**: Solid electrolytes can operate safely at higher voltages than liquids, enabling the use of high-voltage cathodes, which boosts overall energy density.
- **Compatibility with Lithium Metal Anodes**: Lithium metal offers very high specific capacity, but it reacts dangerously with liquid electrolytes and forms dendrites.
- **Dendrite Suppression via Mechanical Resistance**: Solid electrolytes physically block dendrite growth, unlike liquids, which are soft and easily penetrated.
- **Uniform Lithium-Ion Flux**: The solid matrix promotes even lithium plating, reducing hotspots where dendrites usually start.
- **Safety Improvements**: Preventing dendrite penetration stops internal short circuits and thermal runaway risks.

This combination of chemical and mechanical advantages makes solid electrolytes a key enabler for safer, higher-energy–density lithium metal batteries.

On the other hand, the global drive toward sustainable energy has created an urgent demand for improved and safer methods of power storage. This need is particularly evident in the rapidly expanding electric vehicle (EV) market. Today's EV consumers expect vehicles that can travel significantly farther on a single charge-comparable to traditional gasoline cars and recharge at speeds akin to a quick fuel stop, all while maintaining the highest safety standards. Thus, the safety remains a paramount concern for EV batteries to prevent accidents and ensure consumer confidence [1]. However, the current battery technologies often require a trade-off between these critical attributes, as manufacturers struggle to deliver both exceptional performance and extended battery life simultaneously. Additionally, as personal electronics become increasingly advanced and compact, there is a growing need for batteries that are not only more powerful but also smaller, enabling longer-lasting portability. Energy storage really is the linchpin for integrating renewables like solar and wind into the grid effectively. Without reliable storage, the variability of those energy sources can cause instability and limit their usefulness. Solid-state batteries, with their promise of higher energy density, improved safety, and longer lifespans compared to traditional lithium-ion batteries, could be a game-changer in this field (Fig. 1).

1.1.1 Basic Components of Solid-State Batteries

Solid-state batteries indeed operate on the same electrochemical principles as traditional lithium-ion batteries, but the replacement of the liquid electrolyte with a solid one changes many aspects fundamentally. To understand these differences, it's essential to examine the core components of a solid-state battery.

Fig. 1 Key components and features of solid-state batteries [10]

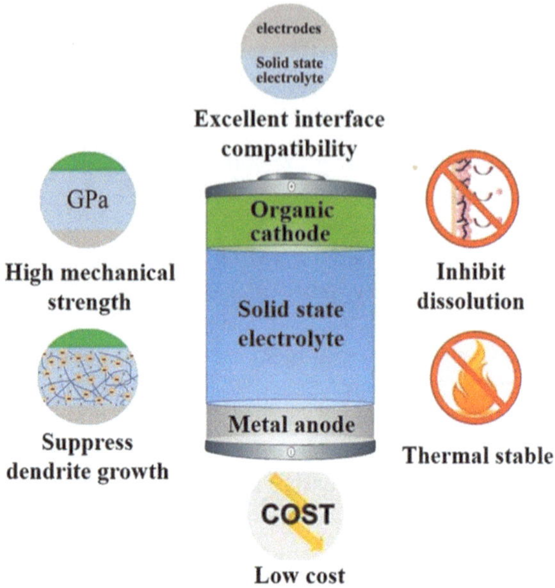

(i) **Anode**

The anode, or negative electrode, donates electrons and stores metal ions while the battery is charging. During discharge, metal ions from the anode are released and migrate toward the cathode, simultaneously releasing electrons into the external circuit. The choice of anode material is especially important because of its unique interactions with the solid electrolyte, which significantly affect battery performance. One of the most exciting prospects for solid-state batteries is their potential to utilize lithium metal as the anode material. Lithium metal offers an exceptionally high theoretical specific capacity of 3860 mAh/g, which far surpasses the 372 mAh/g capacity of graphite, which is currently the standard anode material in most lithium-ion batteries. This significant capacity advantage makes lithium metal highly promising for solid-state battery applications. However, incorporating lithium metal into batteries with liquid electrolytes is challenging due to the formation of needle-like dendrites. These dendrites can penetrate the battery's separator, causing short circuits and serious safety hazards. In contrast, solid-state batteries use solid electrolytes with inherent rigidity and high mechanical strength, which provide a strong physical barrier against dendrite growth. This robust structure not only enhances safety but also makes lithium metal anodes more suitable for use in solid-state battery designs. In addition to lithium metal, the researchers are actively exploring a range of alternative materials for use as solid-state battery anodes. Silicon, for instance, offers an exceptionally high-energy storage capacity (~3579 mAh/g) [11]. However, its significant volume expansion during charging and cycling presents engineering challenges that require careful design to prevent damage to the cell. Meanwhile, carbon-based materials such as graphite and hard carbon share structural similarities with those used in conventional lithium-ion batteries and are also compatible with solid-state systems. The choice of anode material depends on several factors, including the required energy density, cost, cycle life, and compatibility with the selected solid electrolyte.

(ii) **Cathode**

The cathode, or positive electrode, is the source from which metal ions are released during charging and re-intercalated during discharge, driving electrons through the external circuit. A cathode's performance directly influences the battery's voltage, capacity, and overall power output. For solid-state batteries, the ideal cathode material should exhibit high ionic and electronic conductivity to facilitate efficient ion and electron transport. Additionally, it must demonstrate strong mechanical and chemical stability to endure the structural stresses of repeated charge–discharge cycles. Compatibility with the solid electrolyte is also critical, i.e., any chemical reactions between the cathode and electrolyte during cycling can lead to undesirable side reactions, ultimately reducing battery efficiency over time. In many solid-state battery designs, cathode materials are adapted from conventional lithium-ion compounds such as lithium cobalt oxide (LCO), lithium nickel manganese cobalt oxide (NMC), and lithium iron phosphate (LFP) [12]. However, to ensure their effectiveness in solid-state systems, these materials often require surface and structural modifications to enhance their compatibility with solid-state electrolytes.

(iii) **Solid Electrolytes**

The solid electrolyte serves as the heart and soul of a solid-state battery, distinguishing it from conventional batteries that rely on liquid electrolytes. Its primary role is to facilitate the smooth conduction of metal ions between the anode and cathode. At the same time, it must remain chemically inert and electronically insulating to prevent short circuits and maintain stable operation. The substitution of flammable liquid electrolytes in solid-state batteries, which are commonly found in traditional batteries, significantly enhances battery safety by minimizing the risks of thermal runaway, fire, and explosion. Selecting the ideal solid electrolyte material is crucial for battery performance and scalability. A high-performing solid electrolyte must meet several key criteria such as high ionic conductivity [13], chemical stability with both the anode and cathode materials, mechanical robustness to suppress dendrite growth, ease of processing and manufacturability at scale, and thermal and electrochemical stability [6, 7]. Several different types of electrolytes are currently under investigation like oxide-based electrolytes [$Li_7La_3Zr_2O_{12}$ (LLZO)] [14, 15], [$La_{0.5}Li_{0.5}TiO_3$ (LLTO)] [16], sulfide-based electrolytes [$Li_{10}GeP_2S_{12}$ (LGPS)] [17, 18], polymer electrolytes [polyvinylidene fluoride (PVDF)] [19], etc.

(iv) **Interfaces**

Although interfaces are not physical components of a battery like electrodes or electrolytes, they are critically important for its overall performance and longevity. In solid-state batteries, these interfaces are found primarily in three key locations: (i) between the anode and the solid electrolyte, (ii) between the cathode and the solid electrolyte, and (iii) between the individual grains within the solid electrolyte itself. The interfaces between the electrodes and the electrolyte have a profound impact on battery performance. A poor or unstable interface between the anode and the electrolyte can severely hinder the transport of metal ions, leading to high interfacial resistance, which reduces charge and discharge rates, lower power output, and can even promote the growth of dendrites (needle-like structures) that may eventually short circuit the battery [11]. Similarly, inadequate contact at the cathode–electrolyte interface impedes the flow of ions, which reduces power density and limits the effective utilization of the active cathode material. Additionally, both electrodes are subject to volume changes and potential side reactions during repeated charge–discharge cycles, further increasing resistance and degrading performance over time [20]. Interfaces within the solid electrolyte itself also present major challenges. Most solid electrolytes are polycrystalline, composed of numerous microscopic crystal grains. The boundaries between these grains—known as grain boundaries—act as internal interfaces that can hinder ion movement due to their irregular atomic structure. These grain boundaries may also trap impurities or develop secondary phases, further blocking ionic transport and negatively affecting both efficiency and cycle life [21]. The most significant problem associated with these interfaces is their high resistance to ion flow. Such resistance directly translates to lower power density, reduced energy efficiency (as more energy is lost as heat), and slower charging rates. Moreover, poor interfacial contact leads to mechanical stress, unwanted chemical reactions, and structural instability-resulting material degradation and a shorter battery lifespan. Addressing these interfacial challenges is a key focus in the development of next-generation solid-state batteries. This requires innovative solutions through advanced materials science, interface engineering, structural optimization, and improved manufacturing techniques. Enhancing interfacial compatibility and reducing resistance are crucial steps toward safer, more efficient, and longer-lasting solid-state energy storage systems.

(v) **Working Mechanism of Solid-State Batteries**

Fundamentally, all batteries including solid-state batteries operate based on the electrochemical reactions, where ions move between two electrodes: the negative electrode (anode) and the positive electrode (cathode), separated by an electrolyte. The primary function of the electrolyte is to facilitate the movement of specific ions while preventing the internal flow of electrons, which would otherwise cause a short circuit. Instead, electrons flow through the external circuit, generating an electrical current. In a solid-state battery, the distinguishing feature is the use of a solid electrolyte. Unlike the liquid or gel electrolytes found in conventional batteries, this electrolyte is fully solid and is typically composed of ceramic materials (such as oxides, sulfides, or phosphates), polymers, or composite structures. The solid electrolyte serves a dual purpose: It acts as the medium for ion conduction and also physically separates the anode and cathode.

Let's break down the fundamental working mechanism of a rechargeable battery during both discharge and charge cycles, focusing on the generic principles that apply regardless of the specific ion chemistry (Li^+, Na^+, Mg^{2+}, etc.):

During Discharge (Powering a Device)

- **Anode Reaction (Oxidation)**: At the anode (negative electrode), the oxidation of the active material occurs. This process converts the metal into metal ions while releasing electrons to the external circuit. These metal ions are responsible for ionic conduction as they migrate through the solid electrolyte. For example, consider a hypothetical metal anode M. During oxidation, it loses electrons and forms metal ions according to the reaction:

$$\mathrm{M} \rightarrow \mathrm{M}^{n+} + ne^{-}(\text{electrons flow to the external circuit})$$

- **Ion Transport Through Solid Electrolyte**: The metal ions released at the anode now migrate through the solid electrolyte toward the cathode. The solid electrolyte is engineered with a crystal lattice or polymeric structure that contains specialized channels or voids, allowing ions to hop or diffuse efficiently. Crucially, these pathways permit only the movement of ions while blocking electrons, ensuring that electrical energy is delivered through the external circuit.
- **Cathode Reaction (Reduction)**: At the same time, the active material at the cathode receives incoming ions from the electrolyte and electrons from the external circuit. This process triggers a reduction reaction, in which the ions gain electrons and are converted back into their neutral metal form, subsequently depositing onto the cathode. This completes the internal circuit. The difference in electrochemical potential between the anode and cathode drives the spontaneous flow of ions and electrons, generating an electric current to power an external device.

$$\mathrm{M}^{n+} + ne^{-} \rightarrow \mathrm{M}(\text{ions combine with cathode material})$$

During Charge (Recharging the Battery)

- **External Power Application**: When you connect an external power source (like a charger) with a voltage higher than the battery's open-circuit voltage (OCV), it forces current to flow into the battery. This input current drives the electrochemical reactions in the reverse direction compared to what happens during discharge.
- **Cathode Reaction (Oxidation)**: At the cathode, ions that were previously reduced and stored as metal during discharge are now oxidized, releasing into the solid electrolyte and sending electrons into the external circuit—both moving back toward the anode.

$$\mathrm{M} \rightarrow \mathrm{M}^{n+} + ne^{-}(\text{migrating from cathode to anode})$$

- **Ion Transport Through Solid Electrolyte**: Under the influence of the applied electric field, the ions that were liberated from the cathode are moved back toward the anode through the solid electrolyte.
- **Anode Reaction (Reduction)**: At the anode, these incoming metal ions from the solid electrolyte and electrons from the external circuit are absorbed back into the anode material, where they undergo a reduction reaction that essentially "recharges" the anode and stores energy.

$$M^{n+} + ne^{-} \rightarrow M$$

1.1.2 Different Types of Solid-State Electrolytes

The solid electrolytes used in solid-state batteries are broadly classified into following four primary categories and each one exhibits distinct electrochemical and mechanical properties.

Oxide-Based Electrolytes

Oxide-based electrolytes represent a promising class of materials in the development of safer and more stable solid-state batteries. Their inherent chemical robustness and compatibility with established manufacturing processes make them an attractive alternative to conventional liquid electrolytes. Within this category, the several structural types have emerged with unique characteristics. For example, garnet-structured electrolytes such as $Li_7La_3Zr_2O_{12}$ (LLZO) [14, 15] exhibit high lithium-ion conductivity and a wide electrochemical stability window, making them suitable for high-voltage applications. Similarly, perovskite-type electrolytes like $La_{0.5}Li_{0.5}TiO_3$ (LLTO) [16] also offer high ionic conductivity, a critical factor for efficient battery performance. However, their limited electrochemical stability window can restrict compatibility with a broader range of electrode materials. In contrast, NASICON-structured electrolytes offer a favorable combination of good ionic conductivity and mechanical robustness, which supports overall cell integrity [22, 23].

Overall, oxide-based electrolytes provide excellent thermal stability, enabling batteries to operate safely across a wide temperature range. Their stability in ambient air further simplifies manufacturing. Additionally, their relatively high mechanical strength enhances the durability of solid-state batteries. Despite these advantages, the key challenges such as high interfacial resistance and low ionic conductivity at room temperature remain significant barriers to their widespread adoption.

Sulfide-Based Electrolytes

Sulfide-based electrolytes are another class of materials for solid-state batteries, notable for their exceptional ionic conductivity-often comparable to or even exceeding that of conventional liquid electrolytes. This high conductivity enables batteries to charge and discharge at significantly faster rates. Moreover, many sulfide electrolytes are relatively soft, which facilitates better interfacial contact between the electrolyte and electrodes, potentially enhancing overall battery performance. Among the most notable examples is $Li_{10}GeP_2S_{12}$ (LGPS) [17, 18], which exhibits one of the highest reported room-temperature ionic conductivities. Despite this outstanding property, LGPS suffers from high chemical reactivity, posing a significant challenge for practical applications. In contrast, Li_6PS_5Cl, an argyrodite-type electrolyte, offers a more balanced profile, which exhibits good ionic conductivity with improved stability and ease of processing, making it a more viable candidate for real-world use [24, 25]. However, sulfide electrolytes also come with notable challenges. Chief among them is their extreme sensitivity to moisture and air, which necessitates careful handling and controlled environments during manufacturing to prevent material degradation. Additionally, their relatively narrow electrochemical stability window compared to oxide-based electrolytes can limit the range of compatible electrode materials, presenting further obstacles to widespread adoption.

Polymer-Based Electrolytes

Polymer-based electrolytes offer several distinct advantages, particularly in terms of their physical properties. Their inherent flexibility and ease of processing enable the fabrication of thin-film batteries and other flexible energy storage devices. Additionally, the polymers are generally lightweight and mechanically pliable, making them ideal for applications where weight and form factor are critical considerations. The potential for low-cost manufacturing further enhances the appeal of polymer electrolytes. Among the various types, polyethylene oxide (PEO) has emerged as one of the most extensively studied polymer electrolytes [26]. However, despite significant research efforts, PEO continues to suffer from a fundamental drawback such as low ionic conductivity at room temperature. Gel polymer electrolytes present a promising alternative, incorporating liquid electrolytes within a polymer matrix. This hybrid approach seeks to merge the high ionic conductivity of liquid electrolytes with the enhanced safety, mechanical stability, and processability of solid polymers [27, 28]. Nevertheless, the polymer electrolytes still face notable challenges. These include low ionic conductivity at ambient conditions and reduced mechanical strength at elevated temperatures, which can compromise the long-term stability and safety of batteries operating under such environments.

Halide-Based Electrolytes

Halide-based electrolytes represent a rapidly advancing and highly promising frontier in solid-state battery research. They have garnered significant attention for their potential to enable high-performance energy storage systems. One of their most compelling attributes is their exceptionally high ionic conductivity, which in some cases even exceeds that of conventional sulfide-based electrolytes. This superior conductivity opens the door to batteries with higher energy densities and faster charge–discharge capabilities. Notable examples of promising halide electrolytes include Li_3YCl_6, $Li_{10}GeCl_{12}$, and Li_6PS_5Cl, among others [29]. Despite their advantages, the halide electrolytes remain a relatively new area of study. A key challenge lies in their chemical stability particularly their pronounced sensitivity to moisture, which can lead to degradation and compromise performance. Addressing these stability issues will be critical to unlocking the full potential of halide-based solid-state batteries.

2 Lithium-Ion Transportation Mechanism

The movement of lithium ions within solid-state battery materials is a complex process influenced by a combination of structural, atomic, and electronic factors. The favorable interplay among these characteristics is crucial for enabling fast and efficient Li^+ transport, which is essential for the development of high-performance solid-state batteries. Among the various mechanisms governing Li^+ ion transport, a few are briefly discussed below:

2.1 Diffusion Mechanism in Solids

Diffusion in solids is closely related to the presence and movement of defects, which facilitate the movement of atoms. The main diffusion mechanisms in solids can be categorized as follows (Fig. 2):

- **Direct Exchange Mechanism**: This mechanism was proposed in 1896 by Roberts Austen mainly to explain the diffusion of gold in solids [30]. According to this mechanism, it was concluded that the atoms directly swap positions, making it an ideal model in defect free crystalline solids. However, this simple conceptual hypothesis has been ruled out due to requirement of ultra-high activation energy.
- **Interstitial Mechanism**: This mechanism was proposed for light atoms like hydrogen, oxygen, nitrogen, and carbon, which usually occupy the interstitial sites in metals. They jump between interstitial sites without any help of intrinsic point defects, leading to a much higher diffusion coefficient that substitutional atoms [31].

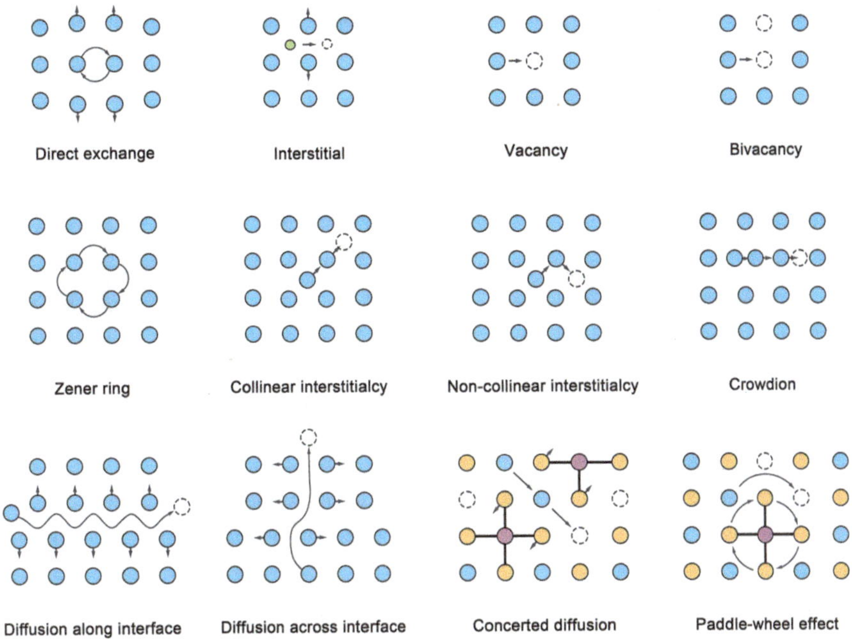

Fig. 2 Different diffusion mechanism in solids [30]

- **Vacancy/Bi-vacancy Mechanism**: Vacancies are an inevitable feature in crystals due to thermal motion. They typically serve as diffusion vehicles in atomic diffusion processes within metals, alloys, and ionic compounds. In addition to single vacancies, paired vacancies known as vacancy pairs can also facilitate diffusion. This mechanism, referred to as the bi-vacancy mechanism, has been proposed in materials such as $LiCoO_2$ and $LiTiS_2$ [31].
- **Zener Ring Mechanism**: The direct exchange mechanism was approached by Zener in 1950 by proposing a ring diffusion mechanism to describe diffusion mechanism in metals [32, 33]. Collective diffusion processes substantially lower the activation energy of ionic dynamics relative to the well-known one-atom hop which has initially been discussed in supercooled liquids and glasses. This is a type of collective diffusion in that several atoms diffuse simultaneously and effectively lowering the energy of activation of the ionic diffusion.
- **Knock-Off Mechanism**: As the name suggests in this mechanism, one atom knocks off its adjacent atom. An interstitial atom will be able to hop to neighboring host lattice position and atom in host position to another interstitial site at the same time. This diffusion mechanism is the so-called knock-off mechanism that can be further classified into two categories according to the fact that the two atoms move collinearly or not: interstitial, collinear, and non-collinear [30]. The Li-ion diffusion in Li_2CO_3 could be well described by the knock-off mechanism [34].

- **Crowdion Mechanism**: A crowdion is a hypothetical structure of interstitials in face-centered cubic metals when one host site is occupied by two atoms. In this mechanism, one of the two atoms of the host site may diffuse into a second host site which is already occupied by an atom. This process of diffusion is referred to as the crowdion mechanism [30].
- **Short-Circuit Diffusion**: Dislocation lines, dislocation walls, grain boundaries, phase boundaries, and 1D and 2D defects are considered as the short circuits to the materials. Diffusion in the short circuits is not in the same way as it happened in lattices due to special atomic environment in the short circuits. The commonly held perception is that diffusion across interfaces is slow compared to those along the interface and this has been proved in solid electrolyte Li_3OCl and cathode material $LiNi_{0.5}Mn_{0.3}Co_{0.2}O_2$ [35].
- **Concerted Diffusion**: Concerted diffusion is a cumulative movement behavior which is made up of two or more atoms. It generally comes in the form of partially occupied atoms in the crystalline lattice. Concerted diffusion leads to a reduction in the number of atoms at high-energy sites and a simultaneous increase of atoms at low-energy sites, so the net effect is a reduced diffusion barrier over that of single-ion migrations [36].
- **Paddle-Wheel Effect**: The paddle-wheel effect is a cooperative movement in which the rotation of groups in crystalline structures drives the movement of the surrounding atoms. It has been predicted in many solid electrolytes with groups of SO_4^{2-}, WO_4^{3-}, CO_3^{2-}, SiF_6^{2-}, AlF_6^{3-}, PO_4^{3-}, PS_4^{3-}, SbS_4^{3-}, BH^{4-}, $XCl_6{}^{2-/3-}$, and BF^{4-} [37].

2.2 *Diffusion in Intercalating-Type Electrodes*

Intercalation and deintercalation reactions in intercalating-type electrodes proceed via solid-state diffusion mechanisms. The kinetics of these diffusion processes play a critical role in determining the rate performance of such electrodes, particularly under challenging conditions such as low temperatures or during fast-charging. The diffusion coefficients, which characterize Li-ion mobility within electrode materials, are influenced not only by the crystallographic diffusion pathways but also by variations in Li-ion concentration within the host structure. This section presents a detailed analysis of Li-ion diffusion in several key intercalating-type electrode materials, including layered graphite, Li_xCoO_2, and $LiMO_2$.

2.2.1 Graphite

The widely used anode material in lithium-ion batteries is graphite, which provides a low average voltage of 0.1 V versus Li/Li^+ and a high theoretical specific capacity of 372 mAh/g, first described by Yazami and Touzain in 1983 [38]. Its large-scale adoption has played a pivotal role in the commercialization of Li-ion battery technology. As a layered allotrope of carbon, graphite serves as a key anode material in various electrochemical energy storage systems, including solid-state batteries. Lithium storage in graphite occurs through a reversible intercalation mechanism, where lithium ions are inserted between the van der Waals-bonded graphene layers, forming graphite intercalation compounds characterized by staging phenomena [39]. Lithium-ion transport within the graphite matrix primarily takes place along the basal planes, facilitated by the expansion of interlayer spacing during lithiation. This diffusion process is influenced by lithium concentration gradients, temperature, and the structural properties of the graphite, which affect the overall electrochemical kinetics of the anode.

2.2.2 Li_xCoO_2

$LiCoO_2$ is a key cathode material in lithium-ion battery technology, featuring an average operating voltage of 3.9 V versus Li/Li^+ and a theoretical specific capacity of 274 mAh/g. $LiCoO_2$ was first synthesized in 1958 by Johnston et al. and brought to energy storage application in 1980 by Mizushima et al.; it has played a central role in the development of Li-ion batteries [40]. However, as with graphite, the complexity of structure of intermediate states during charging and discharging to undergo the chemical reaction possesses a major problem in comprehending exactly the mechanism of lithium-ion transport in $LiCoO_2$. In particular, the phase transition from the layered O3-type $Li_{0.75}CoO_2$ to spinel $Li_{0.5}CoO_2$ further complicates the characterization of these intermediate configurations. To address this complexity, a series of ground-state configurations were determined using cluster expansions, similar to the approach used for graphite. Density functional theory calculations indicate that a di-vacancy diffusion mechanism is energetically favorable for Li^+ transport in Li_xCoO_2. The activation energy in Li_xCoO_2 decreases from 610 to 210 meV as the Li content increases from 0 to 1 [41]. This trend is consistent with results obtained from kinetic Monte Carlo simulations and potentiostatic intermittent titration technique (PITT) measurements. The decrease in activation energy is attributed to changes in the c lattice parameter and the effective valency of Co ions. A diffusion coefficient ranging from 1.19×10^{-13} to 1.27×10^{-11} $cm^2\ s^{-1}$ has been determined using the PITT method. However, $LiCoO_2$ suffers from significant capacity fading during cycling, and the low abundance of Co has motivated the exploration of $LiMO_2$ materials with different transition metals (M) for electrode applications.

2.2.3 $LiMO_2$ Materials

$LiNiO_2$, for example, has a similar theoretical capacity to that of $LiCoO_2$ (275 mAh/g), but the structural changes cause a poor cycling performance. These structural changes involve the movement of Ni^+ from transition metal layers to Li layers, driven by the similar ionic radii of Li^+ and Ni^{2+} ions. $LiMnO_2$ is another cathode material which has gained popularity in research because of the low cost of Mn and comparable theoretical specific capacity (285 mAh/g). However, the practical application of $LiMnO_2$ is hindered by a transition from a layered phase to a spinel phase. To improve the electrochemical properties of $LiMO_2$ materials, a mixing strategy has been proposed, involving combinations of Ni, Co, and Mn (NCM) or Ni, Co, and Al (NCA) [42]. The predicted activation energy in $Li(Ni_xCo_yMn_z)O_2$ depends on the transition metal ions adjacent to mobile Li ions. For example, the predicted activation energies are 357, 430, 545, 510, and 523 meV for Li ions around Ni^{2+}, Ni^{3+}, Ni^{4+}, Co^{3+}, and Mn^{4+}, respectively. It is estimated that the diffusion coefficient will grow from 10^{-9} to 10^{-7} cm^2 s^{-1} as the amount of Ni is increased with no dependency on the state of charge, as is consistent with experimental values [43].

2.3 *Diffusion in Conversion-Type Electrodes*

Compared to the limited capacities of intercalation-type electrodes, alloying and conversion-type electrodes offer significantly higher capacities via conversion reactions. However, accurately determining the crystalline structures involved in these reactions remains a major challenge. As a result, this discussion focuses on lithium (Li) metal and its alloys to examine Li transport mechanisms in conversion-type electrodes. Li metal is a special conversion-type electrode with the maximum theoretical specific capacity of 3860 mAh/g and most negative reduction potential of – 3.04 V versus the standard hydrogen electrode. As with other metallic crystals, bulk diffusion of Li in metals takes place predominantly through vacancy mechanism. While the vacancy migration energy in Li metal is relatively low (55 meV), the high vacancy formation energy (520 meV) at ambient conditions results in a substantial overall activation energy of approximately 575 meV [44]. The diffusion coefficient of Li in the metal phase has been measured in the range of $(1–10) \times 10^{-11}$ cm^2/s, aligning with the experimental values of ~$5–9 \times 10^{-11}$ cm^2/s. This low diffusion rate limits mass transport within Li metal, leading to the morphological changes at solid electrolyte/Li interfaces, such as vacancy accumulation, macroscopic pore formation, and increased interfacial resistance in solid-state batteries. To address these challenges, several strategies have been proposed, including applying external stack pressure, engineering 3D interfaces, and incorporating alloy-based anodes to mitigate dynamic morphological degradation at the Li/solid–electrolyte interface [45].

Alloy anode materials such as Li–Si, Li–Ge, Li–Mg Li–Sn, Li–Sb, Li–Al, Li–In, Li–Zn, Li–Pb, Li–Au, Li–Bi, Li–Ag, and Li–Pt are known to enhance lithium diffusion coefficients compared to pure Li metal. However, this typically comes at the cost of lower theoretical specific capacities (<1000 mAh/g), with notable exceptions being Li–Si (4200 mAh/g) and Li–Mg (3350 mAh/g). Among these, the Li–Mg system is particularly well-suited for studying lithium dynamics due to the formation of a stable β-phase over a wide Li-rich compositional range. Operando galvanostatic electrochemical impedance spectroscopy has revealed intrinsic diffusion coefficients of 2.3×10^{-11} cm^2/s for Li-rich Li-Mg alloy ($Li_{0.9}Mg_{0.1}$) and 0.8×10^{-11} cm^2/s for Li metal. Yet, these values alone do not fully account for the observed differences in Li transport kinetics between alloy and pure Li electrodes. It is now believed that non-equilibrium defect concentrations in both alloy and metal anodes play a significant role in determining lithium diffusivity, helping to explain the broad range of reported diffusion coefficients in Li–Mg alloys (10^{-11} to 10^{-7} cm^2/s). In the case of Li–Sn and Li–In alloys, the migration energies of lithium diffusion vehicles have been calculated using the nudged elastic band method across different Li concentrations. Both interstitial and vacancy-based collective diffusion mechanisms such as knock-off and coordinated exchange have been proposed, with corresponding migration energies ranging from 110 to 490 meV for Li–Sn and 15–260 meV for Li–In [46]. However, the experimental characterization of Li transport in these alloys remains difficult due to their continuous phase transitions during battery operation.

2.4 *Diffusion in Solid Electrolytes*

For these batteries to operate efficiently, Li-ions$^+$ must move easily through the solid electrolyte-which is known as diffusion. This section explores the key factors that influence how lithium ions diffuse through solid electrolytes. The efficiency of Li-ions diffusion is governed by a complex interplay of the material's intrinsic properties, which can be broadly categorized into its structural and dynamic characteristics.

2.4.1 Structural Features and Li Diffusion

The crystal structure of a solid electrolyte, which refers to the arrangement of atoms within it, plays a vital role in facilitating Li-ion movement. Electrolytes with three-dimensional (3D) networks generally exhibit superior Li-ion diffusion due to inter-connected pathways through which Li-ions can easily navigate [47]. Materials with lower dimensional structures, like layered structures, may hinder Li-ion movement in certain directions. The connectivity and openness of these pathways within the structure are critical. Solid electrolytes with 3D frameworks often demonstrate lower activation energies for Li-ion diffusion and higher ionic conductivity [48].

2.4.2 Lattice Dynamics and Li Diffusion

The movement of atoms within a solid electrolyte is not static since they vibrate constantly. These atomic vibrations, or lattice dynamics, also influence Li-ion diffusion. The frequency of atomic vibrations, known as phonon frequency, can affect the energy barrier that Li-ions must overcome to move through the material. Solid electrolytes with lower average phonon frequencies tend to have lower activation energies for Li-ion diffusion [49]. This means that Li ions can move more easily through materials with certain vibrational characteristics.

2.4.3 Electronic Structure and Li Diffusion

The electronic structure of a solid electrolyte, which describes how electrons are arranged within the material, also impacts Li-ion diffusion. Several electronic properties have been identified as important factors. For example, the electronegativity of the constituent atoms, which indicates their ability to attract electrons, influences Li-ion mobility. Higher anionic polarizability, which refers to the ability of negative ions to deform their electron clouds in response to an electric field, generally facilitates Li-ion transport. Conversely, the environments with lower electron density tend to be more favorable for Li-ion diffusion [50].

3 Environmental Advantages of Solid-State Batteries: A Sustainable Energy Future

Solid-state batteries represent a significant advancement in the pursuit of more environmentally sustainable energy storage solutions compared to conventional lithium-ion batteries. Their unique design offers several advantages that help reduce environmental impact. Firstly, solid-state battery eliminates the need for flammable liquid electrolytes, which greatly enhances safety and reduces the risk of thermal runaway. This, in turn, lowers the likelihood of battery failures that could release the harmful substances into the environment. Secondly, solid-state battery offers the potential for higher energy density, enabling the production of smaller and lighter battery packs without compromising energy capacity [3, 4]. This efficiency translates to lower material consumption per unit of stored energy, which is especially valuable benefit in applications such as electric vehicles. By reducing the need for raw material extraction, solid-state battery supports more resource-efficient production.

Moreover, the use of solid electrolytes opens the door to incorporate more abundant and sustainable electrode materials. Combined with the typically longer cycle life and enhanced durability of solid-state batteries, this leads to fewer battery replacements over time, thereby decreasing manufacturing frequency, conserving resources, and minimizing waste. In addition, the simplified architecture of solid-state batteries can allow for more streamlined and potentially less energy-intensive manufacturing processes, further reducing their environmental footprint. Their improved thermal stability also diminishes the need for complex thermal management systems, contributing to overall energy efficiency [6].

In essence, solid-state batteries offer a more sustainable trajectory for energy storage, driven by improved safety, higher energy density, longer lifespan, and the potential use of environmentally benign materials and processes. While continued research and development remain essential, the inherent properties of solid-state batteries suggest a promising reduction in the environmental impact of energy storage technologies, paving the way toward a greener future.

4 Future Trajectories and Commercial Realization of Solid-State Electrolyte Technologies

4.1 Emerging Technological Paradigms

The search for next-generation energy storage solutions extends beyond conventional lithium-ion technologies. Solid-state lithium-air batteries represent a promising avenue for achieving ultra-high-energy densities. These batteries utilize atmospheric oxygen as the active cathode material, potentially surpassing the energy densities of lithium-ion batteries by orders of magnitude [51]. However, the significant challenges remain, particularly in oxygen transport, electrode reversibility, and electrolyte stability [7]. Sodium-based solid-state batteries present a cost-effective alternative, capitalizing on the natural abundance of sodium. Realizing their potential hinges on the development of efficient sodium-ion-conducting solid-state electrolytes. Key challenges include their generally lower ionic conductivity compared to lithium-ion counterparts and the need for compatible electrode materials. Multivalent solid-state batteries such as those based on magnesium or calcium offer the promise of even higher energy densities and enhanced safety [2]. Yet, the advancement of these technologies is constrained by the difficulty of developing solid-state electrolytes that provide both high multivalent ion conductivity and robust electrochemical stability.

4.2 Path to Commercialization

The timeline for the widespread commercialization of solid-state electrolytes depends on overcoming several key technological and economic challenges. Transitioning from laboratory-scale prototypes to large-scale production requires the development of scalable, cost-effective manufacturing processes. This includes optimizing synthesis parameters, designing efficient processing systems, and establishing robust quality control protocols. Collaboration between academia and industry is critical to accelerating this transition. Academic institutions offer deep expertise in fundamental research and materials science, while industry partners contribute advanced manufacturing capabilities and market-driven perspectives. Joint research initiatives, technology transfer agreements, and strategic partnerships are essential to bridging the gap between research and commercial deployment. Additionally, standardizing testing protocols and performance metrics is vital to ensure the reliability and comparability of solid-state electrolytes technologies across the industry. Investment in infrastructure and workforce development will also play a key role in building a resilient supply chain for solid-state electrolytes materials and components.

5 Applications for Solid-State Batteries

Solid-state batteries represent a transformative leap in electrochemical energy storage, offering a wide range of applications across various sectors. Their inherent safety, higher-energy–density potential, and extended cycle life make them a compelling alternative to conventional lithium-ion batteries.

In the electric vehicle (EV) sector, solid-state batteries promise to address key limitations of liquid electrolyte-based systems. By eliminating flammable organic electrolytes, solid-state batteries significantly reduce the risk of thermal runaway, which is a major safety concern in current EV batteries. Additionally, the compatibility of solid electrolytes with high-energy density materials, such as lithium metal anodes, enables substantial improvements in both gravimetric and volumetric energy density, potentially extending EV driving range. Enhanced thermal stability also supports faster charging rates, directly addressing a major consumer concern. Beyond transportation, solid-state batteries offer substantial benefits for consumer electronics. As portable devices continue to shrink in size while demanding more power, the need for safer and more energy-dense batteries becomes critical. The non-flammable nature and high-energy density of solid-state batteries make them ideal for smartphones, laptops, wearables, and medical implants. Their resistance to mechanical stress and ability to function across a wider temperature range further enhance their suitability for these applications (Fig. 3).

Solid-state batteries also hold significant promise for stationary energy storage at the grid level. Their long cycle life and robust safety profile are particularly valuable for large-scale systems required to integrate intermittent renewable sources like solar and wind. Lower maintenance needs and longer operational lifespans can reduce system costs and support a more sustainable energy infrastructure. Moreover, the intrinsic safety of solid-state batteries is especially beneficial in densely

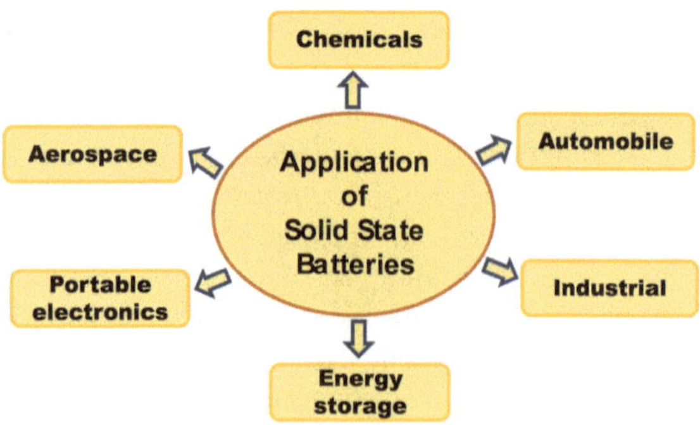

Fig. 3 Different applications of solid-state batteries

populated areas where large energy storage systems are deployed. In the aerospace industry, where weight and safety are paramount, solid-state batteries provide a compelling alternative to traditional batteries. Their high-energy density and non-flammable composition are well-suited for powering electric aircraft and drones, enabling longer flight durations and improved safety. Their ability to operate reliably in extreme temperature conditions also aligns with the rigorous demands of aerospace environments. Finally, the unique properties of solid-state batteries enable their use in specialized applications such as implantable medical devices. Certain solid electrolytes exhibit biocompatibility, while the batteries inherent safety and durability enhance patient outcomes by minimizing risks and reducing the frequency of device replacements.

References

1. Kong L, Wang L, Zhu J, Bian J, Xia W, Zhao R, Lin H, Zhao Y (2021) Configuring solid-state batteries to power electric vehicles: a deliberation on technology, chemistry and energy. Chem Commun 57:12587–12594
2. Trahey L, Brushett FR, Balsara NP, Ceder G, Chenga L, Chiang YM, Hahn NT, Ingram BJ, Minteer SD, Moore JS, Mueller KT, Nazar LF, Persson KA, Siegel DJ, Xu K, Zavadil KR, Srinivasan V, Crabtree GW (2020) Energy storage emerging: a perspective from the joint center for energy storage research. PNAS 117:12550–12557
3. Xiao Y, Wang Y, Bo SH, Kim JC, Miara LJ, Ceder G (2020) Understanding interface stability in solid-state batteries. Nat Rev Mater 5:105–126
4. Arora P, White RE, Doyle M (1998) Capacity fade mechanisms and side reactions in lithium-ion batteries. J Electrochem Soc 145:3647
5. Vetter J, Novák P, Wagner MR, Veit C, Möller KC, Besenhard JO, Winter M, Wohlfahrt-Mehrens M, Vogler C, Hammouche A (2005) Ageing mechanisms in lithium-ion batteries. J Power Sour 147:269–281
6. Bates AM, Preger Y, Torres-Castro L, Harrison KL, Harris SJ, Hewson J (2022) Are solid-state batteries safer than lithium-ion batteries? Joule 6:742–755
7. Janek J, Zeier WG (2016) A solid future for battery development. Nat Energy 1:16141
8. Bhattacharyya R, Key B, Chen H, Best AS, Hollenkamp AF, Grey CP (2010) In situ NMR observation of the formation of metallic lithium microstructures in lithium batteries. Nat Mater 9:504–510
9. Han F, Yue J, Zhu X, Wang C (2018) Suppressing Li dendrite formation in Li_2S–P_2S_5 solid electrolyte by LiI incorporation. Adv Energy Mater 8:1703644
10. Wang S, Lv J, Wang X, Cui H, Huang W, Wang Y (2022) Progress of solid-state electrolytes used in organic secondary batteries. Chem Electro Chem 9:202101005
11. Huo H, Janek J (2022) Silicon as emerging anode in solid-state batteries. ACS Energy Lett 7:4005–4016
12. Minnmann P, Strauss F, Bielefeld A, Ruess R, Adelhelm P, Burkhardt S, Dreyer SL, Trevisanello E, Ehrenberg H, Brezesinski T, Richter FH, Janek J (2022) Designing cathodes and cathode active materials for solid-state batteries. Adv Energy Mater 12:2201425
13. Goodenough JB, Hong HY-P, Kafalas JA (1976) Fast Na^+-ion transport in skeleton structures. Mater Res Bull 11:203–210
14. Allen JL, Wolfenstine J, Rangasamy E, Sakamoto J (2012) Effect of substitution (Ta, Al, Ga) on the conductivity of $Li_7La_3Zr_2O_{12}$. J Power Sour 206:315–319
15. Murugan R, Thangadurai V, Weppner W (2007) Fast lithium ion conduction in garnet-type $Li_7La_3Zr_2O_{12}$. Angew Chem Int Ed 46:7778–7781

16. Cao C, Zhong Y, Zhao L, Seneque H, Shao Z (2023) Enhancing fast-charge capabilities in solid-state lithium batteries through the integration of high $Li_{0.5}La_{0.5}TiO_3$ (LLTO) content in the lithium-metal anode. ACS Appl Mater Interf 15:59370–59379
17. Kamaya N, Homma K, Yamakawa Y, Hirayama M, Kanno R, Yonemura M, Kamiyama T, Kato Y, Hama S, Kawamoto K, Mitsui A (2011) A lithium superionic conductor. Nat Mater 10:682–686
18. Liu D, Zhu W, Feng Z, Guerfi A, Vijh A, Zaghib K (2016) Recent progress in sulfide-based solid electrolytes for Li-ion batteries. Mater Sci Eng, B 213:169–176
19. Wu Y, Li Y, Wanga Y, Liu Q, Chen Q, Chen M (2022) Advances and prospects of PVDF based polymer electrolytes. J Energy Chem 64:62–84
20. Alsac EP, Nelson DL, Yoon SG, Cavallaro KA, Wang C, Sandoval SE, Eze UD, Jeong WJ, McDowell MT (2025) Characterizing Electrode Materials and Interfaces in Solid-State Batteries. Chem Rev 125:2009-2119B
21. Truong E, Patel SV, Liu H, Chen Y, Lacivita V, Zhang C, Oyekunle IP, Ojelade I, Jin Y, Jones BT, Miara LJ, Dravid VP, Gao H, Kim R, Wang Y, Hu Y-Y (2025) Accelerating ion transport in polycrystalline conductors: On pores and grain boundaries. Sci Adv 11:eadt7795
22. Machin S, Morant C, Marquez F (2024) Advancements and challenges in solid-state battery technology: an in-depth review of solid electrolytes and anode innovations. Batteries 10:29
23. Fergus JW (2012) Ion transport in sodium ion conducting solid electrolytes. Solid State Ionics 227:102–112
24. Deiseroth H-J, Kong S-T, Eckert H, Vannahme J, Reiner C, Zaiß T, Schlosser M (2008) Li_6PS_5X: a class of crystalline Li-rich solids with an unusually high Li^+ mobility. Angew Chem 120:767–770
25. Rao RP, Adams S (2011) Studies of lithium argyrodite solid electrolytes for all-solid-state batteries. PSS 208:1804–1807
26. Xi G, Xiao M, Wang S, Han D, Li Y, Meng Y (2021) Polymer-based solid electrolytes: material selection, design and application. Adv Funct Mater 31:2007598
27. Li Z, Fu J, Zhou X, Gui S, Wei L, Yang H, Li H, Guo X (2023) Ionic conduction in polymer-based solid electrolytes. Adv Sci 10:2201718
28. Nie X, Hu J, Li C (2023) Halide-based solid electrolytes: the history, progress, and challenges. Interdiscip Mater 2:365–389
29. Huang L, Zhang L, Bi J, Liu T, Zhang Y, Liu C, Cui J, Su Y, Wu B, Wu F (2024) An insight into halide solid-state electrolytes: progress and modification strategies. Energy Mater Adv 5:0092
30. Fu Z-H, Chen X, Zhang Q (2023) Review on the lithium transport mechanism in solid-state battery materials. WIREs Comput Mol Sci 13:69865
31. Van der Ven A, Ceder G (2000) Lithium diffusion in layered Li_xCoO_2. Electrochem Solid State Lett 3:301
32. Zener C (1950) Ring diffusion in metals. Acta Cryst 3:346–354
33. Donati C, Douglas JF, Kob W, Plimpton SJ, Poole PH, Glotzer SC (1998) Stringlike cooperative motion in a supercooled liquid. Phys Rev Lett 80:2338
34. Shi S, Lu P, Liu Z, Qi Y, Hector LG, Li H, Harris SJ (2012) Direct calculation of li-ion transport in the solid electrolyte interphase. J Am Chem Soc 134:15476–15487
35. Chen B, Xu C, Zhou J (2018) Insights into grain boundary in lithium-rich anti-perovskite as solid electrolytes. J Electrochem Soc 165:A3946–A3951
36. He X, Zhu Y, Mo Y (2017) Origin of fast ion diffusion in super-ionic conductors. Nat Commun 8:15893
37. Zhang Z, Nazar LF (2022) Exploiting the paddle-wheel mechanism for the design of fast ion conductors. Nat Rev Mater 7:389–405
38. Yazami R, Touzain P (1983) A reversible graphite-lithium negative electrode for electrochemical generators. J Power Sour 9:365–371
39. Persson K, Hinuma Y, Meng YS, Ven AV, Ceder G (2010) Thermodynamic and kinetic properties of the Li-graphite system from first-principles calculations. Phys Rev B 82:125416
40. Lyu Y, Wu X, Wang K, Feng Z, Cheng T, Liu Y, Wang M, Chen R, Xu L, Zhou J, Lu Y, Guo B (2021) An overview on the advances of $LiCoO_2$ cathodes for lithium-ion batteries. Adv Energy Mater 11:2000982

41. Wang H, Jang Y-I, Huang B, Sadoway DR, Chiang Y-M (1999) TEM study of electrochemical cycling-induced damage and disorder in $LiCoO_2$ cathodes for rechargeable lithium batteries. J Electrochem Soc 146:473–480
42. Dahn JR, Sacken UV, Juzkow MW, Al-Janaby H (1991) Rechargeable $LiNiO_2$ /carbon cells. J Electrochem Soc 138:2207
43. Wei Y, Zheng J, Cui S, Song X, Su Y, Deng W, Wu Z, Wang X, Wang W, Rao M, Lin Y, Wang C, Amine K, Pan F (2015) Kinetics tuning of Li-ion diffusion in layered $Li(Ni_xMn_yCo_z)O_2$. J Am Chem Soc 137:8364–8367
44. Frank W, Breier U, Elsässer C, Fähnle M (1996) First-principles calculations of absolute concentrations and self-diffusion constants of vacancies in lithium. Phys Rev Lett 77:518
45. Xu R, Liu F, Ye Y, Chen H, Yang RR, Ma Y, Huang W, Wan J, Cu Y (2021) A morphologically stable Li/electrolyte interface for all-solid-state batteries enabled by 3D-micropatterned garnet. Adv Mater 33:2104009
46. Qu J, Xiao J, Wang T, Legut D, Zhang Q (2020) High rate transfer mechanism of lithium ions in lithium–tin and lithium–indium alloys for lithium batteries. J Phys Chem C 124:24644–24652
47. Peled E (1979) The electrochemical behavior of alkali and alkaline earth metals in nonaqueous battery systems—the solid electrolyte interphase model. J Electrochem Soc 126:2047
48. Jun K, Sun Y, Xiao Y, Zeng Y, Kim R, Kim H, Miara LJ, Im D, Wang Y, Ceder G (2022) Lithium superionic conductors with corner-sharing frameworks. Nat Mater 21:924–931
49. Muy S, Schlem R, Shao-Horn Y, Zeier WG (2021) Phonon–ion interactions: designing ion mobility based on lattice dynamics. Adv Energy Mater 11:2002787
50. Monchak M, Hupfer T, Senyshyn A, Boysen H, Chernyshov D, Hansen T, Schell KG, Bucharsky EC, Hoffmann MJ, Ehrenberg H (2016) Lithium diffusion pathway in $Li_{1.3}Al_{0.3}Ti_{1.7}(PO_4)_3$ (LATP) superionic conductor. Inorg Chem 55:2941–2945
51. Nishi Y (2001) Lithium ion secondary batteries; past 10 years and the future. J Power Sour 100:101–106

Fundamentals of Electrochemistry of Solid-State Batteries

Awanish Gupta, Devendra Kumar Rai, and Ashish Bhatnagar

Abstract The solid-state batteries (SSBs) have acquired significant attention due to their improved safety, higher energy density, and superior electrochemical stability, establishing them as a potential replacement for conventional lithium-ion batteries (LIBs). This chapter explores the fundamentals of SSBs, focusing on solid electrolytes (SEs), anode, and cathode materials within SEs. It also examines the electrode–electrolyte interface (EEI) and various strategies to improve the electrochemical properties of SSBs for both LIBs and sodium-ion batteries (SIBs). Additionally, the chapter discusses the latest advancements in oxide-, sulfide-, and polymer-based SEs, along with challenges such as dendrite suppression and interfacial resistance. Various electrochemical measurement techniques used to evaluate SSB performance are also reviewed. The chapter concludes with future perspectives on scalability, commercialization, and material optimization for next-generation energy storage systems.

1 Introduction

This demand for sustainable energy solutions is increasingly driven by the need to fix the climate change issue and reduce dependency on fossil fuels [1, 2]. Therefore, there is an urgent requirement for renewable energy resources such as wind energy, solar energy, and hydropower, but they are unevenly available [3, 4]. Thus, for uninterrupted electricity generation, we require devices called energy storage devices (ESDs), basically batteries. Rechargeable lithium and SIBs are at the core of that. However, the inherent limitations of LIBs/SIBs, such as safety issues, flammability of liquid electrolytes (LEs), and corrosion, highlight the need for innovative alternatives [5–7]. Additionally, during charging, the dendritic growth can lead to

A. Gupta · D. K. Rai · A. Bhatnagar (✉)
Department of Physics and Materials Science and Engineering, Jaypee Institute of Information Technology, Noida, India
e-mail: ashish.bhatnagar@mail.jiit.ac.in

D. K. Rai
Department of Physics, Jaypee University of Engineering and Technology, Guna, India

A. K. Rai et al. (eds.), *Electrochemistry of Next-Generation Solid-State Batteries*, Green Energy and Technology, https://doi.org/10.1007/978-981-95-7603-6_2

short circuits, further compromising battery safety [8]. The electrochemical instability of LEs under extreme conditions such as high voltage or temperature restricts the use of advanced materials like high-capacity cathodes and low-potential anodes (e.g., lithium/sodium metal). The SSBs fulfill this gap and hold promises as the next generation in energy storage technology by providing improved safety, high-energy density, and cycle life [9, 10]. This chapter discusses the fundamental principles of SSBs, their unique advantages, and challenges that must be addressed to realize their full potential [11] by examining emerging materials and innovative designs. The primary challenge of SSB technologies, particularly LIBs and SIBs, is their limited energy density, which hinders their potential for large-scale energy storage applications [8, 12]. The limitation is a constraint to their use in applications demanding high-energy output, such as electric vehicles (EVs) and grid-scale storage.

The primary challenges involved in meeting the necessary parameters to achieve the targeted energy density, including (1) optimizing the thickness of the SE, (2) formulating the cathode to reduce the catholyte content while maximizing electrode thickness, and (3) the use of appropriate anode. These approaches will ensure optimum interface between SE and electrodes and will cease the growth of dendrites responsible for the inferior performance of SSBs as compared to the battery with liquid electrolytes. This chapter presents a wide-reaching discussion of the importance of SSBs in evolving energy storage technologies. It discusses the severe issues current LIBs face and the necessity of new alternatives in SSBs to surmount safety, energy density, and sustainability limitations. Afterward the principles of electrochemistry in SSBs, with particular emphasis on how SEs enhance the performance of batteries, are discussed. The chapter also discusses different types of solid-state electrolytes, interface phenomena in SSBs, some important electrochemical characterization techniques. The chapter ends with a summary of important points and a cautiously optimistic outlook on the potential and future of SSBs.

2 Fundamentals of Electrochemistry in Solid-State Batteries

Electrochemistry involves the conversion of chemical energy into electrical energy and vice versa. Batteries, which store and supply energy as needed, consist of three essential components: a cathode, an anode, and an electrolyte, along with two positive and negative current collectors.

In both conventional and SSBs, the electrode (cathode and anode) releases ions during charging and receives them during discharging. Its material determines the battery voltage and energy density. The electrolyte facilitates ion conduction between the anode and cathode while preventing direct electron flow, ensuring controlled energy transfer. In operation, the movement of positively charged lithium ions through the electrolyte delivers free electrons at the anode, leading to a charge buildup at the positive current collector as shown in Fig. 1. This causes electrical

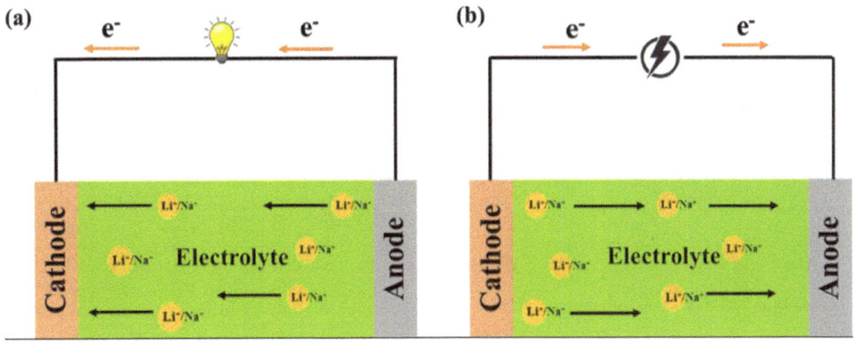

Fig. 1 Solid-state battery component and discharging/charge process

current to flow through an external circuit to a connected device, such as a portable device, smartwatch, cell phone, or computer, before returning to the negative current collector. The electrolyte plays a crucial role in blocking electron flow within the battery while enabling ion transport, preventing short circuits and ensuring stable performance.

2.1 Differences between Liquid and Solid-State Electrolyte Systems

A LE is a solution of ions dissolved in a liquid solvent, allowing for high ionic conductivity because of the free movement of ions, while a SE is a solid material that conducts ions through its crystal lattice. The LE has the issue like leakage, flammable, and corrosion. This issue can resolve by using the SE. Following are the differences between SEs and LEs based on the given category (Table 1).

3 Cathode Materials

The cathode-active material (CAM) is considered the one of the prime components of batteries irrespective of either its LIB/SIB or SSB. These composite cathodes present in the SSBs comprise the CAM and particles of SE in combination with polymeric binders and carbon-based additives. The later materials in above will contribute to enhancing the mechanical and electrical properties of the cathode. Fabrication of high-performing SSB cathodes involves challenges across several length scales, from micro to nano stage level. These challenges arise with regard to the optimization of composite cathodes, CAM particles, and the interface between CAM and SE [15]. Besides microscale optimization of composite cathodes, the CAM itself should be

Table 1 Composition, performance, and safety aspects of liquid versus solid electrolytes in Li/Na-ion batteries [13, 14]

Aspect	Liquid electrolyte	Solid electrolyte
Composition	Made of organic solvents and lithium/sodium salts (e.g., $LiPF_6$, $NaClO_4$, $NaCF_3SO_3$, EC-DMC)	Composed of inorganic compounds (e.g., LiPON, LATP) or polymer materials (e.g., PEO, PVDF-HFP, PMMA) + lithium/sodium salts (e.g., $LiPF_6$, $NaClO_4$, $NaCF_3SO_3$)
Physical state	Liquid, enabling free movement of ions	Solid or semi-solid structure, allowing ions to move within a rigid framework
Ionic conductivity	It offers high ionic conductivity (10^{-3}–10^{-2} S cm^{-1}) with excellent contact between electrodes and electrolytes	Has lower ionic conductivity (10^{-5}–10^{-3} S cm^{-1}), especially at room temperature, and higher interface resistance
Safety	Flammable and volatile, with risks of leaks and fires	Non-flammable and thermally stable, making it safer for use
Stability	It relies on a stable SEI layer to minimize side reactions, which can still occur	Provides better thermal stability but is sensitive to air exposure and interface issues
Energy density	Supports better power density due to efficient ion transport	Can achieve higher energy density by supporting high-voltage materials
Industrialization	Well-developed production processes, highly automated	Requires more complex and less developed manufacturing techniques
Mechanical properties	Can expand uncontrollably during charge and discharge cycles	Maintains more controlled expansion but may face challenges with interface stability

designed to ensure compatibility with the SE in terms of chemical and mechanical properties. It indicates that there shall exist a careful consideration of the materials and processing guidelines; towards optimizing the composite cathode material of the SSBs, as well as the chemical and mechanical properties, for same compatibility with the SE. Although many factors contribute to energy performance, CAM is central to optimization due to the basic primacy it has as key battery parameter. The decrease in the SE content lowers the achievable power density and increases the tortuosity of ionic conduction routes [16–20]. As a result, careful balancing of CAM volume fraction and electrode thickness is necessary for electrode engineering. Fig. 2 provides a glimpse of the tortuosity in liquid and SSBs. In the SSBs, the CAM experiences volume expansion and contraction during cycling, and this affects the performance which leads to battery capacity degradation. In the cathode, volume and morphological changes are primarily associated with the CAM during lithiation/sodiation and delithiation/desodiation, varying with the state of charge (SOC).

Transition metal oxides are generally used as a cathode material; these types of transition metal shows higher valence state when lithium/sodium is removed [21]. The charge neutrality is maintained in transition metal oxides despite large compositional changes; Therefore, the crystal structure of these type of cathodes

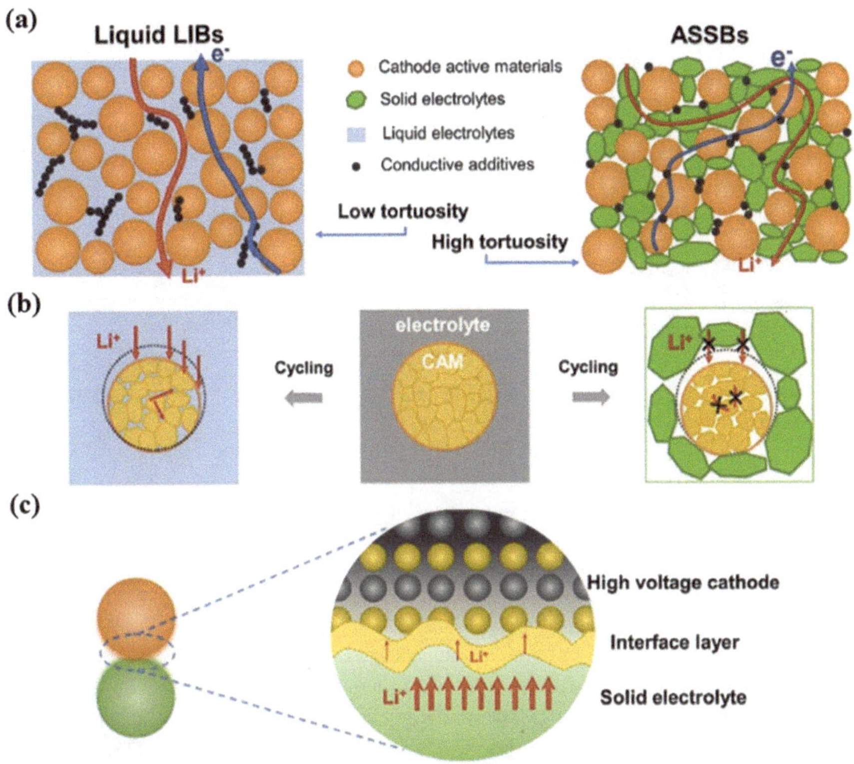

Fig. 2 **a** Comparison of composite cathodes with liquid and solid electrolytes: In a liquid electrolyte system, the electrolyte can permeate the cathode structure and fill its pores, ensuring an evenly distributed charge transport channels, promoting ion transport with minimal tortuosity. In contrast, SEs introduce challenges such as residual porosity and grain boundary resistance, which lead to more convoluted ion transport pathways and significantly higher overall tortuosity. **b** The liquid electrolytes can penetrate freshly created surfaces and adapt to CAM volume changes. The charge transfer between primary particles is disrupted by the intergranular cracks in all SSBs (ASSBs), and the spaces between CAMs and SEs are not electrically or ionically conductive. **c** Illustration of high resistance at the CAM–SE interface. Reprinted with permission from W. Jiang et al., "Design of Composite Cathodes for Sulfide-Based All-Solid-State Batteries," *eTransportation*, 2023, © Elsevier

remains stable over the wide range of compositions. As seen in Fig. 1, during the discharging, the lithium/sodium ions insert into the cathode material and electron release from the anode material. The electron reduces transition metal ions in the cathode to the lower valence. The performance of the cathode is influenced by the microstructure and morphology of the electrode, as these characteristics contribute a major role in the lithium/sodium-ion exchange during charge/discharge process at the EEI. These properties also impact the electrochemical performance of cathode.

The cathode material commonly used in LIBs is $LiCoO_2$ (LCO) because it exhibits a high theoretical capacity and high-energy density [22]. $LiNiO_2$ (LNO) also offers

Table 2 Type of cathode materials for Li-SSBs and advantages, challenges, and capacity

Cathode type	Advantages	Challenges	Capacity	References
Layered oxides (NCM, NCA, LCO)	High-energy density, commercial viability	Volume expansion, electrolyte compatibility issues	200–300 mAh g^{-1}	[27, 28]
Spinel oxides (LMO, LNMO)	Excellent stability and safety	Low electronic conductivity, lower energy density	~148 mAh g^{-1}	[29]
Polyanionic (LFP, $LiMnPO_4$)	Excellent stability and safety	Low electronic conductivity, lower energy density	~170 mAh g^{-1}	[30, 31]
Conversion-type (FeS_2, CuS)	High-capacity, low-cost materials	Huge volume change, slow kinetics	894 mAh g^{-1} for FeS_2 560 mAh g^{-1} for CuO	[32, 33]
Halide-based cathodes	High stability, good conductivity	Expensive, limited availability	150–300 mAh g^{-1}	[34, 35]

a high capacity and is less expensive and less toxic than $LiCoO_2$. Another widely studied cathode material is $LiMnO_2$ (LMO) [23]. In SIBs, transition metal oxides of the form Na_XMO_2 (where M represents Co, Mn, Ti, Ni, Fe, Cr, Al, V, or a mixture of these elements) have been extensively studied [24, 25]. Dezhao Peng et al. [26] studied $Li_{2\text{-}2X}Fe_{1+X}Cl_4$ ($0 \leq x \leq 1/3$) halide-based cathodes for halide-electrolyte all-solid-state batteries (ASSBs), demonstrating high reversibility and stable cycling. Li_2FeCl_4 exhibits no phase change during (de)lithiation, ensuring superior cycling stability (85.1% capacity retention after 5000 cycles at 10C). ASSBs with Li_2FeCl_4 deliver excellent rate performance (122.4 mAh g^{-1} at 0.2C, 91.4 mAh g^{-1} at 10C). Optimizing the Li/Fe ratio enables a maximum capacity of 155.2 mAh g^{-1} ($Li_{4/3}Fe_{4/3}Cl_4$), comparable to commercial $LiFePO_4$ cathodes. These findings highlight the potential of halide cathodes in next-generation ASSBs (Tables 2 and 3).

4 Anode Materials

The anode is a critical component in SSBs, influencing essential performance factors such as energy density, safety, lifespan, and fast-charging capability. As SSBs are poised to revolutionize energy storage by surpassing the limitations of traditional LIBs/SIBs, the focus has shifted to developing advanced anode materials [42]. These materials aim to exploit the benefits of SSBs, including the absence of flammable LEs and the potential for integrating lithium-metal anodes, which have an impressive theoretical capacity of 3860 mAh g^{-1} [43]. For anode materials in SSBs to be ideal,

Table 3 Type of cathode materials for Na-SSBs and advantages, challenges, and capacity

Cathode type	Advantages	Challenges	Capacity	References
Layered oxides ($NaxMO_2$, e.g., $Na_{0.71}CoO_2$, $NaMnO_2$)	High specific capacity, moderate voltage (2.7–3.2 V)	Structural instability, phase transitions, capacity fading	~180–220 mAh g^{-1}	[36]
Polyanionic compounds ($Na_3V_2(PO_4)_3$, $NaFePO_4$)	Excellent cycling stability, strong structural framework	Lower electronic conductivity, limited specific energy	~118 mAh g^{-1}	[37, 38]
Fluorophosphates ($Na_3(VO_{1-x})_2(PO_4)_2F_{1+2x}$)	High operating voltage (~3.8–3.9 V), good cycling stability	Poor electron conductivity, dependence on carbon additives	128 mAh^{-1}	[39]
Hexacyano-metallates ($Na_{0.61}Fe[Fe(CN)_6]_{0.94}$)	High reversible capacity, high voltage (>3.0 V)	Low coulombic efficiency due to crystal water decomposition	~170 mAh g^{-1}	[40]
Organic cathodes (pteridine derivatives, quinone-based compounds)	High sustainability, tunable redox chemistry	Solubility issues, slow kinetics, low electronic conductivity	~70–250 mAh g^{-1}	[36, 41]

they must meet several key requirements. High theoretical capacity is fundamental for maximizing energy storage, while a reduced electrochemical potential comparable to lithium is necessary to achieve high cell voltage and superior energy output. Additionally, efficient electronic conductivity supports faster electron transfer, enhancing the rate at which the battery can charge and discharge. Structural integrity is another crucial attribute, as the anode must endure the physical changes caused by lithium/sodium insertion and removal without degrading, which is essential for long-term battery reliability [44]. Conventionally used anode in SSBs is graphite having a theoretical capacity of 372 mAh g^{-1}. Researchers used alternative anode materials such as silicon, tin, and metal alloys show promise due to their high capacities and compatibility with SEs [45, 46]. The materials like Si, Ge, Sn, and Sb, known for their alloying properties, have been extensively investigated for use in LIBs. The Sb and Ge have considered as more promising anode materials than Sn and Si for SIBs. Additionally, metal oxides and sulfide-based anodes are also popular choices for both LIBs and SIBs [47]. However, significant challenges remain, including managing volume changes during cycling and addressing issues with the SE interphase (SEI). The SEI, a critical layer that forms between the anode and electrolyte, must be strong and stable to inhibit lithium dendrite formation. This is essential not only for ensuring battery safety but also for enabling the use of high-capacity lithium-metal anodes effectively [48]. The shift from LE-based LIBs to solid-state systems introduces additional complexities. The anode–SE interface needs to maintain stability and

Table 4 Type of anode materials for Li-SSBs and capacity, challenges, and strategies for improvement

Anode material	Theoretical capacity (mAh g^{-1})	Challenges	Strategies for improvement	References
Graphite (C)	372	Poor contact with solid electrolytes, limited capacity, risk of lithium plating	Surface modifications, hybrid composites, alternative carbon structures	[49]
Silicon (Si)	4200	Large volume expansion (~300%), SEI formation	Hollow/porous structures, thin films, composite coatings	[50]
Germanium (Ge)	1384	Volume expansion, cost	Nanostructuring, alloying with other elements	[51]
Tin (Sn)	~990	Volume changes, electrode pulverization	Sn-based composites, carbon coatings	[52]
Silicon oxides (SiO_X)	~1000–2000	Irreversible capacity loss, low conductivity	Partial oxidation, composite anodes	[53, 54]
Silicon nitride (SiN_X)	~700	Limited capacity, interfacial resistance	Doping, composite formulations	[55]
Tin sulfide (SnS)	~800–1100	Conversion reaction irreversibility	Use of solid electrolytes to stabilize cycling	[56, 57]
Silicon-carbon composites	~1000–3500	Trade-off between capacity and stability	Carbon coating, nano-engineering	[58, 59]

conductivity during battery operation to ensure efficient ion transport. Any disruption at this interface could lead to performance degradation. Therefore, the careful selection and optimization of anode materials are paramount. These materials must not only be compatible with the SE but also foster the creation of a durable interface that supports reliable battery function (Tables 4 and 5).

5 Electrolytes

The SEs contribute a significant role in shaping the outcome of SSBs and overcoming the challenges of LEs like leakage, flammability, and explosives. Their significance extends across key parameters such as ionic conductivity, thermal and electrochemical stability, and compatibility with electrode materials. Inorganic electrolytes and PEs are the classifications of SEs. Each category is characterized by features that impact battery functionality. Fig. 3 shows the advantages of the SEs.

Table 5 Type of anode materials for Na-SSBs and reaction mechanism, capacity, and voltage range

Anode material	Reaction mechanism	Capacity (mAh g^{-1})	Voltage range (V)	References
Hard carbon	Na^+ insertion	~ 300–335	0.1–2.5	[60, 61]
Tin (Sn)	Alloying reaction	847	<0.7	[62]
Antimony (Sb)	Alloying reaction	~660	~0.3	[63]
Phosphorus (P)	Alloying reaction	~2596	~0.5	[64]
MoS_2	Intercalation + conversion	~240	0.12–0.94	[65]
$Na_2Ti_3O_7$	Intercalation	~177	~0.13	[66]
$Na_2Ti_6O_{13}$	Intercalation	~190	0.01–2.5	[67]
$Na_3V_2(PO_4)_3$	Intercalation	~113	0.3–1.6	[68]
$NaZr_2(PO_4)_3$	Intercalation	~150	0.01–2.5	[69]

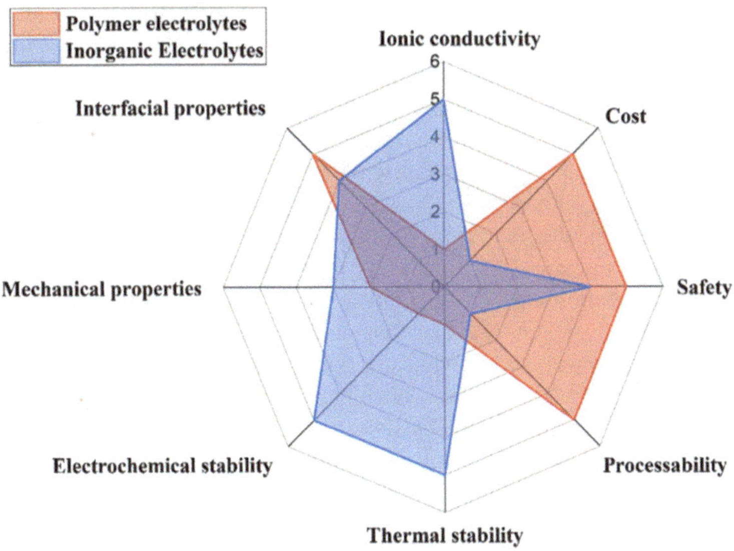

Fig. 3 Types of solid electrolytes and comparative properties of solid electrolytes

5.1 Inorganic Solid Electrolytes

Inorganic SEs, also known as ceramic electrolytes, contain crystalline, partially crystalline (glass-ceramics), and amorphous glasses, all of that exhibit Li^+/Na^+ conductivity. These electrolytes possess several important properties, including high ionic conductivity, electrochemical stability, mechanical strength, and safety. A distinctive feature of inorganic SEs is that they act as single-ion conductors, with a lithium transference number close to unity, except in the case of halides. The ion transport mechanism in these electrolytes occurs through vacancies or interstitial sites, whereas in LEs, ions move freely within the solvent.

Inorganic SEs are generally classified into oxides and sulfides. Li-hydrides and Li-halides have certain drawbacks, due to issues like instability against cathode materials or poor ionic conductivity, making them less extensively studied. Examples include Li-hydrides (e.g., $LiBH_4$, Li_3AlH_6, $Li_2BH_4NH_2$) and Li-halides (e.g., $Li_{1.8}N_{0.4}Cl_{0.6}$, Li_2CdCl_4, Li_3YCl_6, $Li_3InBr_3Cl_3$).

5.1.1 Oxide Solid Electrolytes

The oxide SEs offer high stability against air and elevated temperatures, making them more suitable for large-scale industrial production. Despite having lower ionic conductivity compared to sulfide electrolytes, oxide-based electrolytes exhibit sufficient lithium-ion mobility to match the performance of LIBs. However, their widespread adoption faces hurdles, primarily due to their mechanical properties. The hardness and brittleness of oxide electrolytes hinder the formation of dense and conductive interfaces without resorting to high-temperature sintering, which can lead to resistive reaction layers at the electrode/electrolyte junctions. Ongoing research aims to overcome these challenges by optimizing processing techniques and materials interfaces to enhance performance while retaining the advantages of oxide electrolytes. These electrolytes are mainly classified into perovskite, NASICON-type, LISICON-type (structure shown in Fig. 4), garnet, and LiPON groups, each having a distinct structure. Omid Sharifi et al. [70] investigated Ga-doped $Li_7La_3Zr_2O_{12}$ (LLZO) SEs (structure shown in Fig. 4) for solid-state LIBs using the sol-gel combustion method. The research revealed that doping with 0.2 mol of Ga effectively stabilized the cubic phase, achieving the maximum overall ionic conductivity of 5.85×10^{-3} S cm^{-1} at 20 °C, along with a grain boundary activation energy (Ea) of 0.39 eV. Santosh Kumar Pal et al. [71] investigated NASICON-type $Na_3Zr_2Si_2PO_{12}$ SEs doped with Sc^{3+} and Yb^{3+} to enhance Na-ion conductivity for SIBs. Their study found that replacing 16.5 at.% of Zr^{4+} with Sc^{3+} increased room-temperature (RT) ionic conductivity from 0.61×10^{-3} to 0.96×10^{-3} S cm^{-1}, and further substituting 11.11 at.% of Sc^{3+} with Yb^{3+} raised it to 1.62×10^{-3} S cm^{-1}. This improvement was attributed to microstructure refinement, elimination of ZrO_2 impurities, and lattice expansion, making the material promising for solid-state SIBs. The figure shows the Na-ion migration pathway.

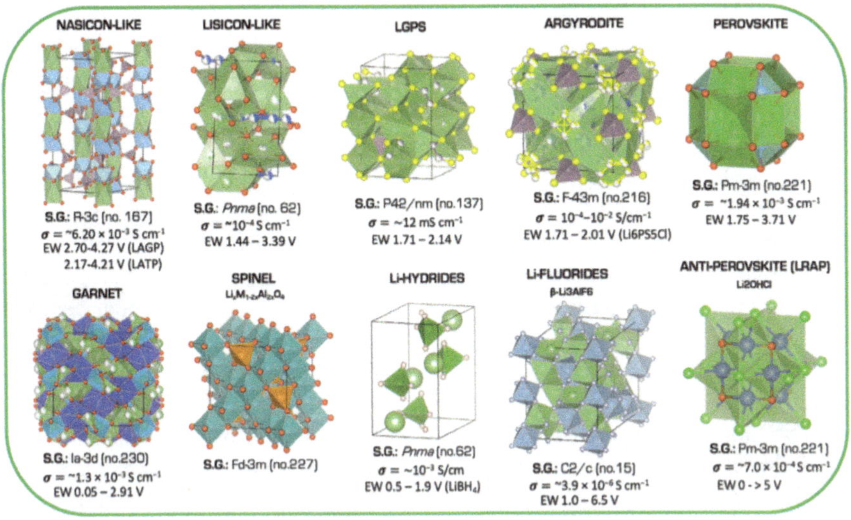

Fig. 4 Structural types, ion transport, and electrochemical window (EW) properties of inorganic solid electrolytes. Reprinted with the permission from Ref. N. V. Kireeva et al., "Modeling ionic conductivity and activation energy in garnet-structured solid electrolytes: The role of composition, grain boundaries and processing," *Eur Polym J.*, 2016 © Elsevier

5.1.2 Perovskite-Type Electrolytes

The perovskite-type materials (structure shown in Fig. 4) are characterized by a general formula of ABX_3, with corner-shared BX_6 octahedra forming a three-dimensional framework, providing a highly flexible lattice structure for ion conduction. The well research perovskite-type electrolyte is $Li_{3x}La_{2/3}$-$xTiO_3$ (LLTO). Because this shows the highest ionic conductivity 1×10^{-3} S/cm with $x = 0.1$ [27]. Lithium/sodium ions move stepwise through interconnecting pathways in the lattice, positioned next to the A-site defect, which is again surrounded by oxygen atoms [72]. The major structural features affecting the size of diffusion bottlenecks, and hence the activation energy for ion transport, include octahedral tilting and variation in lattice volume. Although these materials provide benefits such as chemical stability and compatibility with various battery architectures, challenges related to grain boundary resistance along with issues about interfaces with lithium metal electrodes require immediate improvements. Perovskite-type SEs represent a vast promise for safer and more powerful energy storage systems, as this research continues to further optimize their structural and electrochemical properties. Frederick Z.T. Yang et al. [73] studied the perovskite-type sodium-ion SE ($Na_{1/2-x}La_{1/2}$-$xSr_{2x}ZrO_3$) and reported a highest ionic conductivity of 1.025×10^{-5} S cm^{-1} at RT. Maycol F. Mena et al. [74] investigated the electrochemical performance of perovskite-type $Li_{0.3}La_{0.57}TiO_3$ (LLTO) SEs by optimizing the synthesis parameters of a sol-gel process and found ionic conductivity of 0.54×10^{-5} S cm^{-1} at 30 °C.

5.1.3 Sulfide-Based Electrolytes

Sulfide electrolytes exhibit high ionic conductivity ranging from 10^{-4} to 10^{-3} S cm^{-1} at RT due to the lower electronegativity of sulfur and the larger radius of sulfur ions. The weaker bonding strength of Li-S compared to Li-O allows for more free moving lithium ions in sulfide electrolytes. Structurally, the sulfide electrolytes can be categorized into glass, glass-ceramics, and crystalline types. Glassy sulfides like lithium phosphorus sulfide (LPS) demonstrate ion conductivity around 10^{-4} S cm^{-1}, with higher ionic conductivities compared to crystalline sulfides due to the absence of grain boundary resistance. Addition of lithium halide or ortho-oxo salts enhances ionic conductivity to around 10^{-3} S cm^{-1}. Glass-ceramics exhibit high ionic conductivities attributed to superionic phases within the glass. Crystalline sulfides like Li_6PS_5X and LISICON show promising ionic conductivities up to 10^{-2} S cm^{-1} at RT [74, 75]. The $Li_{10}GeP_2S_{12}$ type SEs shows high ionic conductivity >10^{-2} S cm^{-1} at RT and is able to compete with LE. Despite their high conductivity, sulfide electrolytes face challenges from interfacial issues, including side reactions with electrodes, poor electrolyte–electrode contact, and Li dendrite growth. Strategies like interlayer materials and cathode coatings are employed to mitigate these challenges and improve interfacial stability, enhancing the overall performance of SSBs utilizing sulfide electrolytes. Meirong Li et al. [76] investigated $Li_{3.25}Ge_{0.25}P_{0.75}S_4$ (LGPS)-based electrolytes and found that a lithium-sulfur full cell using an LGPS electrolyte exhibited excellent rate performance, delivering 1183 mAh g^{-1} at 0.2 C and 719 mAh g^{-1} at 0.5 C. Additionally, the cell maintained a capacity of 588 mAh g^{-1} after 50 cycles at 0.1 C. Similarly, sulfide-based electrolytes are also used in SSBs. Han Nguyen et al. [75] studied the Na_3PS_4 SE for sodium SSBs and found that it exhibits high ionic conductivity of ~0.2 mS cm^{-1}. Additionally, a sodium SSB using Na_3PS_4 delivered a capacity of 185 mAh g^{-1} on the first discharge and demonstrated excellent rate performance with a TiS_2 cathode and a $Na_{15}Sn_4$ anode.

5.2 *Polymer Electrolytes*

Compared with inorganic SEs, polymer electrolytes (PEs) provide numerous benefits, including flexibility, ease of processing, and cost-effectiveness [75, 77]. These characteristics facilitate additional advantages, such as close electrode/electrolyte contact, uniform coatings, and the ability to fabricate continuous components using roll-to-roll techniques. The PEs exhibit numerous advantages over LEs and inorganic SEs, including improved safety, enhanced flexibility, processability, and better ability to buffer electrode volume expansion during charge/discharge cycles. In PEs, dendrite growth can be minimized or suppressed under certain conditions, especially in solvent-free formulations. Polyethylene oxide (PEO)-based PEs were initially investigated by Fenton et al. in 1973, demonstrating their ionic conductivity as PEO hosts facilitate ion migration through alkali salts [78]. Subsequent research over the past two decades has focused on developing new PEs, understanding ionic transfer

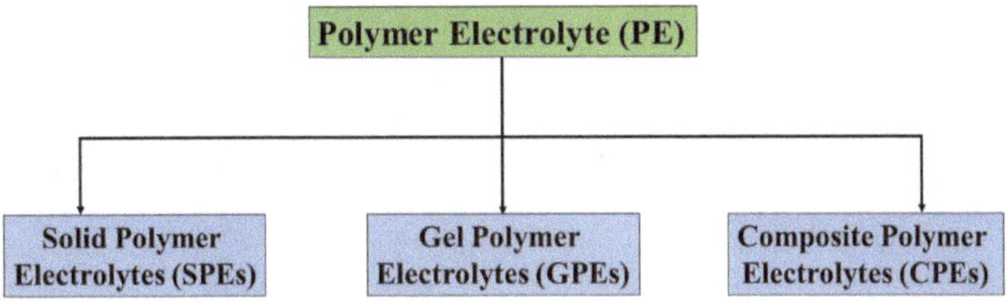

Fig. 5 Classification of PE based on their composition

mechanisms, and optimizing physical and chemical properties, as well as the EEI [79–81]. Initially, solid polymer electrolytes (SPEs), also known as dry SPEs, utilized polymer matrices and salts. However, SPEs suffered from drawbacks such as poor ionic conductivity and low transference numbers. For preparing PEs, PEO and its derivatives have emerged as optimal candidates due to their ability to dissolve lithium salt ions through ether oxygen coordination within the PEO matrix. Over the past two decades, advancements in LIB commercialization have centered around dry SPEs, aiming to enhance ionic conductivity, ion transfer numbers, and EEIs. Other polymer electrolyte systems, including different polymer hosts, salts, polymer-in-salt systems, and single-ion polymer systems, have also been developed. Progress in PEs has led to strategies such as the incorporation of ionic liquids and fillers to improve ionic conductivity. The development of gel polymer electrolytes (GPEs) is the result of this progress, which combine solid-like strength with liquid-like properties to achieve high ionic conductivity, enhanced transference numbers, and improved EEIs. GPEs can achieve ionic conductivities up to 10^{-3} S cm^{-1}, two orders of magnitude higher than SPEs, albeit with some compromise in mechanical strength. This drawback can be mitigated by incorporating fillers into GPEs, enhancing both their electrochemical and mechanical properties. Based on composition, PEs are classified (Fig. 5) into three main types:

5.2.1 Solid Polymer Electrolytes (SPEs)

Based on their composition, SPEs consist of a polymer matrix and a dissolved lithium or sodium salt. These PEs facilitate ion transport through ionic conductivity. Polymers serve as excellent hosts for dissolving salts while providing electrochemical stability. Among the earliest investigated SPEs, PEO attracted significant attention. However, to enhance ionic conductivity, researchers have explored other polymers with lower glass transition temperatures (Tg), such as polypropylene oxide (PPO) [82, 83], poly[bis(methoxy-ethoxy-ethoxy)phosphazene] (MEEP) [6, 84, 85], and polysiloxane (PSi) [86–88]. These polymers exhibit an amorphous nature at ambient temperatures but generally suffer from poor ionic conductivity. Poly(acrylonitrile) (PAN)-based SPEs exhibit an ionic conductivity of approximately 10^{-3} S cm^{-1} at RT, an ionic transference number of 0.6 [89]. However, they suffer from poor EEI stability.

Similarly, poly(methyl methacrylate) (PMMA) has been explored for LIBs due to its promising properties, but it lacks mechanical strength. Research has also been conducted on poly(vinyl chloride) (PVC), but these electrolytes generally exhibit poor ionic conductivity and limited compatibility with lithium metal anodes. SPEs are ion-rich systems where ion transport occurs primarily through the amorphous regions of the polymer, facilitated by polymer chains segmental motion. The efficiency of ion transference in PEs is primarily governed by two factors: the amorphous nature of the polymer matrix and its glass transition temperature (Tg). Enhancing ionic conductivity in SPEs can be achieved by increasing the amorphous nature of the polymer or reducing its Tg. This can be accomplished through copolymerization, crosslinking, or polymer blending (Fig. 6).

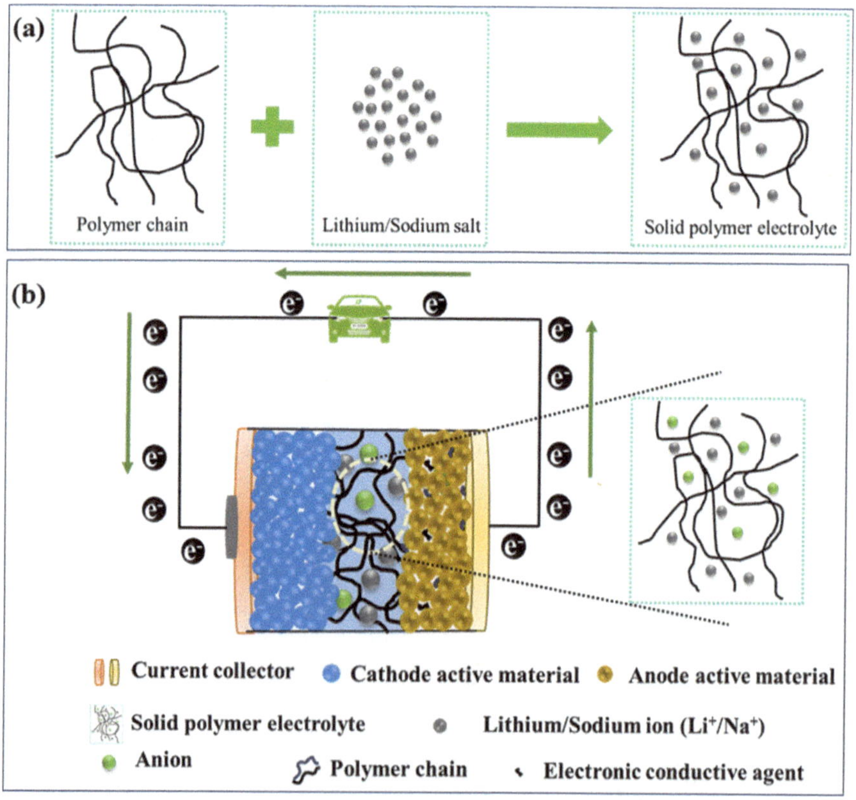

Fig. 6 **a** Conceptual representation of a solid polymer electrolyte, **b** schematic representation of a lithium/sodium solid polymer electrolyte battery during the discharge process [90, 91]

5.2.2 Gel Polymer Electrolyte (GPE)

The GPE was first introduced by Feuillade and Perche in 1975 [92]. It is also known as plasticized GPE, where the addition of a plasticizer results in a gel-like polymer electrolyte or a PE swollen in a LE. GPE exhibits properties of both solid and LEs, offering improved EEI, high ionic conductivity (10^{-3}–10^{-2} S cm^{-1}), good safety, and excellent chemical, mechanical, and electrochemical properties. The presence of a plasticizer in the polymer matrix reduces crystallinity, thereby enhancing segmental motion, which facilitates ion transport [93]. Figure 9 illustrates that how the plasticizer impacts on the polymer chain and improves the lithium/sodium transport. Shenhua Song et al. [94] studied on PVDF-HFP/$LiBF_4$/EC + PC based GPE which exhibited high ionic conductivity (8.1×10^{-4} S cm^{-1}), a wide ESW (4.6 V), and excellent mechanical properties (Young's modulus: 36.8 MPa, breaking strain: 320%). When used in a Li-ion battery, it delivered a high initial discharge capacity of 165.1 mAh g^{-1} at 0.1 C, with only 4.8% capacity fading after 120 cycles, demonstrating superior cycling stability compared to other PVDF-HFP-based GPEs. Wu et al. [95] studied a GPE using PVDF-HFP as polymer and 1 M $NaClO_4$ salt (in solvent EC: DEC: DMC = 1: 1: 1 v/v as plasticizer). The GPEs show the ionic conductivity 6.0×10^{-4} S cm^{-1} at RT. Fig. 7 illustrates that how the plasticizer impacts on the polymer chain and improves the lithium/sodium transport. (a) Ion mobility in SPE is improved through interactions between polymer chains and the plasticizer, which enhances polymer chain flexibility. (b) Ion mobility is further enhanced due to the presence of the plasticizer, as lithium cations exhibit a greater affinity for polar solvents than for polymer side chains [96].

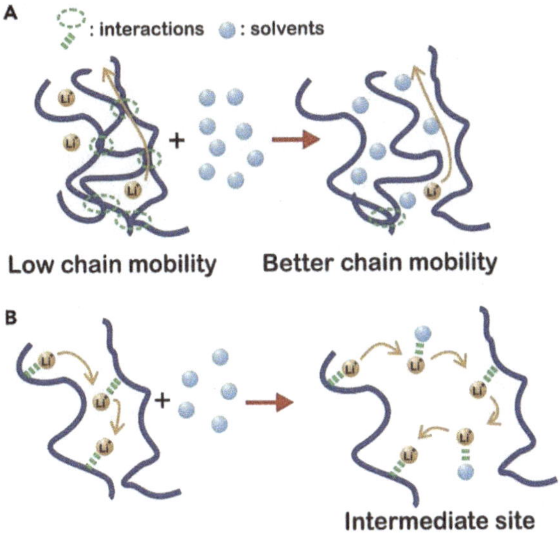

Fig. 7 Structural changes due to solvent incorporation impacting Li^+ ion mobility in polymer electrolytes [96]

5.2.3 Composite Polymer Electrolytes (CPEs)

In CPEs, inorganic fillers are incorporated into a polymer containing a dissolved salt. These fillers perform an essential function in improving the electrochemical and mechanical properties of the electrolyte. The concept of CPEs emerged as a strategy to overcome the limitations of conventional SPEs, such as poor ionic conductivity and low mechanical stability. The ionic conductivity of CPEs in the range 10^{-7} S cm^{-1} to 10^{-4} S cm^{-1} [97, 98]. In GPEs, plasticizers improve ionic conductivity and the transference number but reduce mechanical strength. The incorporation of fillers such as SiO_2, TiO_2, Al_2O_3, or MgO into SPEs/GPEs enhances mechanical strength, transport properties, and electrochemical performance. Additionally, active fillers like Li_3N, $LiAlO_2$, $NaAlO_2$, and NASICON-type materials further improve electrochemical properties, ionic conductivity, and the transference number. Qingya Guo et al. [99] prepared a CPE by incorporating 2 wt% modified $Li_{6.4}La_3Zr_{1.4}Ta_{0.6}O_{12}$ (LLZTO) filler into PEO, demonstrating an enhanced ionic conductivity of 1.1×10^{-4} S cm^{-1}, which is twice that of unmodified LLZTO/PEO and 20 times higher than pure PEO electrolyte. The CPE also exhibited high oxidation stability (4.8 V), an ionic transference number of 0.34, and improved interface stability with the lithium anode. As a result, $LiFePO_4$|Li ASSBs using this CPE achieved a capacity of 152.3 mAh g^{-1} with 90.35% retention over 500 cycles at 0.5 C (60 °C), indicating excellent cycling stability. Figure 8a illustrates the solid-state battery, while Fig. 8b demonstrates the device application by powering an LED light.

6 Interfacial Phenomena in Solid-State Batteries

The "interfacial phenomena" in the SSBs refers the various physical and chemical process at the boundaries between the SE and the electrodes. This phenomenon is also known as EEI. These processes can affect ionic conductivity, interfacial resistance, and electrochemical stability at these contact points, as well as lead to interphase formation, space-charge layers, lithium dendrite growth, and changes in contact quality as a result of volume variations during cycling. All of these processes can limit the lifespan of the battery and impede effective ion transport [100]. The EEI of SSBs can be understood through the SE interface (SEI) and cathode–electrolyte interface (CEI) formed on the anode and cathode sides, respectively. The SEI formation is depended on the physical–chemical properties of anode materials like particle size, pore size, basal-to-edge-plane ratio, crystallinity degree, and surface composition [101, 102]. The EEI in SSBs can be understood with the help of the diagram shown in Fig. 9.

The main strategies to mitigate EEI issues include the use of interfacial protection layers, which can help reduce large interfacial resistance and prevent Li/Na dendrite growth. Important modifications to the SE are necessary to improve the cycling stability and interfacial conductivity of Li/Na metal anodes. Additionally,

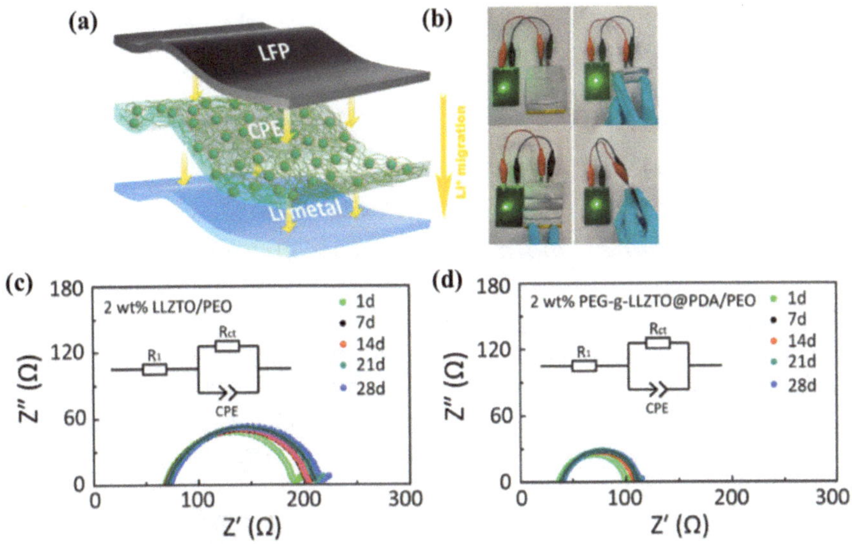

Fig. 8 **a** Schematic representation of the ASSB structure and **b** an illustration demonstrating the battery's application in powering LED., **c** Nyquist plot of Li|2 wt% LLZTO/PEO|Li, **d** Nyquist plot of Li|2 wt% PEG-g-LLZTO@PDA/PEO|Li symmetric cells after various storage times at RT. Reprinted with permission from Q. Guo et al. "Poly(ethylene glycol) brush on Li6.4La3Zr1.4Ta0.6O12 toward intimate interfacial compatibility in composite polymer electrolyte for flexible all-solid-state lithium metal batteries," *J Power Sources*, 2021, © Elsevier

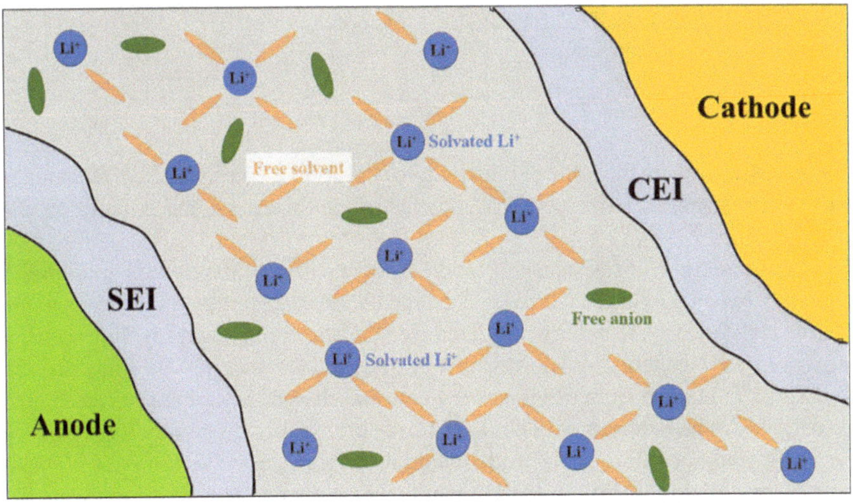

Fig. 9 Illustration of electrode–electrolyte interfaces (EEIs), including the solid–electrolyte interface (SEI) on the anode side and the cathode–electrolyte interface (CEI) on the cathode side

the modification of anode and cathode materials is essential for enhancing the EEI, electrochemical properties, and overall cycling stability [103].

7 Electrochemical Characterization Techniques

7.1 Electrochemical Impedance Spectroscopy (EIS)

EIS is a highly valuable technique for analyzing ionic conductivity, charge transfer resistance, and interfacial properties in electrochemical systems, particularly in ASSBs. The impedance spectrum obtained from EIS can be divided into three distinct regions: high-frequency bulk electrolyte resistance, mid-frequency charge transfer at interfaces, and low-frequency Warburg diffusion. For ionic conductivity measurements, the electrolyte is sandwiched between stainless steel (SS) electrodes, and a small voltage is applied across the cell while performing a frequency sweep from 1 Hz to 10^6 Hz. The resulting graph, which plots real impedance against imaginary impedance, is known as a Nyquist plot or Cole-Cole plot. From this plot, the bulk resistance (R_b) of the SE is determined. R_b arises due to the ion flow through the interior of the electrolyte material. It reflects how difficult for ions (Li^+/Na^+) to move through the solid lattice of the electrolyte. This resistance is mainly influenced by three factors: (1) availability of vacant sites and the energy barrier for hopping from one site to another. (2) Concentration of charge carriers. (3) Temperature. The following formula is used to compute the ionic conductivity (σ):

$$\sigma = \frac{\mathrm{d}}{R_b \times A}.$$

Here, σ represents the ionic conductivity of the solid electrolyte (SE), R_b denotes the bulk resistance, d corresponds to the electrolyte thickness, and A indicates the contact area.

The charge transfer resistance of the SSBs occurs due to the lithium/sodium-ion transfer between the SE to electrode. The high charge transfer resistance leads to the internal resistance in the battery which reduced the power output and cycling stability and increased the energy loss during the charging and discharging. By analyzing the Nyquist plot obtained from EIS measurements, the charge transfer resistance (R_{ct}) of the battery can be determined. The R_{ct} depends on the surface morphology, particle size, and decomposition of electrode materials. Figure 8c and d show the Nyquist plots of the cell. In these plots, the semicircle that starts on the X-axis represents the R_b, denoted as R_1. The point where the semicircle intersects the X-axis corresponds to the R_{ct} of the cell.

7.2 Cyclic Voltammetry (CV)

Cyclic voltammetry is among the most widely used electrochemical methods for battery. It finds uses in the investigation of oxidative and reductive processes. This technique employs the measurement of current flow generated during a redox reaction. It addresses changing voltage between the electrodes. The CV applies to inferring redox properties, charge–discharge mechanisms, and the electrochemical stability of materials. The reversible, quasi-reversible, and irreversible behavior in an electrochemical system can be discerned using cyclic voltammetry.

A CV is a nitrogenous "duck-shaped" plot originating from the CV experiment with solution reactants (Fig. 10). The scan rate dependence of peak heights in cyclic voltammograms enables one to monitor the predominant control mechanism for electron transfer. For cathode materials like $LiFePO_4$, CV evaluates their redox behavior under practical conditions at slow scan rates, mimicking real battery operations, with temperature-dependent studies showing that higher temperatures favor ion diffusion and charge transfer. Similarly, for anode materials, CV reveals SEI formation via irreversible peaks and distinguishes charge storage mechanisms, providing a pathway for better alternatives to graphite.

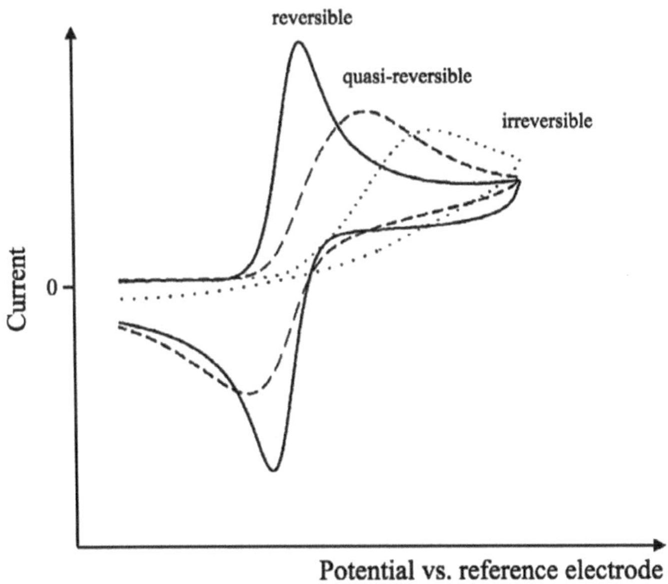

Fig. 10 Representative cyclic voltammogram showing reversible **a**, quasi-reversible **b**, and irreversible **c** behavior in electrochemical systems [104]. Reprinted with permission from J. Izquierdo and C. Kranz, "Electrochemical techniques for investigating redox active macromolecules," *Eur Polym J.*, 2016 © Elsevier

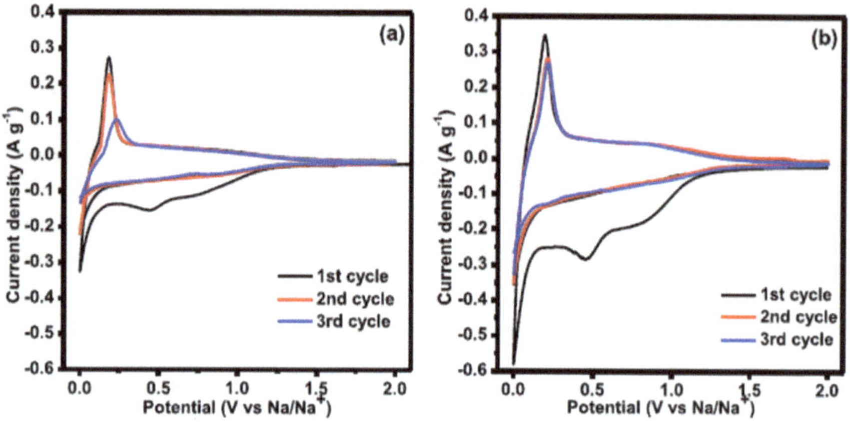

Fig. 11 CV responses of cell; **a** EHC/PE/Na and **b** EHC/LE/Na. Reprinted with permission from S.A. Hashmi et al., "Hard carbon anode derived from pre-treated bio-waste sugarcane bagasse for high-capacity sodium-ion battery fabricated with biodegradable porous polymer electrolyte," *J Energy Storage*, 2024 © Elsevier

In general, this makes CV a crucial driver for understanding reaction kinetics, ion diffusion, and material stability that play an important role in optimizing SSBs components and enhancing their performance.

The oxidation/reduction behavior of all electrochemical cells (EHC/PPE/Na and EHC/LE/Na) was investigated by CV responses in the potential range of 1 mV–2 V at a scan rate of 0.1 mV s^{-1}, as shown in Fig. 11a and b. The ethanol-soaked hard carbon (EHC) powder electrodes exhibit two prominent irreversible cathodic peaks at approximately 0.8 V and 0.5 V with both the LE and porous polymer electrolyte (PPE). The peak around 0.8 V corresponds to the decrease of oxygen-containing groups adsorbed on the electrode surface, while the peak near 0.5 V is attributed to electrolyte decomposition and the subsequent formation of the SE interphase (SEI) at the EHC electrolyte interface [105]. These irreversible cathodic peaks are significantly diminished in the following cycles, indicating substantial capacity loss after the first cycle.

7.3 *Galvanostatic Charge–Discharge (GCD)*

GCD is a technique to measure the capacity of the battery. In this technique, a constant current applied for charging/discharging the cell and voltage of the cell monitored over the time to analyze the charge/discharge behavior of the cell. Figure 12a and b shows the GCD curves of the cell using a PPE (poly-ε-caprolactone + $NaClO_4$ + EC: PC) and a LE (1 M $NaClO_4$ + EC: PC (1:1 v/v)), with ethanol-soaked hard carbon (EHC) as the anode. From the GCD curves, it is evident that the cell using electrolyte as the polymer electrolyte delivers a capacity (212 mAh g^{-1}) comparable

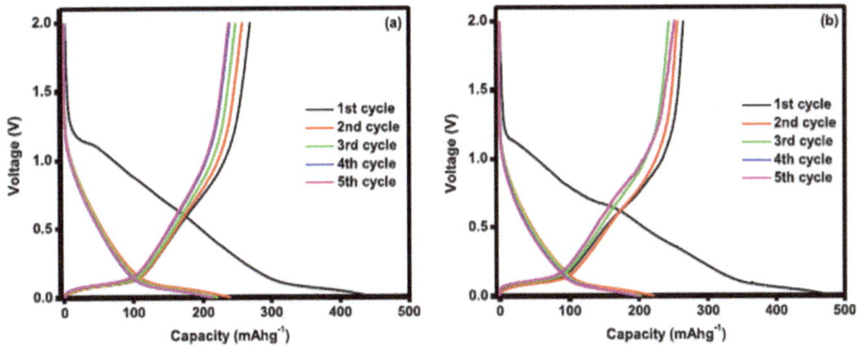

Fig. 12 Initial GCD profiles of all the cells recorded at current rate of 0.05 A g^{-1}; **a** EHC/PE/Na and **b** EHC/LE/Na. Reprinted with permission from S.A. Hashmi et al., "Hard carbon anode derived from pre-treated bio-waste sugarcane bagasse for high-capacity sodium-ion battery fabricated with biodegradable porous polymer electrolyte," *J Energy Storage*, 2024 © Elsevier

to that of the cell using the LE (241 mAh g^{-1}). This demonstrates the potential of SSBs to achieve similar performance to conventional LE-based batteries [105].

8 Conclusion

In recent years, rapid progress has been made in the development of Li/Na-ion SSBs, with increased efforts in both academic and industrial sectors aimed at enhancing their electrochemical performance. However, compared to conventional LEs, there are several challenges that must be addressed before these batteries can be widely adopted for industrial applications. These challenges include reducing bulk resistance, minimizing grain boundary resistance at the interface between SEs and electrodes, and improving the long-term stability of SE structure and interface. The interfacial properties of SSBs play a crucial role in their performance, primarily due to Li/Na-ion diffusion within the SE bulk. In practical applications, the stability and electrochemical compatibility of the interface between the SE and electrode are essential to minimize interfacial resistance. Future research on SSBs should focus on optimizing SEs to reduce interfacial resistance and enhance conductivity/ ionic transport. Additionally, the performance of SSBs is significantly influenced by the characteristics of the anode and cathode materials. Therefore, it is necessary to develop anode and cathode materials that exhibit suitable and high theoretical capacities when combined with SEs.

Acknowledgments The authors gratefully acknowledge the financial support received from the MOE-SPARC Program (Project No. SPARC/2019-2020/P2531/SL) and the Anusandhan National Research Foundation (ANRF) through the Core Research Grant (CRG) Scheme (File No. CRG/ 2023/003048).

References

1. Hwang JY, Myung ST, Sun YK (2017) Sodium-ion batteries: present and future. Chem Soc Rev 46:3529–3614
2. Shalu SK, Chaurasia SK, Singh SC (2013) Thermal stability, complexing behavior, and ionic transport of polymeric gel membranes based on polymer PVdF-HFP and ionic liquid, [BMIM][BF_4]. J Phys Chem B 117:897–906
3. Prajapati AK, Bhatnagar A (2023) A review on anode materials for lithium/sodium-ion batteries. J Energy Chem 83:509–540
4. Chaudhary ML, Patel R, Bhatnagar A, Gupta RK (2025) MXene-derived nanostructures: pioneering anodes for lithium-ion and sodium-ion batteries. J Mater Chem A 13(35), 28718–28748
5. Zhao W, Yi J, He P, Zhou H (2019) Solid-state electrolytes for lithium-ion batteries: fundamentals, challenges and perspectives. Electrochem Energy Rev 2:574–605
6. Chayambuka K, Mulder G, Danilov DL, Notten PHL (2018) Sodium-ion battery materials and electrochemical properties reviewed. Adv Energy Mater 8:1–49
7. Joshi A, Mishra DK, Singh R, Zhang J, Ding Y (2025) A comprehensive review of solid-state batteries. Appl Energy 386:125546
8. Pasta M, Deshpande V, Doerrer C et al (2020) 2020 roadmap on solid-state batteries. J Phys Energy 2:32008
9. Machín A, Morant C, Márquez F (2024) Advancements and challenges in solid-state battery technology: An in-depth review of solid electrolytes and anode innovations. Batteries 10:29
10. Richards WD, Miara LJ, Wang Y, Kim JC, Ceder G (2016) Interface stability in solid-state batteries. Chem Mater 28:266–273
11. Yu X, Chen R, Gan L, Li H, Chen L (2023) Battery safety: from lithium-ion to solid-state batteries. Engineering 21:9–14
12. Hansen S, Quiroga-González E, Carstensen J, Adelung R, Föll H (2017) Size-dependent physicochemical and mechanical interactions in battery paste anodes of Si-microwires revealed by fast-Fourier-transform impedance spectroscopy. J Power Sources 349:1–10
13. Perea A, Dontigny M, Zaghib K (2017) Safety of solid-state Li metal battery: solid polymer versus liquid electrolyte. J Power Sources 359:182–185
14. Chen S, Wen K, Fan J, Bando Y, Golberg D (2018) Progress and future prospects of high-voltage and high-safety electrolytes in advanced lithium batteries: from liquid to solid electrolytes. J Mater Chem A 6(25), 11631–11663
15. Minnmann P, Strauss F, Bielefeld A et al (2022) Designing cathodes and cathode active materials for solid-state batteries. Adv Energy Mater 12:2201425
16. Kaiser N, Spannenberger S, Schmitt M, Cronau M, Kato Y, Roling B (2018) Ion transport limitations in all-solid-state lithium battery electrodes containing a sulfide-based electrolyte. J Power Sources 396:175–181
17. Shi T, Tu Q, Tian Y et al (2020) High active material loading in all-solid-state battery electrode via particle size optimization. Adv Energy Mater 10:1902881
18. Minnmann P, Quillman L, Burkhardt S, Richter FH, Janek J (2021) Editors' choice-quantifying the impact of charge transport bottlenecks in composite cathodes of all-solid-state batteries. J Electrochem Soc 168:40537
19. Park J, Kim KT, Oh DY et al (2020) Digital twin-driven all-solid-state battery: unraveling the physical and electrochemical behaviors. Adv Energy Mater 10:2001563
20. Zhang W, Weber DA, Weigand H et al (2017) Interfacial processes and influence of composite cathode microstructure controlling the performance of all-solid-state lithium batteries. ACS Appl Mater Interfaces 9:17835–17845
21. Zuo W, Innocenti A, Zarrabeitia M, Bresser D, Yang Y, Passerini S (2023) Layered oxide cathodes for sodium-ion batteries: storage mechanism, electrochemistry, and techno-economics. Acc Chem Res 56:284–296
22. Lyu Y, Wu X, Wang K et al (2021) An overview on the advances of $LiCoO_2$ cathodes for lithium-ion batteries. Adv Energy Mater 11:2000982

23. Langdon J, Manthiram A (2021) A perspective on single-crystal layered oxide cathodes for lithium-ion batteries. Energy Storage Mater 37:143–160
24. Kanwade A, Gupta S, Kankane A et al (2022) Transition metal oxides as a cathode for indispensable Na-ion batteries. RSC Adv 12:23284–23310
25. Sayahpour B, Hirsh H, Parab S, Nguyen LHB, Zhang M, Meng YS (2022) Perspective: design of cathode materials for sustainable sodium-ion batteries. MRS Energy Sustain 9:183–197
26. Peng D, Li R, Xu K, Si R, Zhang Z, Hu YS (2025) A low-strain lithium cathode material $Li_{2-2x}Fe_{1+x}Cl_4$ for halide-based all-solid-state batteries. ACS Energy Lett 10:1421–1429
27. Stramare S, Thangadurai V, Weppner W (2003) Lithium lanthanum titanates: a review. Chem Mater 15:3974–3990
28. Meng X, Wang J, Li L (2023) Layered-oxide cathode materials for fast-charging lithium-ion batteries: a review. Molecules 28:4007
29. Bhuvaneswari S, Varadaraju UV, Gopalan R, Prakash R (2019) Structural stability and superior electrochemical performance of Sc-doped $LiMn_2O_4$ spinel as cathode for lithium ion batteries. Electrochim Acta 301:342–351
30. Nwachukwu IM, Nwanya AC, Ekwealor ABC, Ezema FI (2024) The potentials of $LiMnPO_4$ cathode material for aqueous Li-ion batteries: An investigation into solid state and green chemistry approaches. Appl Surf Sci Adv 19:100537
31. Wani TA, Suresh G (2021) A comprehensive review of $LiMnPO_4$ based cathode materials for lithium-ion batteries: current strategies to improve its performance. J Energy Storage 44:103307
32. Tran DT, Dong H, Walck SD, Zhang SS (2015) Pyrite FeS_2-C composite as a high capacity cathode material of rechargeable lithium batteries. RSC Adv 5:87847–87854
33. Hosseini SM, Varzi A, Ito S, Aihara Y, Passerini S (2020) High loading CuS-based cathodes for all-solid-state lithium sulfur batteries with enhanced volumetric capacity. Energy Storage Mater 27:61–68
34. Liang J, Li X, Kim JT et al (2023) Halide layer cathodes for compatible and fast-charged halides-based all-solid-state Li metal batteries. Angew Chemie 135:e202217081
35. Xu J, Pollard TP, Yang C et al (2023) Lithium halide cathodes for Li metal batteries. Joule 7:83–94
36. Xiang X, Lu Y, Chen J (2017) Advance and prospect of functional materials for sodium ion batteries. Acta Chim Sin 75:154–162
37. Xu C, Zhao J, Yang C, Hu YS (2023) Polyanionic cathode materials for practical Na-ion batteries toward high energy density and long cycle life. ACS Cent Sci 9:1721–1736
38. Chari A, El Ouardi K, Tayoury M et al (2022) $Na_3CoZr(PO_4)_3$ as high-voltage cathode material for sodium-ion batteries. J Power Sources 548:232046
39. He J, Tao T, Yang F, Sun Z, Huang H (2022) Phase-manipulated hierarchically core-shell $Na_3(VO_{1-x}PO_4)_2F_{1+2x}(0\leq x\leq 1)@Na_3V_2(PO_4)_3$ and its synergistic effect with conformally wrapped reduced graphene oxide framework towards high-performance cathode for sodium-ion batteries. Mater Today Phys 27:100813
40. Liao JY, Hu Q, Zou BK, Xiang JX, Chen CH (2016) The role of potassium ions in iron hexacyanoferrate as a cathode material for hybrid ion batteries. Electrochim Acta 220:114–121
41. Wilkinson D, Bhosale M, Amores M, Naresh G, Cussen SA, Cooke G (2021) A quinone-based cathode material for high-performance organic lithium and sodium batteries. ACS Appl Energy Mater 4:12084–12090
42. Goriparti S, Miele E, De Angelis F, Di Fabrizio E, Zaccaria RP, Capiglia C (2014) Review on recent progress of nanostructured anode materials for Li-ion batteries. J Power Sources 257:421–443
43. Xu W, Wang J, Ding F et al (2014) Lithium metal anodes for rechargeable batteries, energy. Environ Sci 7:513–537
44. Liang C, Gao M, Pan H, Liu Y, Yan M (2013) Lithium alloys and metal oxides as high-capacity anode materials for lithium-ion batteries. J Alloys Compd 575:246–256
45. Kamali AR, Fray DJ (2010) Review on carbon and silicon based materials as anode materials for lithium ion batteries. J New Mater Electrochem Syst 13:147–160

46. Kama AR, Fray DJ (2011) Tin-based materials as advanced anode materials for lithium ion batteries: a review. Rev Adv Mater Sci 27:14–24
47. Kim Y, Ha KH, Oh SM, Lee KT (2014) High-capacity anode materials for sodium-ion batteries. Chem-Eur J 20:11980–11992
48. Toki GFT, Hossain MK, Rehman WU, Manj RZA, Wang L, Yang J (2024) Recent progress and challenges in silicon-based anode materials for lithium-ion batteries. Ind Chem Mater 2:226–269
49. Asenbauer J, Eisenmann T, Kuenzel M, Kazzazi A, Chen Z, Bresser D (2020) The success story of graphite as a lithium-ion anode material–fundamentals, remaining challenges, and recent developments including silicon (oxide) composites. Sustain Energy Fuels 4:5387–5416
50. Liao D, Kuang X, Xiang J, Wang X (2018) A silicon anode material with layered structure for the lithium-ion battery. J Phys Conf Ser 986:12024
51. Loaiza LC, Monconduit L, Seznec V (2020) Towards germanium layered materials as superior negative electrodes for Li-, Na-, and K-ion batteries. Batter Supercaps 3:417–426
52. Mou H, Xiao W, Miao C, Li R, Yu L (2020) Tin and tin compound materials as anodes in lithium-ion and sodium-ion batteries: a review. Front Chem 8:141
53. Ashuri M, He Q, Shaw LL (2023) Silicon oxides for Li-ion battery anode applications: toward long-term cycling stability. J Power Sources 559:232660
54. Chen T, Wu J, Zhang Q, Su X (2017) Recent advancement of SiO_x based anodes for lithium-ion batteries. J Power Sources 363:126–144
55. Nemaga AW, Lai SY, Nguyen T, Ulvestad A, Foss CEL (2025) Unveiling the advantages of silicon nitride nanoparticle anodes for enhanced cyclic stability, rate performance, and mechanical robustness. ACS Omega 10:2608–2615
56. Tripathi AM, Mitra S (2014) Tin sulfide (SnS) nanorods: structural, optical and lithium storage property study. RSC Adv 4:10358–10366
57. Jingjing W, Chao L, Jianfeng M et al (2015) Solid-state fabrication of SnS_2/C Nanospheres for high-performance sodium ion battery anode. ACS Appl Mater Interfaces 7:11476–11481
58. Boorboor Ajdari F, Asghari P, Molaei Aghdam A et al (2024) Silicon solid state battery: the solid-state compatibility, particle size, and carbon compositing for high energy density. Adv Funct Mater 34:2314822
59. Xiao Z, Wang C, Song L, Zheng Y, Long T (2022) Research progress of nano-silicon-based materials and silicon-carbon composite anode materials for lithium-ion batteries. J Solid State Electrochem 26:1125–1136
60. Müller C, Wang Z, Hofmann A et al (2023) Influences on reliable capacity measurements of hard carbon in highly loaded electrodes. Batter Supercaps 6:e202300322
61. Wu C, Yang Y, Zhang Y et al (2023) Hard carbon for sodium-ion batteries: progress, strategies and future perspective. Chem Sci 15:6244–6268
62. Palaniselvam T, Babu B, Moon H et al (2021) Tin-containing graphite for sodium-ion batteries and hybrid capacitors. Batter Supercaps 4:173–182
63. He J, Wei Y, Zhai T, Li H (2018) Antimony-based materials as promising anodes for rechargeable lithium-ion and sodium-ion batteries. Mater Chem Front 2:437–455
64. Ramireddy T, Xing T, Rahman MM et al (2015) Phosphorus–carbon nanocomposite anodes for lithium-ion and sodium-ion batteries. J Mater Chem A 3:5572–5584
65. Khan MA, Kang YM (2016) Synthesis and electrochemical performance of colloidal MoS_2 nanosheets as an anode material in sodium ion battery. J Energy Storage 7:252–257
66. Pan J, Wang N, Lv D, Dong W, Yang J, Huang F (2021) Layered structure $Na_2Ti_3O_7$ as a promising anode material for sodium-ion batteries. Adv Energy Sustainability Res 2:2000095
67. Zhu L, Yin X, Pan C et al (2022) Facile synthesis of nanorods $Na_2Ti_6O_{13}$ as anode materials for high-performance sodium ion batteries. J Alloys Compd 906:164306
68. Zhou M, Chen Y, Zhang Q et al (2019) $Na_3V_2(PO_4)_3$ as the sole solid energy storage material for redox flow sodium-ion battery. Adv Energy Mater 9:1901188
69. Ji Y, Honma T, Komatsu T (2023) Formation of sodium ion conductive $NaZr_2(PO_4)_3$ composite via liquid phase sintering method with sodium disilicate glass. Solid State Ionics 395:116213

70. Sharifi O, Golmohammad M, Soozandeh M, Mehranjani AS (2023) Improved Ga-doped $Li_7La_3Zr_2O_{12}$ garnet-type solid electrolytes for solid-state Li-ion batteries. J Solid State Electrochem 27:2433–2444
71. Pal SK, Saha R, Kumar GV, Omar S (2020) Designing high ionic conducting NASICON-type $Na_3Zr_2Si_2PO_{12}$ solid-electrolytes for Na-ion batteries. J Phys Chem C 124:9161–9169
72. Aguesse F, Roddatis V, Roqueta J et al (2015) Microstructure and ionic conductivity of LLTO thin films: influence of different substrates and excess lithium in the target. Solid State Ionics 272:1–8
73. Yang FZT, Peterson VK, Schmid S (2021) Composition and temperature dependent structural investigation of the perovskite-type sodium-ion solid electrolyte series $Na_{1/2-x}La_{1/2-x}Sr_{2x}ZrO_3$. J Alloys Compd 863:158500
74. Mena MF, Vásquez FA, Florentin O et al (2025) Electrochemical performance enhancement of perovskite-type $Li_{0.3}La_{0.57}TiO_3$ ceramic electrolyte by controlling synthesis parameters. J Eur Ceram Soc 45:116972
75. Zhang Q, Cao D, Ma Y, Natan A, Aurora P, Zhu H (2019) Sulfide-based solid-state electrolytes: synthesis, stability, and potential for all-solid-state batteries. Adv Mater 31:1901131
76. Li M, Frerichs JE, Kolek M et al (2019) Solid-state lithium-Sulfur battery enabled by Thio-LiSICON/polymer composite electrolyte and sulfurized Polyacrylonitrile cathode. Adv Funct Mater 30:1910123
77. Balo L, Gupta H, Singh VK, Singh RK (2019) Flexible gel polymer electrolyte based on ionic liquid EMIMTFSI for rechargeable battery application. Electrochim Acta 230:123–131
78. Fenton DE (1973) Complex of alkali metal ions with poly (ethylene oxide). Polymer (Guildf) 14:589
79. Berthier C, Gorecki W, Minier M, Armand MB, Chabagno JM, Rigaud P (1983) Microscopic investigation of ionic conductivity in alkali metal salts-poly (ethylene oxide) adducts. Solid State Ionics 11:91–95
80. Gorecki W, Andreani R, Berthier C et al (1986) NMR, DSC, and conductivity study of a poly (ethylene oxide) complex electrolyte: $PEO(LiClO_4)_x$. Solid State Ionics 18:295–299
81. Tarascon JM, Gozdz AS, Schmutz C, Shokoohi F, Warren PC (1986) Performance of Bellcore's plastic rechargeable Li-ion batteries. Solid State Ionics 86:49–54
82. Vachon C, Labreche C, Vallee A, Besner S, Dumont M, Prud'Homme J (1995) Microphase separation and conductivity behavior of poly (propylene oxide)-lithium salt electrolytes. Macromolecules 28:5585–5594
83. Andersson D, Carlsson P, Engberg D et al (1999) Modelling of segmental dynamics in polymer electrolyte $PPO-LiClO_4$, by surface fitting of quasi-elastic neutron scattering data. Phys B Condens Matter 266:126–130
84. Blonsky PM, Shriver DF, Austin P, Allcock HR (1984) Polyphosphazene solid electrolytes. J Am Chem Soc 106:6854–6855
85. Blonsky PM, Shriver DF, Austin P, Allcock HR (1986) Complex formation and ionic conductivity of polyphosphazene solid electrolytes. Solid State Ionics 18:258–264
86. Hooper R, Lyons LJ, Mapes MK, Schumacher D, Moline DA, West R (2001) Highly conductive siloxane polymers. Macromolecules 34:931–936
87. Snyder JF, Ratner MA, Shriver DF (2003) Ion conductivity of comb polysiloxane polyelectrolytes containing oligoether and perfluoroether sidechains. J Electrochem Soc 150:A1090
88. Kang Y, Lee W, Suh DH, Lee C (2003) Solid polymer electrolytes based on cross-linked polysiloxane-g-oligo (ethylene oxide): ionic conductivity and electrochemical properties. J Power Sources 119:448–453
89. Wang SH, Kuo PL, Hsieh CT, Teng H (2014) Design of poly (acrylonitrile)-based gel electrolytes for high-performance lithium ion batteries. ACS Appl Mater Interfaces 6:19360–19370
90. De Cachinho Cordeiro I, Li A, Lin B, Ma D, Xu L, Eh A, Wang W (2023) Solid polymer electrolytes for zinc-ion batteries. Batteries 9(7):343
91. Gupta A, Bhatnagar A, Rai DK (2026) Impact of 2D layered VS 2 on the electrochemical properties of gel polymer electrolytes for sodium-ion battery applications. Nanoscale 18(17), 9425–9443

92. Feuillade G, Perche P (1975) Ion-conductive macromolecular gels and membranes for solid lithium cells. J Appl Electrochem 5:63–69
93. Long L, Wang S, Xiao M, Meng Y (2016) Polymer electrolytes for lithium polymer batteries. J Mater Chem A 4:10038–10069
94. Song S, Wang J, Tang J, Muchakayala R, Ma R (2017) Preparation, properties, and Li-ion battery application of EC+ PC-modified PVdF-HFP gel polymer electrolyte films. Ionics (Kiel) 23:3365–3375
95. Yang YQ, Chang Z, Li MX, Wang XW, Wu YP (2015) A sodium ion conducting gel polymer electrolyte. Solid State Ionics 269:1–7
96. Foran G, Mankovsky D, Verdier N et al (2020) The impact of absorbed solvent on the performance of solid polymer electrolytes for use in solid-state lithium batteries. Iscience 23:101597
97. Leš K, Jordan CS (2020) Ionic conductivity enhancement in solid polymer electrolytes by electrochemical in situ formation of an interpenetrating network. RSC Adv 10:41296–41304
98. Liu W, Liu N, Sun J et al (2015) Ionic conductivity enhancement of polymer electrolytes with ceramic nanowire fillers. Nano Lett 15:2740–2745
99. Guo Q, Xu F, Shen L et al (2021) Poly(ethylene glycol) brush on $Li_{6.4}La_3Zr_{1.4}Ta_{0.6}O_{12}$ towards intimate interfacial compatibility in composite polymer electrolyte for flexible all-solid-state lithium metal batteries. J Power Sources 498:229934
100. Liu M, Song A, Zhang X et al (2025) Interfacial lithium-ion transportation in solid-state batteries: challenges and prospects. Nano Energy 136:110749
101. An SJ, Li J, Daniel C, Mohanty D, Nagpure S, Wood DL III (2016) The state of understanding of the lithium-ion-battery graphite solid electrolyte interphase (SEI) and its relationship to formation cycling. Carbon 105:52–76
102. Li Y, Liu M, Feng X et al (2021) How can the electrode influence the formation of the solid electrolyte Interface? ACS Energy Lett 6:3307–3320
103. Wang J, Chen L, Li H, Wu F (2023) Anode interfacial issues in solid-state Li batteries: mechanistic understanding and mitigating strategies. Energy Environ Mater 6:1–21
104. Izquierdo J, Kranz C (2016) Electrochemical techniques for investigating redox active macromolecules. Eur Polym J 83:428–449
105. Mohit SA (2024) Hashmi, "hard carbon anode derived from pre-treated bio-waste sugarcane bagasse for high capacity sodium-ion battery fabricated with bio-deFgradable porous polymer electrolyte,". J Energy Storage 83:11069

Key Electrode Materials for Solid-State Batteries

Sarathkumar Krishnan, Mayank K. Singh, and Dhirendra K. Rai

Abstract Solid-state batteries (SSBs) are considered a next-generation energy storage technology due to their potential to enhance safety and increase energy density as compared to conventional liquid-electrolyte-based lithium-ion batteries (LIBs). This chapter explores the key electrode materials and properties that are essential to the performance of SSBs, emphasizing their critical role in determining specific capacity, rate performance, and stability. This chapter starts with an introduction to electrode processes, outlining the fundamental mechanisms of ion transport and electron transfer. The discussion progresses to the critical properties and requirements for electrode materials that are suitable for SSBs. With a major focus on lithium-ion batteries, the chapter explores the advancements and possible challenges in cathode materials, particularly intercalation and conversion-type electrode materials, which offer significant improvements in energy density and cyclic stability. On the anode side, the chapter investigates the potential of metal anodes, carbon materials, and alloy anodes, particularly highlighting their performance and the problems they face in practical applications. The chapter also discusses the broader challenges faced by SSB technology with anodes, including poor ionic conductivity, interface compatibility, and dendrite formation. Finally, we expand our scope to other alkali-metal ion batteries, such as sodium (Na^+) and potassium (K^+), discussing their unique advantages and future research directions. Through a comprehensive review of recent advancements and ongoing research challenges, this chapter provides valuable insights into the key materials and strategies that will drive the future of solid-state battery technology, paving the way for building state-of-the-art energy storage technologies.

Keywords Solid-state batteries · Lithium metal anodes · Intercalation cathodes · Solid electrolytes · Electrode–electrolyte interfaces

S. Krishnan · M. K. Singh · D. K. Rai (✉)
Sustainable Energy and Environmental Materials (SEEM) Lab, Department of Metallurgical Engineering and Materials Science, Indian Institute of Technology Indore, Simrol, Indore, India
e-mail: dkrai@iiti.ac.in

A. K. Rai et al. (eds.), *Electrochemistry of Next-Generation Solid-State Batteries*, Green Energy and Technology, https://doi.org/10.1007/978-981-95-7603-6_3

1 Introduction to the Electrode Process

Solid-state batteries (SSBs), which represent a paradigm leap in battery technology, have garnered a substantial amount of interest because of the growing demand for efficient and sustainable energy storage systems. In contrast to traditional liquid-electrolyte batteries, SSBs utilize solid electrolytes, which provide improved safety, enhanced thermal stability, and the possibility of providing larger energy densities. As a result of these characteristics, SSBs are prospective candidates for a diverse array of applications that include renewable energy storage, portable electronics, and electric vehicles (EVs) [1, 2]. The development of novel electrode materials that are capable of satisfying the critical requirements of this cutting-edge technology is vital to the performance and success of solid-state batteries. The materials used for electrodes in solid-state batteries are essential for promoting ion transport and ensuring effective electrochemical reactions. The intricate relationship between the electrode and solid electrolyte plays a crucial role in determining the overall performance of the battery, which includes energy density, cycling stability, and power output. As a result, a comprehensive understanding of the essential mechanisms that regulate ion transport in solid electrodes and their interfaces is required [3, 4]. In addition, it is necessary to address critical obstacles to designing high-performance electrode materials. These issues include attaining high ionic and electronic conductivity, maintaining structural stability during long-term charge/discharge cycling, and reducing interfacial resistance and side reactions with solid electrolytes [5].

The recent advancements in cathode materials have played a crucial role in facilitating the progress of SSBs [3, 6]. Particularly, layered oxides, spinel compounds, and polyanionic materials exhibit significant potential owing to their capacity to provide high energy densities and work effectively at increased voltage levels. Nevertheless, the adaptation of these materials for solid-state systems presents a range of new complexities. This includes the necessity to enhance their compatibility with solid electrolytes, as well as addressing challenges such as phase transitions and the decomposition of electrolytes [7, 8]. In a similar way, the selection of anode materials significantly impacts the overall performance of SSBs. Despite lithium metal being recognized as an optimal anode owing to its substantial theoretical capacity, it is imperative to address the considerable challenges posed by dendrite formation and the constraints of interfacial stability, which continue to hinder its practical application [9]. Research is currently being conducted on alternative anode materials, such as alloy-based and conversion-type electrodes, to address and overcome these limitations. Various strategies, including interface engineering, material doping, and the advancement of composite structures, are currently being actively explored to address these challenges existing in electrode materials [10]. Consequently, it improves the performance and reliability of SSBs, thereby advancing their potential for commercial viability. The investigation of alternative alkali-metal ion systems, including sodium-ion and potassium-ion batteries, which are gaining significant interest beyond the realm of lithium-ion chemistry due to its vast abundance and affordability [11]. Nevertheless, the transition toward these chemistries requires the fabrication of specialized electrode materials capable of accommodating the larger ionic radii and distinct electrochemical properties of sodium and potassium ions [12]. Addressing these challenges could open new pathways for the widespread adoption of solid-state batteries in various applications.

The future of solid-state batteries (SSBs) largely depends on advancements in material design and engineering methods. Recent developments in computational modeling, in situ characterization techniques, and scalable fabrication methods are expected to accelerate the search for next-generation electrode materials. By addressing current limitations and utilizing emerging technologies, SSBs have the potential to transform the energy storage landscape, thereby meeting the growing demands of a sustainable and electrified world. This chapter explores the essential characteristics of electrode materials, providing an in-depth analysis of their various types and features, the challenges faced, and the mechanisms driving the progress of solid-state battery technology, as well as offering future perspectives.

2 Mechanism and Electrode Reactions

The operation of all-solid-state lithium-ion batteries (SSLBs) is similar to that of conventional lithium-ion batteries (LIBs) [13]. The schematics of the SSLBs and LIBs are shown in Fig. 1. The primary difference is the substitution of liquid electrolytes with solid electrolytes. During the discharge process, Li atoms oxidize and lose electrons at the anode. The resulting Li^+ ions then migrate from the anode to the cathode through the solid electrolyte. Meanwhile, the electrons left behind at the anode travel through an external circuit to reach the cathode, where they reduce the redox-active centers of the cathode material, aided by the incoming Li^+ ions from the anode. Since the overall free energy change during this discharging process (the oxidation of Li and the reduction of the cathode material) is negative, energy is generated in the form of electricity, which is used to power various devices. During charging, the process reverses: the redox-active centers on the cathode are oxidized, and Li^+ ions migrate from the cathode back to the anode, where they are reduced back to Li within the anode material [6, 14]. Unlike during discharge, the total free energy change during charge (the oxidation of the cathode material and the reduction of Li^+) is positive. Therefore, electrical energy must be supplied to drive this process.

Solid electrolytes fulfill two essential functions: they facilitate the transportation of lithium ions by utilizing the chemical potential difference between the electrodes while simultaneously providing an insulating barrier for electrons, thereby ensuring their passage through the external circuit. In contrast to liquid electrolytes, solid electrolytes significantly improve safety by reducing the chances of flammability and leakage. Ion migration and electron flow are facilitated by the free energy difference between the cathode and anode, which is the driving force behind the thermodynamic operation of SSLBs [15]. From a kinetic perspective, the effectiveness of ion transport relies on surpassing activation energy barriers, which are caused by the characteristics of the solid electrolyte and the interfaces between the electrode and electrolyte

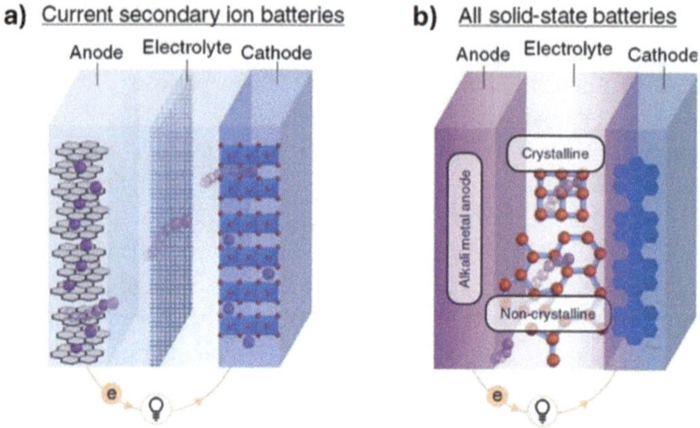

Fig. 1 Schematic representation of liquid and solid-state batteries. Adopted under creative commons CC BY 4.0 license from Ref. [13]

[16]. The solid-state design presents opportunities for increased safety and enhanced energy densities by enabling access to high voltage; however, it also brings various challenges, including the need to optimize ionic conductivity and reduce interfacial resistance. These factors highlight the necessity of customized materials and designs to attain the best performance in SSLBs.

Cathode Reactions in LIBs

Cathode reactions in batteries can be classified into two main types: (i) displacement (or conversion) reactions and (ii) insertion (or intercalation) reactions. In displacement reactions, the quantities of different phases within the electrode microstructure change, while the composition of each individual phase remains the same. On the other hand, the insertion reactions involve compositional changes within one or more phases due to the insertion or removal of atomic species from their crystal structures. These two types of reactions exhibit distinct thermodynamic and kinetic characteristics, which significantly influence their performance in battery systems. The voltage profiles represent a significant distinction between these types of reactions. In displacement reactions, the electrode potential remains constant throughout the reaction due to the absence of additional degrees of freedom. This leads to a distinct flat voltage plateau [17] observed during the discharge or recharge process, as illustrated schematically in Fig. 2. This characteristic is advantageous for applications that necessitate a stable and predictable voltage output. In insertion reactions, the incremental addition or extraction of atomic species into the crystal lattice of the host material results in variations in the electrode potential corresponding to the concentration of the inserted species. As a result, batteries that utilize insertion-reaction electrodes generally display sloped voltage-capacity curves throughout the discharge and charge processes, even when operating under optimal conditions. This variation indicates the continual adjustment of chemical potential throughout the progression of the reaction. The variations in voltage behavior highlight the essential

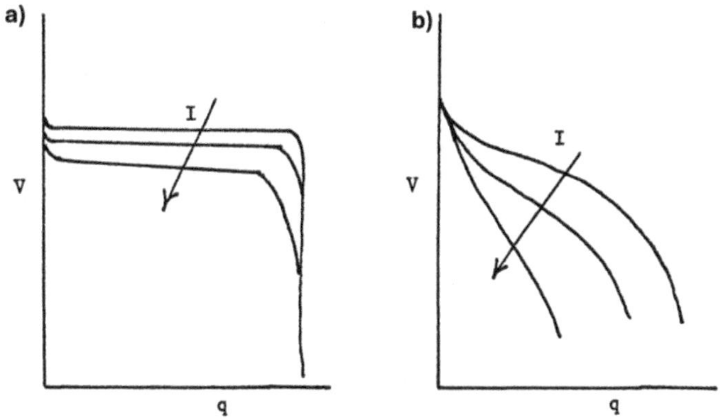

Fig. 2 Typical discharge curves at varying current densities for **a** displacement reaction and **b** insertion reactions (right). Adopted with permission from Ref. [17]

differences between displacement and insertion reactions. It is important to understand these mechanisms in order to optimize electrode materials and tailor their performance for targeted energy storage applications.

The two types of cathode reactions can be further understood from the following examples. In conversion-type electrodes, a single compound undergoes a reaction that results in the formation of entirely different phases. For example, lithium–sulfur (Li–S) batteries [18] involve a conversion-type cathode reaction as given below:

$$S_8 + 16Li^+ + 16e^- \underset{\text{charging}}{\overset{\text{discharging}}{\longleftrightarrow}} 8Li_2S$$

In this process, solid sulfur (S_8) reacts with lithium to form lithium sulfide (Li_2S). During discharge, the amount of S_8 decreases while the amount of Li_2S increases, and the reverse occurs during charging. However, the stoichiometries of the two phases remain unchanged. In insertion processes, ions are reversibly incorporated or extracted from the crystal lattice of the host material without modifying its overall structure. This phenomenon is prevalent in intercalation-type electrodes. The cathode reaction in lithium-ion batteries involves the insertion of Li^+ into a layered transition metal oxide cathode, such as lithium cobalt oxide ($LiCoO_2$), which can be expressed as follows [19]:

$$LiCoO_2 \underset{\text{discharging}}{\overset{\text{charging}}{\longleftrightarrow}} Li_{1-x}CoO_2 + xLi^+ + xe^-$$

In this process, Li^+ ions are intercalated into the cathode during discharge, while they are extracted during charging. It implies that as charging or discharging progresses, the Li content in the $LiCoO_2$ cathode continuously changes. Therefore, the stoichiometry of cathode material depends on the extent of charging or discharging [19, 20].

Anode Reactions in LIBs

Similar to cathodes, anodes in LIBs also show various mechanisms for lithium storage. The primary types of anode mechanisms include intercalation, alloying, conversion, and plating. The simplest anode system is made from lithium metal itself, which offers the highest theoretical capacity. Here, Li^+ undergoes plating on the Li metal anode during the charging process and stripping during discharging. However, severe dendrite formation on lithium-metal anodes raises safety concerns, limiting their use in modern devices. Currently, the most widely used anodes are of the intercalation type. In these systems, Li^+ reversibly intercalates into and de-intercalates from layered host materials (graphite or lithium titanate (LTO)) during the charging and discharging processes, respectively. The reaction illustrating the (de)intercalation of lithium in a graphite anode is shown below:

$$C_6 + xLi^+ + xe^- \underset{\text{charging}}{\overset{\text{discharging}}{\longleftrightarrow}} Li_xC_6$$

The researchers have shown the potential of alloying and conversion-type anodes (such as Sn, Ge, As, and transition metal oxides), which have specific advantages; however, significant volume change during the charge–discharge cycle is the key issue that needs to be addressed before they can be widely employed.

3 Challenges with Cathodes and Anodes with Solid Electrolytes

The discussion in this section is mainly focused on Li-ion-based SSBs. In liquid-electrolyte batteries, the fluid characteristics of the electrolyte facilitate its ability to effectively wet and infiltrate the active materials within the composite electrode, thereby promoting rapid ion transport for electrochemical reactions. Conversely, SSBs encounter considerable obstacles stemming from insufficient interaction between the solid electrolyte (SE) and the composite electrode (the active electrode materials blended with solid electrolytes through a binder and carbon additive). The interface between solid phases in solid-state batteries introduces an ion transfer resistance that is significantly greater than that observed in liquid-electrolyte batteries. The elevated interfacial resistance observed at the cathode in SSBs is attributable to a variety of factors, as discussed below. Tackling these challenges is essential for enhancing the interface in solid-state batteries and elevating their overall electrochemical performance. These issues are more detailed in the following sections.

3.1 Cathode Microstructure and Charge Transport

The microstructure of the cathode in SSBs must be carefully designed to achieve optimal ionic and electronic conductivities while ensuring a high concentration of cathode active material (CAM) [21]. Achieving this balance is crucial for reducing internal cell resistance (to several tens of Ω cm^2) and improving both energy and power density. An ideal composite cathode should have a CAM proportion close to 100%, as a higher CAM content is necessary to increase specific capacity and, thus, energy density (Fig. 3) [22]. However, this also reduces the amount of solid electrolyte (SE), which increases the tortuosity of the ionic channels and limits power density. Experimental studies typically use about 50 vol% CAM to ensure ionic percolation, although simulations suggest that an ideal CAM composition should range from 60 to 70 vol%, depending on porosity and distribution uniformity. Advanced manufacturing techniques and liquid additives that reduce porosity and promote uniform particle distribution could enable commercial SSBs to achieve a CAM fraction of up to 70 vol%. Unlike cathodes that use liquid electrolytes, where porosity assists electrolyte infiltration, porosity in solid-state battery cathodes is not beneficial (Fig. 3). It decreases the interface between the SE and CAM and, thus, hinders ionic and electronic conductivity, resulting in a significant rise in overvoltage due to inadequate charge transport pathways. Therefore, minimizing porosity is essential for enhancing effective charge transport in solid-state battery cathodes.

Solid electrolytes also have specific challenges due to their microstructure, which includes bulk regions and grain boundaries, each with unique ionic transport characteristics. Poorly functioning grain boundaries can increase the effective tortuosity of ionic transport beyond what would be expected based on phase distribution alone. This results in irregular and poorly defined transport properties at the microscale, negatively affecting the overall performance of the cathode [23, 24]. Therefore, the microstructures of both SE and CAM particles play a crucial role in determining the electrochemical performance of solid-state batteries (SSBs). To enhance these microstructures and promote efficient charge transport, simulation studies paired with experimental research are essential for developing high-performance solid-state battery cathodes [25, 26].

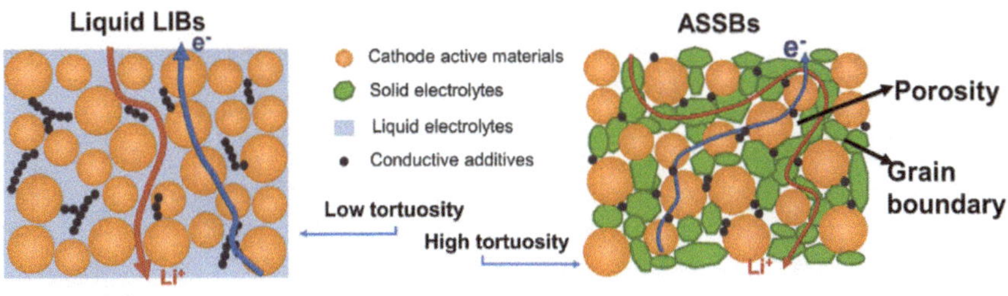

Fig. 3 Low-tortuous traditional liquid LIBs are due to the proper wetting property of liquid, and high tortuous all-SSBs are due to poor solid–solid contact and the presence of porosity. Adopted with permission from Ref. [22]

3.2 Poor Physical Contact

The inadequate interfacial contact between solid electrolytes and cathode materials poses a significant challenge in the development of solid-state batteries (SSBs). The rigid characteristics of solid electrolytes hinder the creation of conformal interfaces with cathode active materials, resulting in increased interfacial resistance and poor ion transport. Advanced imaging techniques, especially the 3D reconstruction methods developed through scanning electron microscopy (SEM) analysis, have revealed the presence of micropores and inefficient interfaces that considerably impair SSB performance. Recently, Choi et al. demonstrated that a composite cathode comprising 29% solid electrolyte contains a substantial amount of micropores (approximately 15% by volume), which obstruct Li-ion transport. Only about 23% of the total interface was shown to be effective for ionic or electronic conduction (Fig. 4) [27], highlighting the inefficiency of the interface. This inefficiency is attributed to the presence of more micropores and improper dispersion of conductive additives, which lead to inferior interfacial properties.

While layered oxide cathodes have the potential for high energy density, their complex shapes and susceptibility to cracking exacerbate these challenges. For instance, $LiNi_{0.6}Co_{0.2}Mn_{0.2}O_2$ increases the difficulties of achieving effective ionic and electronic conduction due to its intricate morphology. Li^+ percolation relies on interparticle interfaces, which are often disrupted by voids or cracks. Strauss et al. [28] found that approximately 30% of the cathode material remains inactive at low current densities due to insufficient conductive contact, as confirmed by the operando X-ray diffraction study. They observed that establishing adequate ionic pathways

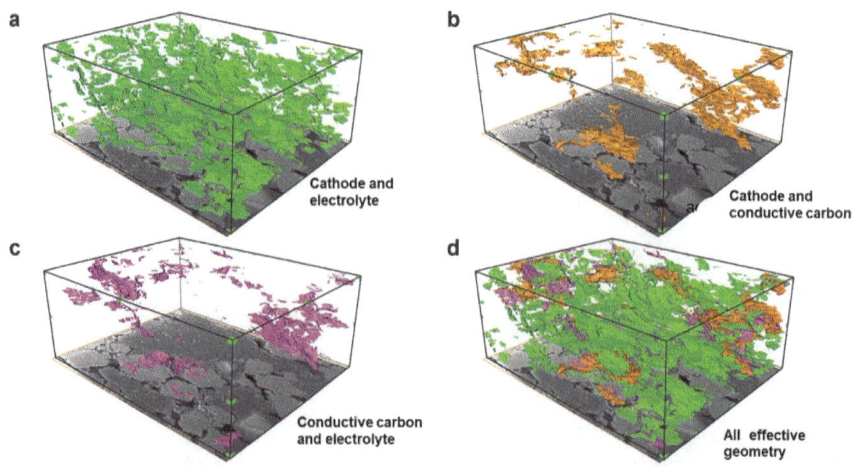

Fig. 4 3D reconstructed images of the two-phase boundary (TPB) of two components in the composite cathode: **a** CAM and SE, **b** CAM and conductive carbon, **c** conductive carbon and SE, and **d** all effective geometry (23%). Adopted with permission from Ref. [27]

requires a significant mass fraction of solid electrolyte (over 20%), which consequently reduces the practical energy density compared to liquid-electrolyte-based batteries (which typically contain around 2% electrolyte).

The mechanical stress caused by volume changes in cathode materials during cycling worsens interfacial contact issues, leading to delamination, cracking, and the formation of voids [29, 30]. This results in "dead zones" where the active material loses electrical connectivity, which diminishes the effective capacity and energy density of the battery. The challenges are further complicated by the need to balance high cathode mass loading with efficient pathways for ion and electron transport. Future research should focus on optimizing the morphology of the cathode to reduce voids and cracks, enhancing solid electrolytes to improve ionic conductivity, and refining fabrication techniques to create stable, low-resistance interfaces. Addressing these challenges will enable solid-state batteries (SSBs) to reach their full potential.

3.3 *Thermal Interdiffusion*

Thermal interdiffusion presents a significant challenge in solid-state batteries, especially at the cathode-electrolyte interface. High-temperature processing techniques, including hot pressing and sintering, are frequently employed to enhance contact and establish continuous ionic pathways [31]. However, elevated temperatures facilitate elemental migration across interfaces, resulting in the creation of undesirable interphases characterized by poor ionic and electronic conductivity.

In oxide-based solid electrolytes and cathode systems, transition metal ions such as Ni, Co, and Mn from the cathode can diffuse into the electrolyte, while elements from the electrolyte, including Li, O, and dopants, migrate into the cathode [32, 33]. The mutual diffusion modifies the chemical composition and structure of both materials, resulting in resistive layers that impede Li^+ transport and increase interfacial resistance. For instance, the interphase diffusion between $LiCoO_2$ and $Li_7La_3Zr_2O_{12}$ (LLZO), induced by pulsed laser deposition at 700 °C, results in the formation of La_2CoO_4, a phase characterized by restricted ionic conductivity, which leads to an interfacial resistivity of up to 2600 Ω cm^2. As shown in Fig. 5, the TEM images from sections A to B, particularly the intensity map and EDS results, clearly show that there is an unidentified layer between the CAM and SE, which is due to the formation of La_2CoO_4 [34]. Aluminum contamination in garnet electrolytes similarly results in phase decomposition, significantly decreasing ionic conductivity by two orders of magnitude. Interfacial reactions degrade battery performance and introduce mechanical stresses resulting from mismatches in lattice parameters and thermal expansion coefficients, which can lead to delamination and cracking. Researchers are developing strategies to reduce thermal interdiffusion. These strategies include using diffusion barrier layers, applying surface coatings to cathode particles, forming core–shell structures, and utilizing chemically stable solid electrolytes. Addressing thermal interdiffusion is crucial for enhancing the longevity, efficiency, and commercial viability of SSBs, particularly in high-temperature and fast-charging applications.

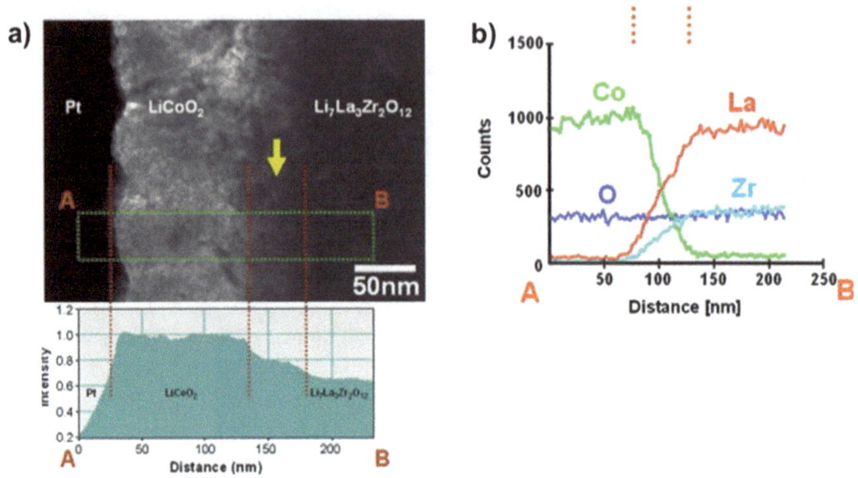

Fig. 5 Thermal interdiffusion: **a** TEM image and intensity map, and **b** EDS line profile. The red-dot line indicates the reaction area. The figure was adopted with the permission from Ref. [34]

3.4 Space-Charge Layer Formation

The space-charge layer (SCL) is a nanoscale interfacial region defined by the directional migration of charge carriers at heterojunctions. It was initially conceptualized in solid-state ionics to elucidate the anomalous conductivity observed at the LiI/Al_2O_3 interface [35, 36]. The formation of SCL at cathode/SE interfaces, driven by electrochemical potential gradients, results in lithium-ion depletion. This occurs due to the high electrostatic potential of the cathode, which reduces the concentration of Li^+ in the surface region of the electrolyte. The depletion establishes a diffusion barrier that diminishes ionic conductivity and rate performance, particularly more pronounced in sulfide electrolytes with weak Li^+ interactions. Takada et al. noted similar Li^+ removal during the initial charging of $LiCoO_2$ and $LiMn_2O_4$ in sulfide electrolytes, associating voltage slopes with SCL effects and layer thicknesses of ~10 nm (Fig. 6). The depleted layer thickness was calculated from the potential versus capacity slope because the curve suggests the number of Li-ions stripped, the amount of Li-ions in the electrolyte, and the interface area [37]. Through density functional theory (DFT) calculations, Haruyama et al. [38] suggested that the formation of SCL at equilibrium is mainly due to the surface adsorption of Li^+ from the SE to CAM. The effects of SCL are particularly noticeable in high-voltage electrodes. For instance, using 5 V-class $LiCr_{0.05}Ni_{0.45}Mn_{1.5}O_{4-\delta}$ results in significant interfacial resistance that exceeds 10^7 Ω cm^2 in LiPON-based thin-film batteries. In contrast, moderate resistance is observed in 4 V-class cathodes [39]. Gittleson and El-Gabaly [40] demonstrated Li^+ transfer at $LiCoO_2$/LiPON interfaces using XPS and indicated valence-state shifts that differ from those observed in oxide/sulfide systems, presumably due to increased vacancy energy levels in $LiCoO_2$.

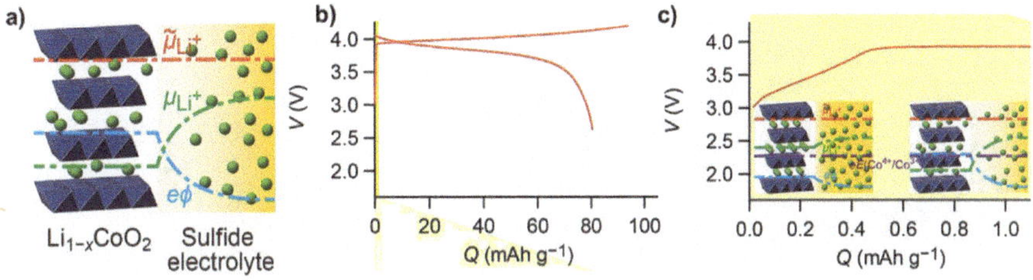

Fig. 6 **a** Schematic illustration of $Li_{1-x}CoO_2$/sulfide SE interface, **b** first charge/discharge curve, and **c** magnified image of first charge curve from the capacity range of 0–1 mAh/g. Adopted with permission from Ref. [37]

Quantifying the effects of SCL formation is challenging due to the low elemental sensitivity of lithium in conventional techniques and the limitations associated with nanoscale confinement. Electrochemical impedance spectroscopy (EIS) can be used to measure the interfacial resistance caused by SCL formation; for example, a resistance of approximately 10^4 Ω has been observed for $LiCoO_2/Li_{3.25}Ge_{0.25}P_{0.75}S_4$ [41]. However, overlapping semicircles in Nyquist plots complicate the separation of contributions from the cathode and the electrolyte, necessitating alternative methods for accurate quantification. A better understanding of the critical role of SCL in interfacial dynamics, along with an emphasis on advanced interfacial engineering, will contribute to the development of high-performance solid-state batteries.

3.5 Chemo Mechanics (Electro-Chemo-Mechanical) and Their Effects

In inorganic SSBs, the solid-state composition of all components and their limited volume within the cell housing significantly affect performance when alterations in volume and morphology occur during cycling. The changes experienced by the cathode active materials (CAM) during lithiation and de-lithiation depend on the state of charge (SOC). Intercalation-type CAMs typically undergo modest volume changes, usually in the range of a few percent, while preserving their crystal structure. In contrast, conversion-type CAMs can experience substantial volume changes, with some electrodes exceeding 100%. This can lead to anisotropic stresses and mechanical damage along various crystallographic axes. The chemo-mechanical effects—such as interparticle stresses, expansion and contraction of the CAM, and strain-induced damage—are critical factors that influence both performance and degradation, as well as interface deterioration in SSBs [42].

Changes in volume within the CAM can lead to interparticle cracking, the formation of pores, and reduced contact between the CAM and SE (Fig. 7). These issues can result in increased resistance, diminished ionic and electronic percolation, and a decline in capacity over extended cycling periods. Unlike liquid electrolytes

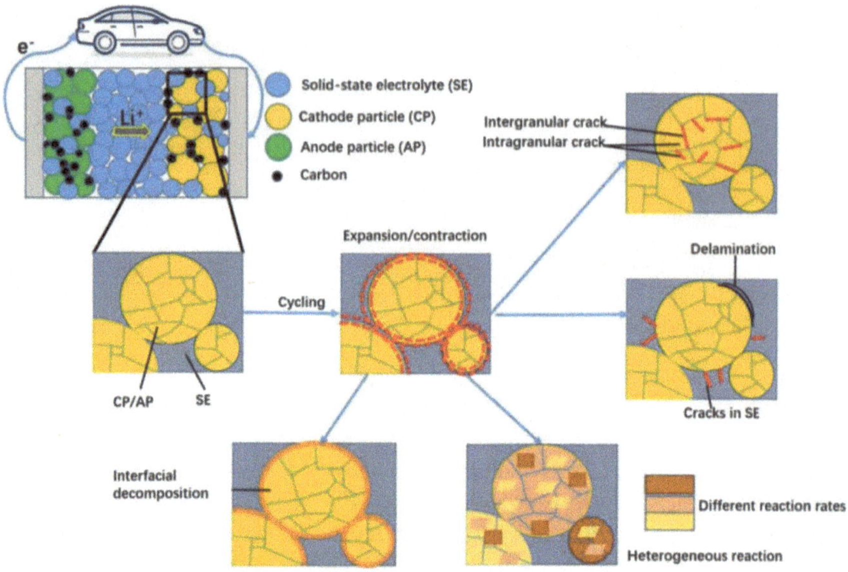

Fig. 7 Pictorial representation of chemo-mechanical effect in the SSBs. Adopted under creative commons CC BY 4.0 license from Ref. [42]

(LEs), SEs cannot penetrate these cracks or pores, which accelerates the degradation process. For example, Ni-rich NMC or conversion-type CAMs often aggravate these problems, leading to increased impedance. In laboratory cells, external pressures ranging from a few to several tens of MPa are applied to mitigate these effects; however, such pressures are not practical for large-scale applications. Therefore, it is essential to develop material combinations, including malleable thiophosphate-based solid electrolytes (such as LPS or LPSCl), that can effectively accommodate volume changes and reduce mechanical stress. This approach would enhance the durability and performance of solid-state batteries.

The percentage volume changes that occur due to lithium insertion and extraction in various cathode materials—whether they are intercalation-based or conversion-based—during cycling are illustrated in Fig. 8. The olivine-structured $LiFePO_4$ undergoes a reversible transition to $FePO_4$ at a stable voltage plateau of 3.5 V relative to Li^+/Li, with a volumetric change of 6.81% [43, 44]. In contrast, spinel $LiMn_2O_4$ experiences two cubic phase transitions at 3.96 and 4.07 V, leading to the transformation into λ-Mn_2O_4 and resulting in a volume change of approximately 6% [45]. While these structural transformations are less pronounced than those in conversion-based materials, they also pose challenges for SSLBs, where the stability of the electrode–electrolyte interfaces is crucial for long-term cycling stability. Various strategies are being investigated to reduce electro-chemo-mechanical degradation. These include the addition of buffer layers, the development of engineered composite electrodes, and the use of advanced electrolyte formulations to minimize stress accumulation and maintain interfacial integrity.

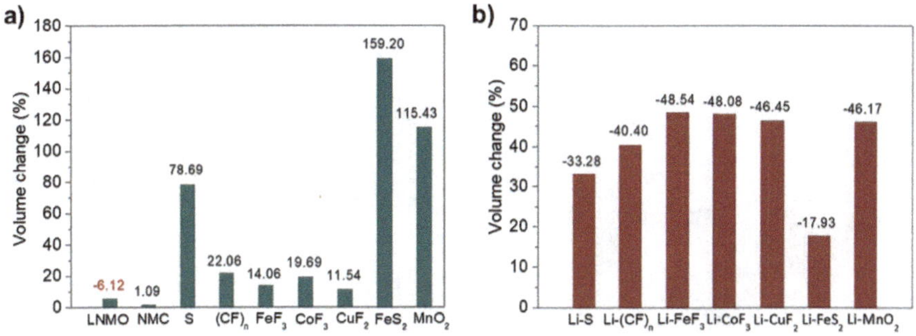

Fig. 8 **a** Theoretical volume change of the Li/non-Li-cathodes and **b** theoretical volume change of the cell containing Li-cathodes. Adopted with permission from Ref. [46]

3.6 Undesirable Side Reactions

In SSB design, ensuring electrochemical compatibility among the cathode, anode, and SE poses a significant challenge. Extensive research has been conducted on the electrochemical stability [47–49] window (ESW) of various SEs to determine the voltage range in which they remain stable, avoiding oxidation or reduction (Fig. 9) [50]. Preliminary investigations using cyclic voltammetry (CV) and linear sweep voltammetry (LSV) indicate a wider ESW range for a variety of cathodes in SSBs. However, DFT calculations predict significantly narrower stability windows than those observed experimentally [47–49]. This discrepancy can be attributed to the kinetic sluggishness of decomposition reactions, which were not accounted for in the DFT calculations. The kinetics of interfacial reactions are so sluggish that an additional overpotential is necessary to induce decomposition, thereby extending the observed stability window beyond its thermodynamic limit. For example, DFT calculations indicate that the upper voltage stability limit of $Li_7La_3Zr_2O_{12}$ (LLZO) is 2.91 V; however, experimental measurements demonstrate stability extending to 4.0 V, which is ascribed to kinetic effects. The combination of high-voltage oxide cathodes with SEs having low ESW is incompatible, leading to interfacial side reactions. This results in the formation of an ineffective interfacial layer at the cathode-electrolyte interface, which can hinder charge transfer and ultimately reduce overall performance. Comprehending and managing these interfacial interactions is crucial for the development of high-performance, long-cycle-life SSBs.

The interaction between carbon additives and solid electrolytes, alongside the cathode-electrolyte interface, represents a significant concern in terms of side reactions. Carbon additives are frequently employed in composite cathodes to enhance electronic conductivity, given that numerous cathode materials exhibit low intrinsic conductivity. Research indicates that carbon can enhance undesirable electrolyte decomposition, adversely affecting SSB performance. Zhang et al. [51] examined $LiCoO_2$ cathodes incorporating various carbon additives in conjunction with the $Li_{10}GeP_2S_{12}$ solid electrolyte. The presence of carbon resulted in a pronounced

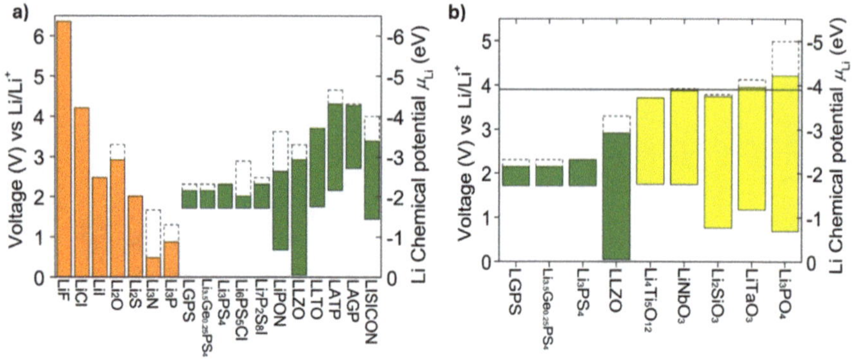

Fig. 9 **a** ESW of various SEs and **b** commonly used coating layers. The solid bar is the potential window, and the unfilled dotted bar is the maximum potential required for full de-lithiation. Adopted under creative commons CC BY 4.0 license from Ref. [50]

plateau in the charge profile (Fig. 10), elevated overpotential during initial charging, and significant electrolyte decomposition. This led to increased interfacial resistance, swift capacity degradation, and low Coulombic efficiency, underscoring the significance of electrochemical compatibility in composite electrodes.

4 Key Properties Required for the Electrode Materials

An optimal cathode for all-solid-state batteries must meet certain requirements, as mentioned below (Fig. 11), to tackle the specific challenges presented by solid-state architectures [52–54].

- It is crucial for the material to exhibit high electronic and ionic conductivity to establish a continuous and stable dual-transport network. Efficient electron transport minimizes resistive losses during charging and discharging, while rapid ion migration ensures uniform lithium diffusion at the cathode-electrolyte interface. Unlike liquid-electrolyte systems, which have continuous ionic pathways, SSBs require the cathode material to possess inherent ion-conducting properties or to maintain close contact with a solid electrolyte phase.
- The composition of components in the cathode needs careful optimization. A well-balanced composition ensures that the cathode contains enough reversibly active Li^+ ions per unit mass, which is essential for achieving high capacity while maintaining structural and electrochemical stability. The active material (such as NMC or $LiCoO_2$) should make up the majority of the composite cathode's mass fraction to optimize energy density. However, it is also important to include enough conductive additives (like carbon) to enhance electronic conductivity, as well as solid electrolyte phases (such as sulfide or oxide electrolytes) to facilitate Li^+ transport. Using too much active electrode material can create stress on

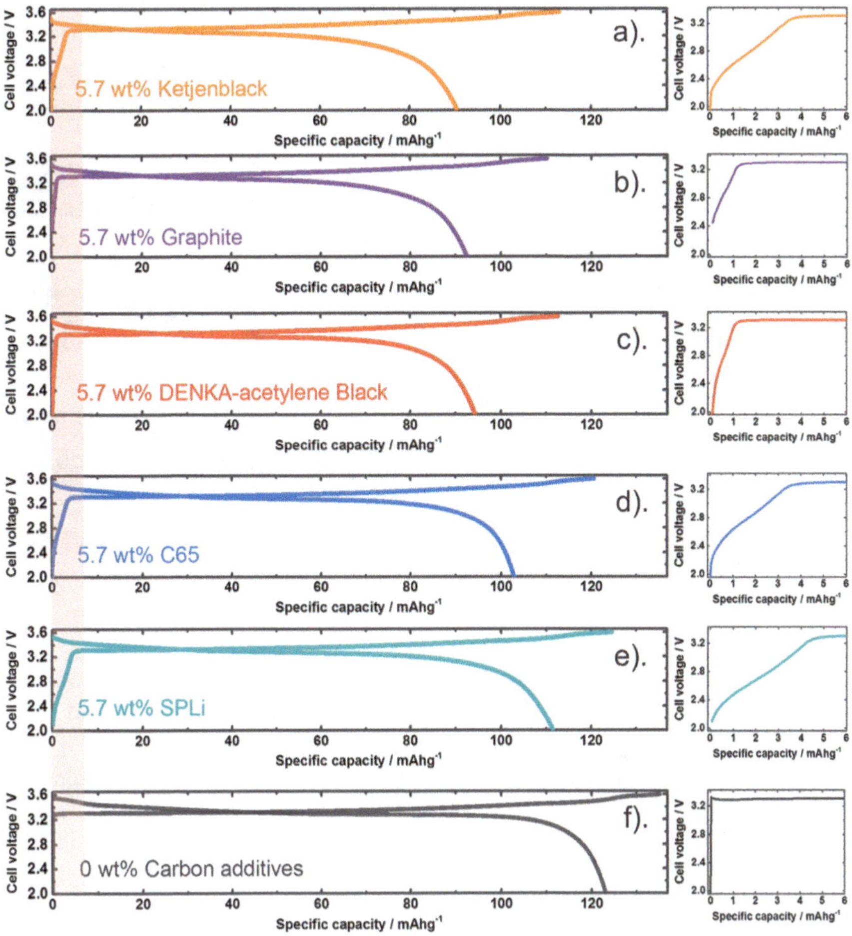

Fig. 10 a–f First charge/discharge curves with and without carbon additives (Left). Right side images are magnified images of charge/discharge curves in the range of 0–6 mAh/g. The slope before the plateau suggests the occurrence of side reactions. Adopted with permission from Ref. [51]

the composite structure during cycling, while its insufficiency can lead to faster capacity degradation.

- The electrode must exhibit mechanical and electrochemical stability throughout cycling. Volume changes during lithiation and de-lithiation can lead to interfacial delamination or microcracks in rigid solid–solid interfaces, disrupting ion and electron pathways. A robust electrode design should incorporate materials that exhibit minimal lattice strain, such as spinel or olivine structures, or utilize nanocomposites to mitigate mechanical stress. The cathode architecture must enhance percolation networks, which are interconnected pathways for ion conduction.

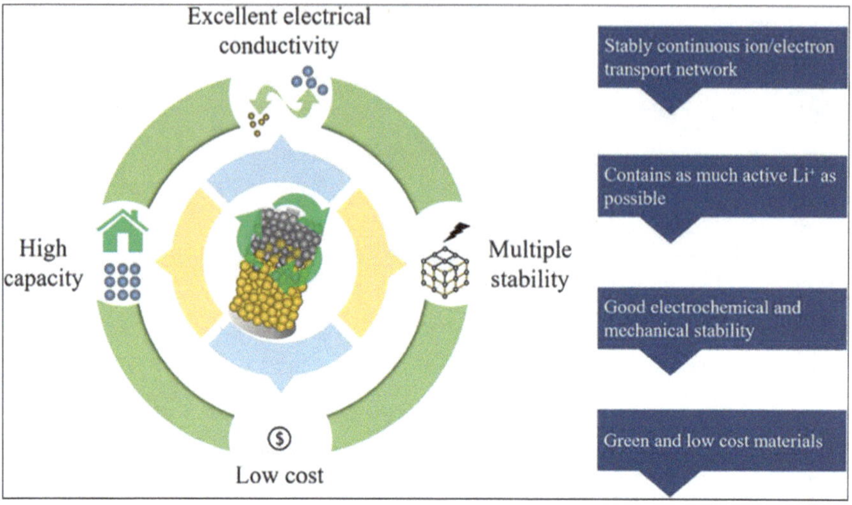

Fig. 11 Key properties required for cathode active materials. Adopted with permission from Ref. [52]

- Thermal stability is a crucial factor because localized Joule heating in SSBs can worsen interfacial degradation. The cathode must be able to withstand phase transitions or decomposition at elevated voltages (>4 V versus Li/Li^+) and high temperatures. It is essential for the cathode material and solid electrolyte to be compatible in order to minimize chemical reactions at the cathode-electrolyte interface, such as sulfurization or oxidation, which can lead to increased impedance. To address these issues, the common strategies include surface coatings, such as coating of $LiNbO_3$ on layered oxides, and hybrid cathodes that combine active materials with ion-conductive polymers or sulfide-based electrolytes.
- Scalability and cost-effectiveness are critical factors that determine the practical viability of battery materials. While materials such as lithium nickel manganese cobalt oxide (NMC) and lithium-rich manganese-based oxides offer high capacity, they require specialized processing to minimize interfacial resistance in solid-state configurations. Recent advancements in nanostructuring techniques, including the use of core–shell particles and porous frameworks, along with modern manufacturing methods like 3D printing and atomic layer deposition, are enabling the development of cathodes with enhanced conductivity, stability, and manufacturability.

5 Electrode Materials for Solid-State Lithium Batteries (SSLBs)

5.1 *Cathode Materials*

So far, the research on cathodes has focused on material selection, charge transport, composite interfaces, and framework stability. The selection of cathode active materials (CAM) was initially focused on transition metal oxides, which facilitate ion storage and release through redox reactions. Subsequent research has progressively transitioned from examining general materials to emphasizing specific elements and structures.

Based on their internal ion pathway structures, cathodes can be categorized as either having an intrinsic IP (intercalation or conversion) or a composite IP (combination of intercalation and conversion). Figure 12 illustrates that current intrinsic-IP-based CAMs for SSBs can be classified into two primary categories [3, 52]. The first type comprises intercalation compounds (>3.5 V) that store lithium ions via intercalation mechanisms such as layered structures (e.g., LiM_xO_2, where M = Ni, Co, and Mn), spinel structures (e.g., $LiMn_2O_4$), and olivine structures (e.g., $LiFePO_4$). Layered structures, including $LiNi_xMn_yCo_{1-x-y}O_2$ (NMC) and $LiCoO_2$ (LCO), are frequently utilized in high-energy systems. Whereas, spinel and olivine structures such as $LiMn_2O_4$ (LMO) and $LiFePO_4$ (LFP) are favored for high-power applications. The second category comprises conversion materials with high specific capacities (>500 mAh/g) such as metal chalcogenides, metal disulfides, and sulfur.

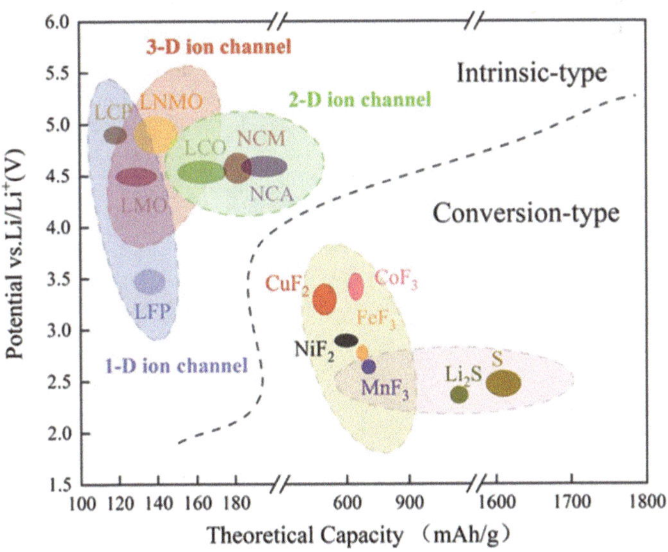

Fig. 12 Different types of cathode materials. Adopted with permission from Ref. [52]

Intercalation compounds are currently the primary focus of research and represent the most promising candidates, as solid-state pouch cells using these materials have the potential to exceed 300 Wh/kg at cell-level energy density [3, 55]. Conversely, conversion materials encounter considerable challenges, especially in enhancing the active material content in the cathode to exceed 80% in order to match the energy density of intercalation compounds.

5.1.1 Olivine Structured CAM (1D Ion Pathway)

Olivine structure-based CAMs are characterized by linear or 1D ion pathways (IPs) and are expressed as $LiMPO_4$ (M = Fe, Mn, Co, etc.). In these materials, Li^+ ions occupy designated lattice sites within the crystal structure and migrate through microscopic defects, establishing linear or networked Li^+ transport pathways. $LiFePO_4$ (LFP, Fig. 13a) is extensively utilized in conventional liquid-electrolyte lithium-ion batteries because of its superior cycling stability. Nonetheless, despite this benefit, LFP has not achieved widespread adoption in SSBs for two primary reasons. The development of SSLBs aims to achieve high power and high capacity, necessitating that all components maintain stability across a broad voltage range. The optimal operating voltage of LFP is 3.5 V, while the lithium-ion transport potential is approximately 3.45 V [56]. As a result, its performance exceeding 4 V is inadequate, rendering it less appropriate for high-energy–density SSLBs utilizing inorganic solid electrolytes. Nevertheless, due to the limited electrochemical stability window of solid polymer electrolytes (SPEs), LFP continues to be a suitable choice in SPE-based systems [57].

The second limitation of LFP is its low electronic conductivity, which ranges from 10^{-9} to 10^{-10} S/cm at room temperature [59]. To address this issue, the incorporation of conductive additives is necessary, resulting in a decreased proportion of

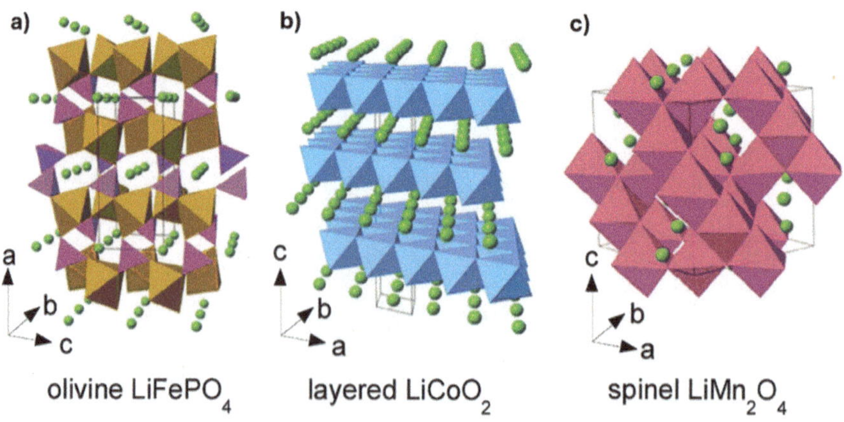

Fig. 13 Crystal structures: **a** olivine $LiFePO_4$, **b** layered $LiCoO_2$, and **c** spinel $LiMn_2O_4$. Adopted under creative commons CC BY 4.0 license from Ref. [58]

CAMs in the electrode. Moreover, research indicates that carbon additives (detailed in Sect. 3.1), although enhancing conductivity, may compromise the long-term cycling performance of SSBs. Consequently, LFP and analogous materials are utilized less frequently in SSLBs than cathodes with high voltage or capacity.

Future research should concentrate on strategies to improve the electrochemical stability and conductivity of these materials under high-voltage conditions. Methods like element doping and surface coating can enhance the stability of IPs. Beyond LFP, $LiCoPO_4$ and $LiNiPO_4$ cathodes have been examined to enhance their high-voltage stability [60], thereby increasing their suitability for next-generation SSLBs.

5.1.2 Intercalation or Layered Structures (2D Ion Pathway)

Intercalation-based CAMs typically exhibit lower specific capacities; however, they can achieve high energy densities that exceed 300 Wh/kg. This is primarily due to their elevated discharge voltages, making them valuable for practical applications. Generally, these materials are composed of two-dimensional structures that operate based on an intercalation mechanism. Such electrode materials facilitate lithium-ion diffusion in a planar configuration (2D ion pathway) during charge and discharge cycles, allowing for the reversible insertion and extraction of ions through SEs. Layered cathodes, including $LiCoO_2$ (LCO) and $LiNi_xCo_yMn_zO_2$ (NMCs with diverse compositions), exhibit an ordered α-$NaFeO_2$-type layered structure characterized by a hexagonal crystal system. These materials have the general formula $LiMO_2$ (where M stands for Co, Ni, Mn, Al, etc.), and they have R-3 m symmetry. Layered cathodes are recognized for their high energy densities, positioning them as promising candidates for next-generation batteries. However, their application in solid-state lithium batteries (SSLBs) presents significant challenges, particularly regarding interfacial compatibility and stability. Many solid-state electrolytes, especially solid polymer electrolytes (SPEs), struggle to maintain stability when paired with these cathodes at high voltages.

In $LiCoO_2$ (Figs. 13b and 14a), lithium ions are capable of reversible intercalation and deintercalation, which maintains structural integrity and enhances high-rate capability. When the charging voltage surpasses 4.6 V, the overlapping Co-t_{2g} and O-2p bands may allow oxidation of oxide (O^{2-}), leading to the formation of peroxide species (O_2^{2-}) and eventually oxygen release from the lattice (Fig. 14b). This may lead to possible structural failure [58]. Additionally, $LiCoO_2$ experiences several phase transitions during the deintercalation of lithium ions. The phase transformations, along with volume fluctuations during cycling, lead to mechanical degradation of secondary particles, which further aggravates interfacial contact problems.

$LiNiO_2$ has a theoretical discharge capacity of 275 mAh/g, while its reversible specific capacity remains closely above 180 mAh/g. Due to the comparable ionic radii of Ni^{2+} and Li^+, Ni^{2+} ions preferentially occupy Li^+ sites during the charge and discharge processes. This leads to structural phase transitions that restrict cycling stability. $LiMnO_2$ presents a high theoretical capacity of 285 mAh/g; however, its

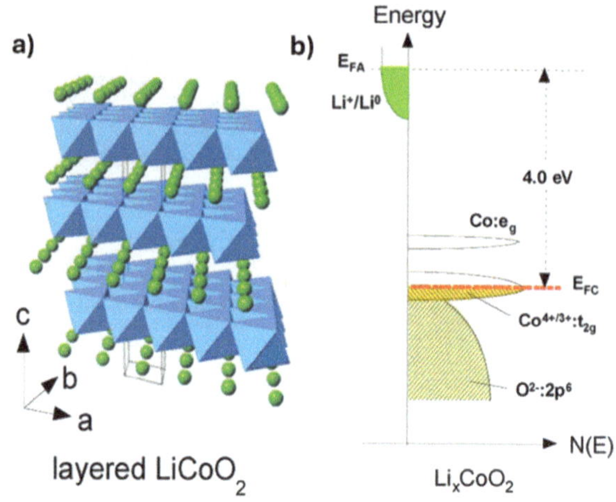

Fig. 14 **a** Crystal structure, and **b** energy versus density of states diagram of $LiCoO_2$. Adopted under creative commons CC BY 4.0 license from Ref. [58]

cycling performance is compromised by its layered rock salt structure. Li^+ deintercalation in $LiMnO_2$ induces a phase transformation to a spinel-type structure ($LiMn_2O_4$), facilitating the migration of Li^+ into Mn^{3+} layers, which ultimately causes capacity decay [61].

Layered intercalation phases typically consist of alternating edge-sharing LiO_6 and MO_6 octahedra, resulting in an "O3-type" structure characterized by an "ABCABC…" stacking sequence (Fig. 15a). During cycling, particularly under high-voltage conditions, excess de-lithiation can change this structure into a non-layered spinel structure. This transition leads to vertical collapse and irreversible structural changes. Liu et al. [62] demonstrated that nickel-rich layered oxides can shift from an ordered structure to a disordered state due to cation mixing, which occurs because of the similar ionic sizes of Ni^{2+} and Li^+ (Fig. 15b). In normal ordered layered structures, transition metal (TM) ions are ideally positioned at well-defined 3a sites, while Li ions occupy 3b sites. The effects of diffusion, alongside the comparable atomic sizes of TM and Li ions, result in the migration of TM ions into Li vacancies. Consequently, this creates disordered interphase structures marked by decreased interlayer spacing and increased diffusion barriers for Li^+ ions. The migration of TM ions is more pronounced at elevated voltages, which leads to greater structural instability and capacity degradation.

Transformations from an ordered to a disordered layered structure or from a layered structure to a disordered spinel-like phase are challenging to prevent completely. While framework stabilization and transition metal ion doping have shown promise in mitigating these effects, further research is needed to identify optimal stabilization techniques and improve long-term performance in SSLBs. For instance, partially substituting Co with Mn and Ni has led to the development of $LiNi_{1-y-z}Mn_yCo_zO_2$ (NMC) materials. These materials effectively address the limitations associated with $LiCoO_2$, $LiNiO_2$, and $LiMnO_2$. In NMC cathodes, Ni^{2+} and Co^{3+} enhance electrochemical activity, leading to a discharge capacity exceeding 180

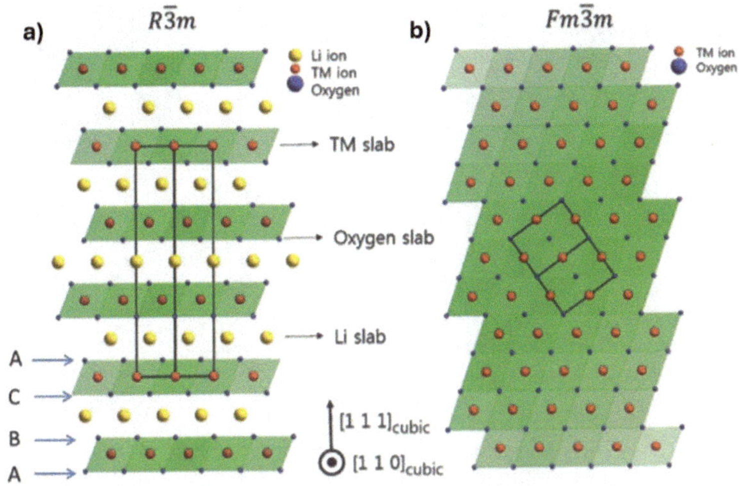

Fig. 15 Layered oxide: **a** an ordered structure, R-3 m, also called an O3-type structure, and **b** disordered structure due to cation disorder. Adopted with permission from Ref. [62]

mAh/g, while Mn^{4+} serves as a structural stabilizer, maintaining the integrity of the crystal lattice. Figure 16a–c illustrates that Co^{3+} presence markedly decreases Li^+/Ni^{2+} mixing and cation disorder, thereby maintaining the perfect crystal structure of the NMC materials. However, Co-rich structures induce irreversible phase transformations due to continuous activation of unstable oxygen redox reaction [63]. Moreover, Co-free and *ca.* 3–7% cation disorder yields several benefits for NMC cathodes, such as elevated volumetric energy density, an extensive operational voltage range, and economic viability.

Two significant challenges emerge when layered cathodes, particularly LCO and NMC cathodes, are combined with SEs. At elevated voltages, SEs experience electrochemical oxidation when in contact with NMC, resulting in the formation of a high-resistance cathode-electrolyte interphase (CEI) that hinders ionic conductivity.

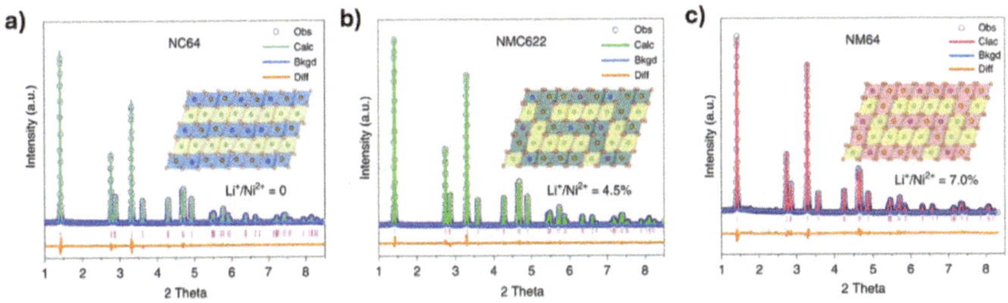

Fig. 16 **a** XRD and Rietveld refinement of different Li^+/Ni^{2+} disorder ratios: **a** NC64, **b** NMC622, and **c** NM64, where Li^+/Ni^{2+} disorder ratios are 0, 4.5, and 7.0%, respectively. Adopted with permission from Ref. [63]

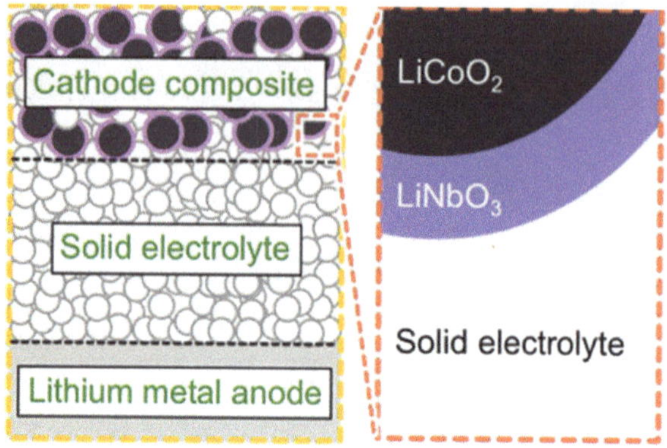

Fig. 17 Schematic illustration of protection of $LiCoO_2$ CAM by $LiNbO_3$ buffer layer. Adopted with permission from Ref. [64]

Secondly, substantial volume changes during cycling result in mechanical degradation and a loss of contact among secondary particles, which further diminishes the long-term stability of the material. It is essential to address these issues to optimize NMC materials for solid-state battery applications.

The incorporation of buffer layers or surface coatings is essential for stabilizing the CEI at elevated voltage. For example, a typical LCO cathode was coated with a $LiNbO_3$ buffer layer, which is a widely utilized buffer layer to stabilize the CEI, thereby mitigating the interfacial side reactions, as demonstrated in Fig. 17. Multiple deposition techniques (both chemical and physical) and in situ surface modification methods have been established to reduce direct contact between CAMs and SEs. These techniques improve interfacial compatibility and increase the effective contact area between the cathode and SE, resulting in enhanced electrochemical performance.

Despite these advancements, the scalability and uniformity of these coatings continue to pose significant challenges, particularly in the context of large-scale solid-state battery manufacturing. To tackle these problems, researchers have a series of coating materials made of ceramics ($LiNbO_3$, Li_3PO_4, Li_2O–ZrO_2, AlF_3, and $Li_4Ti_5O_{12}$) and polymers (polyimide, polyacrylonitrile, poly(3,4-ethylenedioxythiophene), and poly((4-vinylbenzyl)trimethylammonium bis(trifluoromethanesulfonylimide))). These coatings enhance the long-term performance of batteries by improving interfacial stability and reinforcing the structural durability of multilayer cathodes, including NMC.

Recently, Liang et al. [65] reported a gradient coating strategy for $LiNi_{0.8}Mn_{0.1}Co_{0.1}O_2$ (NMC811) particles utilizing thiophosphate salts ($Li_3P_{1+x}O_4S_{4x}$). In this method, Li_3PO_4 was deposited on the surface of the electrode material by atomic layer deposition (ALD) technique, and then in situ vulcanization was performed in order to produce a gradient coating of $Li_3P_{1+x}O_4S_{4x}$. The precise regulation of coating thickness and morphology significantly reduced structural degradation during cycling. The Solid-State Lithium Battery (SSLB) exhibited stable cycling performance, exceeding 250 cycles at a current density of 0.178 mA/cm^2, while achieving 80% capacity retention. However, some challenges remain unresolved, such as layer cracking during extended cycling and high production costs. Moreover, interfacial reactions between CAM and SE also resulted in undesirable side reactions, generating byproducts like phosphates and sulfates/sulfites that impair battery performance. Walther et al. [66] proposed a Li_2CO_3/$LiNbO_3$ coating for $LiNi_{0.6}Mn_{0.2}Co_{0.2}O_2$ (NCM622) cathodes to address this issue. The investigation demonstrated that the coating possessed a core–shell structure, with an average thickness of around 15 nm, consisting of an amorphous $LiNbO_3$ core and a Li_2CO_3 outer shell. The dual-layer coating significantly inhibited the formation of oxygenated phosphorus and sulfur compounds at the cathode/solid electrolyte interface, thereby enhancing the battery's initial Coulombic efficiency and minimizing capacity fading. Consequently, the battery's long-term cycle stability was much improved.

Element doping can substantially modify the intrinsic properties of materials, aiding in the maintenance of their structural integrity and reducing capacity loss during cycling. Doping ions can typically be classified into cations and anions according to their charge properties. Doping NMC materials with ions including Na^+, K^+, Mg^{2+}, Al^{3+}, La^{3+}, Zr^{4+}, and Nb^{5+} can effectively mitigate phase transitions. Particularly, Al^{3+} has been demonstrated to enhance the bond strength between transition metal (TM) ions and oxygen atoms, thereby reducing cation mixing and improving the structural stability of NMC materials [67, 68]. An additional effective approach to enhance the TM–O bond strength involves the doping of oxygen sites with anions or polyanions. While doping can improve the electrochemical performance of cathodes [69], it is essential to optimize the uniformity of doping, as this directly influences the homogeneity of the electrode material.

In addition to phase transitions, layered CAMs also undergo considerable volume changes during the (de-)lithiation processes. This results in contact loss between CAMs and SEs and can lead to rupture in the secondary particles of CAMs [70], as illustrated in Fig. 18. Although secondary particle rupture is not a significant concern in liquid LIBs, where liquid can readily infiltrate new cracks, it poses challenges in SSLBs because of the stiff characteristics of the SEs. The dense solid nature of materials in solid-state batteries frequently leads to inadequate contact between the cathode material and the solid electrolyte throughout the cycling process.

Strauss et al. [71] proposed that adjusting the molar ratio of Ni and Co in NMC materials can affect anisotropic volume changes (Fig. 19). The rise in nickel content results in an expansion of the lattice parameter, attributed to the greater ionic radii of Ni^{2+} (0.69 Å) and Ni^{3+} (0.56 Å) compared to Co^{3+} (0.55 Å) and Mn^{4+} (0.53 Å). In a similar manner, the lattice parameter c exhibits an increase; however, there is no discernible trend that correlates this with ion size. In contrast to LCO and NCM811 (>5%), these CAMs exhibit relative unit-cell volume variations of <1% when charged to 4.5 V against Li^+/Li. Nevertheless, they demonstrate significant volume contraction at elevated voltages. The anisotropic lattice parameter changes, which include expansion along the c-axis and contraction along the a-axis, are balanced at approximately $x(Li) \approx 0.45$. Consequently, these materials are commonly referred to as "quasi-zero-strain" electrode materials.

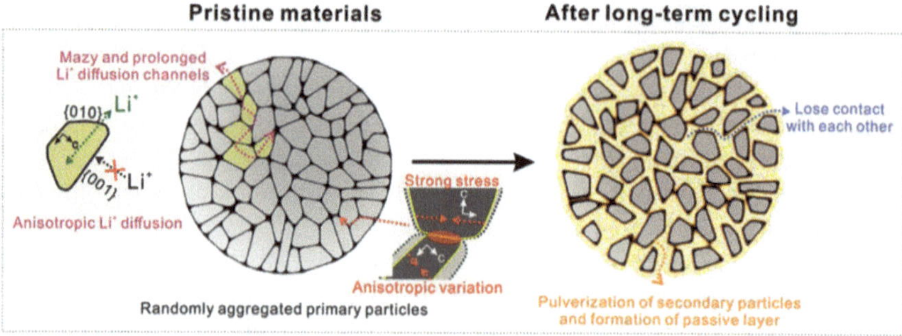

Fig. 18 Schematic depiction of NCM811 in pristine condition (left) and after cycling (right). Adopted with permission from Ref. [70]

Fig. 19 Molar ratio of Ni and Co versus volume change of CAMs. Adopted with permission from Ref. [71]

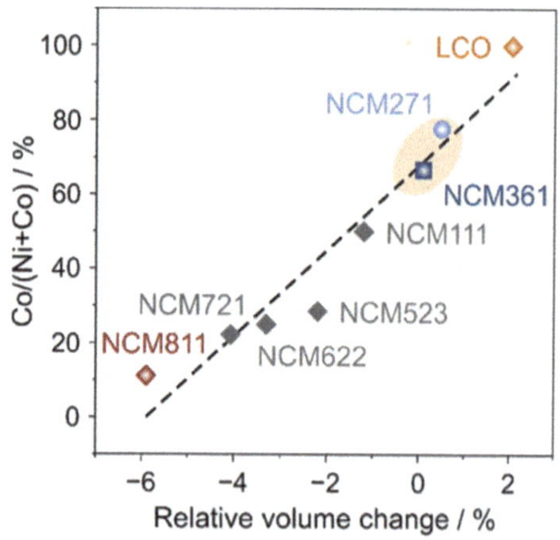

5.1.3 Spinel Structures (3D Ion Pathways)

Spinel-type cathodes such as $LiMn_2O_4$ (LMO, Fig. 13c) and $LiNi_{0.5}Mn_{1.5}O_4$ (LNMO) are favored for their cost-effectiveness, low toxicity, and ability to withstand high voltages. LMO, characterized by its cubic spinel structure, has been the subject of research for more than four decades. The material crystallizes in the Fd-3 m space group, exhibiting a lattice parameter of a = 8.2449 Å at room temperature. In this configuration, Mn ions are situated in octahedral positions, forming MnO_6 octahedra (16d sites), whereas lithium ions occupy tetrahedral (8a sites) positions. This configuration establishes a three-dimensional pathway (8a–16d–8a–16d) for Li^+ migration, facilitating rapid ion transport. Yamada et al. [72] observed that LMO exhibits reduced activation energy and diminished resistance variations during cycling when compared to LCO thin electrodes. Nonetheless, its application is constrained by quick capacity degradation and reduced specific capacity due to severe cathode material dissolution. This is because the Jahn–Teller (JT) distortion in Mn^{3+} triggers its disproportionation into Mn^{2+} and Mn^{4+}. While Mn^{4+} is stable, Mn^{2+} tends to dissolve in electrolytes, causing structural damage to the cathode material.

The introduction of cationic disorder in LMO by enabling excess Li at the Mn site and Li/Mn exchange has the potential to enhance the cycle stability of LMO (Fig. 20) [73]. This disorder interferes with the symmetric arrangement of Mn^{3+}, inhibiting cooperative JT distortion (CJTD) by reducing the MnO_6 elongation, resulting in enhanced material capacity. In pristine LMO, CJTD leads to the phase transition of LMO due to the simultaneous distortion of each JT center, leading to uniform stretching of the MnO_6 octahedron along the c-axis. As a result, capacity degradation continues to occur due to the formation of cracks during cycling and structural changes resulting from the disproportionation of Mn^{3+} in the pristine LMO. Hunter [74] suggested that Mn^{3+} can be transformed into Mn^{2+} and Mn^{4+} through the action of hydrofluoric acid ($2Mn^{3+} \rightarrow Mn^{4+} + Mn^{2+}$), resulting in an elevated concentration of Mn^{2+} in liquid-electrolyte LIBs during cycling. The internal ion pathways in spinel cathodes play a critical role in determining their performance. Element doping and surface modifications are widely utilized methods to improve stability. Although diffusion occurs at a slower rate in solid-state systems, which mitigates cathode dissolution, it is essential to further investigate the side reactions between LMO-based cathodes and SEs.

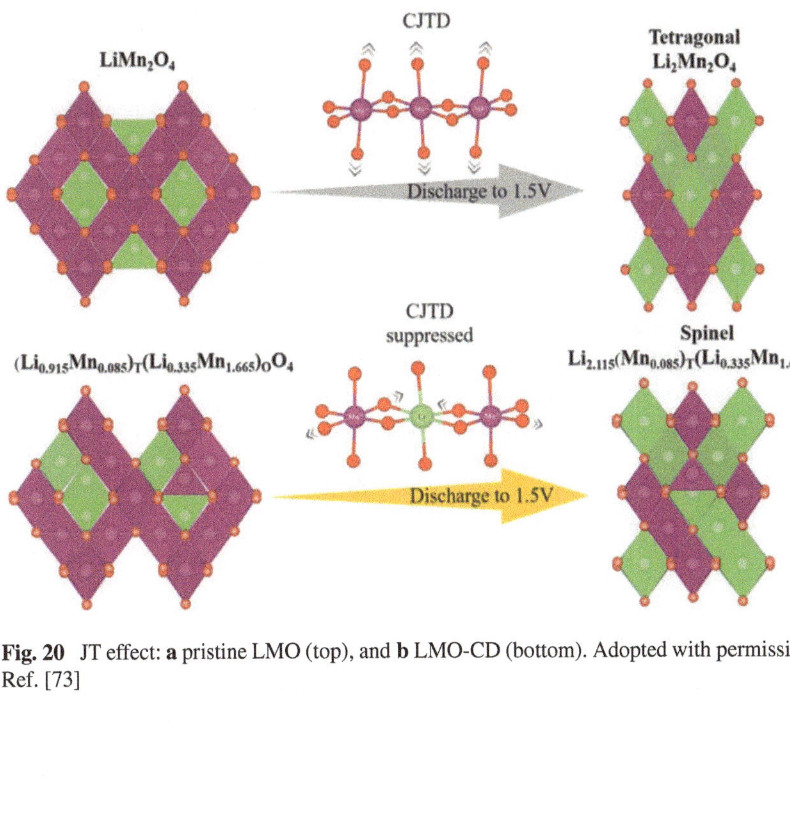

Fig. 20 JT effect: **a** pristine LMO (top), and **b** LMO-CD (bottom). Adopted with permission from Ref. [73]

5.1.4 Conversion-Based Electrode Materials

Conversion-type cathode materials (CAMs) undergo significant phase transformations during cycling, distinguishing them from intercalation and insertion materials that maintain their structural integrity. These conversion-type electrode materials usually exhibit a high specific capacity, often ranging from several hundred mAh/g. However, they also experience considerable volume changes, with alterations typically ranging from tens to more than 100%. Furthermore, the degradation of electrolytes during cycling results in the development of thicker surface films, frequently causing <75% of initial Coulombic efficiencies (CEs). The development of these films, in conjunction with the formation of nanoscopic and amorphous structures, results in notable voltage hysteresis, thereby diminishing overall energy efficiency. The issue becomes significant when employing SEs that possess narrow electrochemical stability windows.

A significant challenge for conversion-type CAMs in SSLBs is their inadequate electronic and ionic conductivity. This restricts the development of conductive networks within the CAM, impeding charge transport. To resolve this issue, it is essential to nanosize the CAM material and incorporate significant quantities of conductive additives, which may lead to a decrease in specific capacity.

The development of various fluoride and sulfide-based cathodes for the solid-state battery is in the infancy stage [75–77]. Transition metal fluorides have been reported to show the highest redox potentials, e.g., CuF_2 attaining 3.4–3.7 V relative to Li^+/Li [78, 79]. Although such a conversion reaction exhibits partial reversibility, it leads to the formation of intermediate and amorphous phases. Also, FeS_2 (pyrite) exhibits a theoretical specific capacity of 894 mAh/g and a cell voltage of 1.84 V. However, experimental results often deviate from these ideal metrics, similar to what is observed in lithium-ion batteries that use liquid electrolytes.

Sulfur-based conversion-reaction CAMs have garnered significant interest due to their theoretical capacity of 1675 mAh/g. The reaction mechanism in solid-state lithium-sulfur batteries differs from that in traditional liquid-electrolyte batteries (LIBs). In sulfur-based SSLBs, during discharge, Li^+ ions move from the anode to the cathode, breaking the –S–S– bonds of S_8 and forming lithium sulfide (Li_2S) at the cathode. During the charging process, Li_2S is oxidized, and the Li^+ ions return to the anode, where they get deposited as metallic lithium or incorporated into the anode material. This solid-to-solid process stands in contrast to lithium-sulfur batteries with liquid electrolytes, where, in addition to Li_2S, intermediate lithium polysulfides (LiPSs:Li_2S_{2-8}) are also formed at the cathode. These LiPSs can cause a "shuttling effect," which leads to material loss and decreased cycle efficiency. By combining sulfur with ceramic-based solid electrolytes, it is possible to create a direct conversion pathway to Li_2S. This approach eliminates the formation of LiPSs and establishes specific design criteria for the cathodes in solid-state lithium-sulfur batteries (SSLSBs) (Fig. 21) [80]. In response to the limitations associated with single-structured intrinsic ion pathways, composite ion pathways (CCIPs) are attracting increasing attention. These pathways are established through the integration of various CAMs, SEs, or specialized structures. The CCIPs can be categorized into three types such as 1D–3D, 1D/3D–2D, and 2D–2D. The hybrid pathways integrate the benefits of various CAMs, enhancing capacity, power, stability, and safety in comparison to individual CAMs [52].

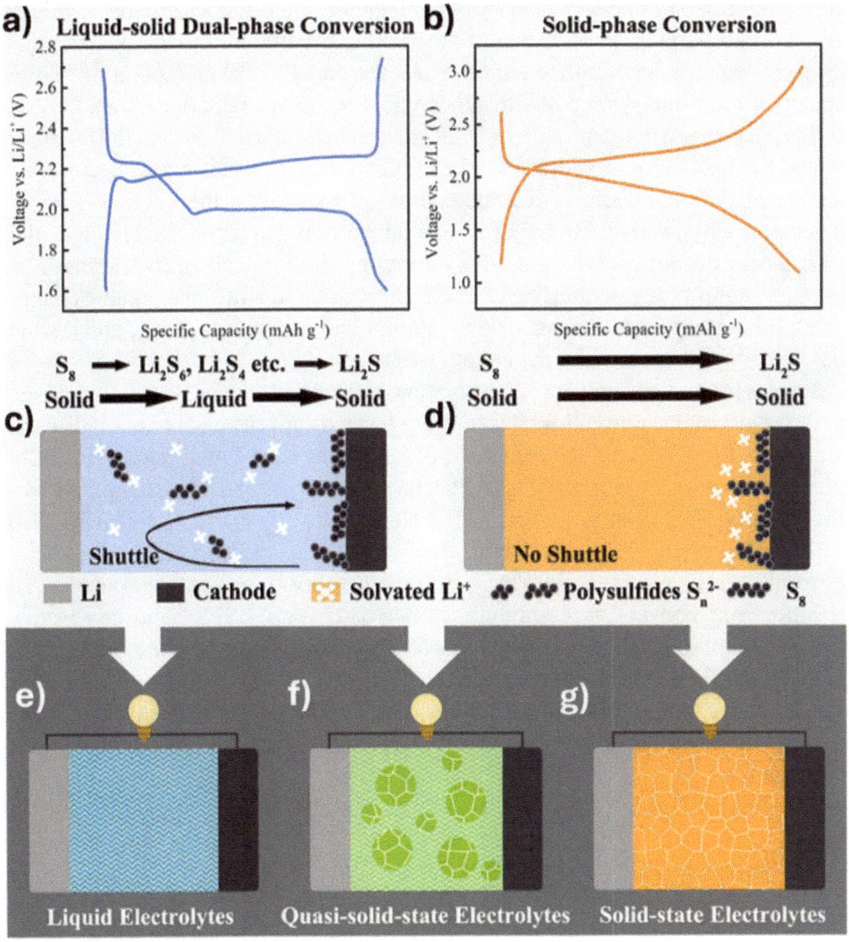

Fig. 21 Pictorial representation of liquid–Li–S and solid Li–S batteries: **a** and **b** charge/discharge profiles, **c** and **d** reaction mechanisms in both battery systems, and **e** and **f** schematic models in different electrolyte systems. Adopted with permission from Ref. [80]

5.2 Anode Materials

5.2.1 Properties Required for the Anode Materials

The anodes in SSBs play a critical role in influencing energy storage capacity, safety, lifespan, and charging speed. In contrast to traditional batteries that rely on flammable liquid electrolytes, SSBs utilize solid electrolytes, enabling the incorporation of high-capacity lithium-metal anodes with a theoretical capacity of 3860 mAh/g[4]. In SSBs, the optimal anode must possess a high theoretical capacity to maximize energy storage per unit mass. It is essential for the material to exhibit a low electrochemical potential in comparison to lithium in order to attain a high cell voltage and enhanced energy output. High electronic conductivity is crucial for effective electron transfer during operation and for improving the rate capability. Furthermore, structural stability is essential, as the anode is required to withstand volume changes during lithium intercalation and deintercalation without serious degradation, which could otherwise compromise the battery's lifespan.

Besides lithium and graphite, investigations are being conducted on materials such as silicon, tin, and various alloys due to their elevated capacity and compatibility with solid electrolytes [10, 81]. Nonetheless, these materials encounter issues, including volume expansion during the charging/discharging process and the development of an unstable solid electrolyte interphase (SEI). The stability of the SEI is critical in SSLBs to inhibit lithium dendrite formation, which poses risks of short circuits and safety concerns, thereby playing a vital role in the functionality of lithium-metal anodes. The transition from conventional liquid-electrolyte batteries to SSBs featuring advanced anodes presents significant complexities. It is essential for the interface between the anode and the solid electrolyte to maintain stability and conductivity over the entire lifespan of the battery to facilitate efficient ion transport. Comprehensive selection and optimization of materials are essential to ensure stability and improve battery performance.

5.2.2 Selection of Anode Materials for SSLBs

Lithium Anode: Lithium metal is widely studied for solid-state batteries due to its high capacity of 3860 mAh/g and negative potential of −3.04 V. However, challenges such as unwanted side reactions, an unstable solid electrolyte interphase (SEI), and lithium dendrite formation hinder practical use. These issues not only reduce efficiency but also raise safety concerns, especially in sulfide-based solid-state batteries (SSBs). The growth of lithium dendrites can lead to internal short circuits and battery failure, while the instability of lithium-metal anodes complicates their commercial application in solid-state lithium batteries (SSLBs). To improve the stability and performance of lithium-metal anodes (LMAs), strategies like surface engineering and protective layers are being explored to minimize side reactions and dendrite formation.

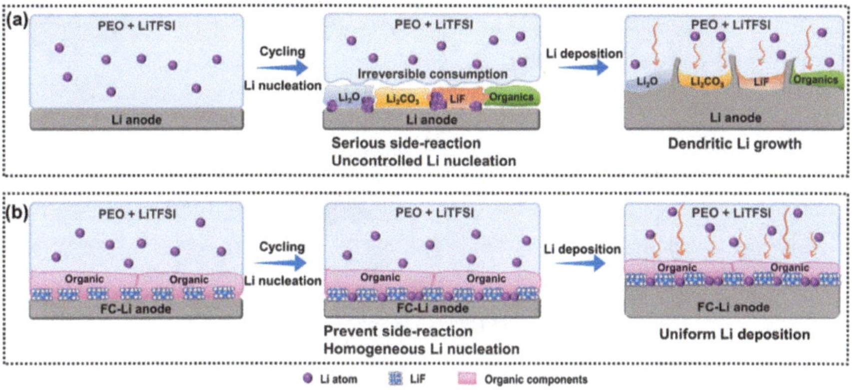

Fig. 22 Cartoon representation of pristine Li anode (**a**), and FC coated Li anode (**b**). Adopted under creative commons CC BY-NC-ND 4.0 license from Ref. [82]

Zhang et al. [82] synthesized a fluorinated lithium salt coating characterized by a soft–rigid gradient structure via an in situ reaction involving lithium metal and fluorinated carboxylic acid (heptafluorobutyric acid). In this reaction, lithium metal with high activity facilitates a single-electron transfer process, resulting in the cleavage of C–F bonds from the carboxylic precursor and the formation of a LiF-enriched layer (Fig. 22). Concurrently, fluorinated organic lithium salts accumulate partially on the outer layer (formation of gradient structure), which is attributed to the strong binding energy (>400 kJ mol^{-1}) and the short bond length (1.4 Å) of C–F bonds. The resultant FC coating effectively stabilizes the LMA/polyethylene oxide (PEO) interface, promotes uniform lithium deposition, and improves structural stability.

The application of a liquid metal (GaInSn) coating on a lithium-metal anode, achieved through an ex situ brushing technique [83] (Fig. 23), establishes a robust liquid–solid–liquid interphase that effectively inhibits dendrite formation. Liquid metals exhibit self-healing properties and a strong affinity for lithium, enhancing interfacial stability, accommodating volume fluctuations, and minimizing interfacial resistance. Moreover, the distinct wetting characteristics of LMA enhance the contact between the lithium-metal surface, leading to a reduction in interfacial resistance and an improvement in stability.

Wang et al. developed a 20 μm electroplated lithium (eLi) anode featuring a dual-layered interphase. The plating bath consists of lithium bis(fluorosulfonyl)imide (FSI), $LiNO_3$, polyethylene oxide, fluoroethylene carbonate, and dimethyl ether. The interphase formed during the electroplating consists of a SEI that is enriched with N/F-based compounds alongside a composite-polymer interphase (CPI) that incorporates lithium salts. This configuration improves ionic conductivity, mitigates

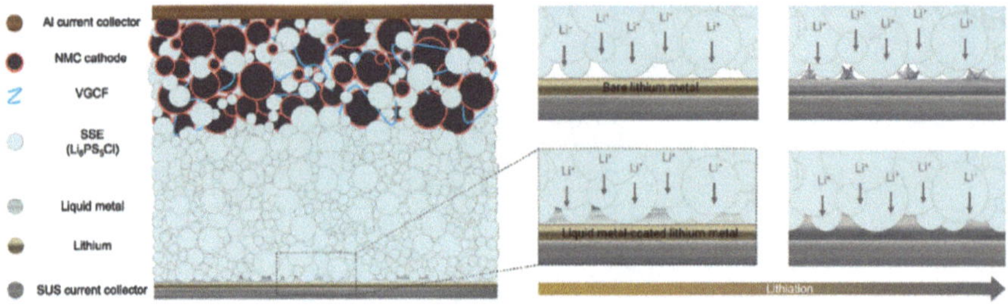

Fig. 23 Cartoon depiction of liquid metal coated Li anode (left), and electrode reaction in pristine Li anode and coated anode (right). Adopted under Creative Commons CC BY 4.0 license from Ref. [83]

dendrite formation, and preserves interfacial contact. The enhanced interfacial reaction kinetics of eLi facilitate the decomposition of FSI^-/NO^{3-} anions in the electrolyte, resulting in the generation of LiF, Li_2O, and Li_3N within the SEI. The lithiophobic characteristics of the SEI, combined with its favorable mechanical properties, effectively inhibit dendrite penetration [84].

Graphite Anode: Graphite has emerged as the most commercially viable anode material for LIBs, attributed to its cost-effectiveness, widespread availability, and consistent performance characteristics. Even though it has a lower theoretical capacity of 372 mAh/g, it continues to be a practical option for immediate application in SSLBs due to its proven high-rate capability and cycle stability observed in conventional liquid LIBs. Graphite anodes exhibit limited effectiveness in SSLBs, primarily due to considerable polarization [85]. This phenomenon arises from two key factors: (i) the increased contact resistance between graphite and sulfide-based SEs, and (ii) the reductive decomposition of the sulfide SE on the graphite surface at lithiation potential. Optimizing the particle size of sulfide solid electrolytes to reduce contact resistance can enhance interface stability; however, this may also lead to accelerated reductive decomposition of the SEs due to increased interaction with graphite. Consequently, attaining optimal performance necessitates proper adjustment of critical structural parameters, including particle size, porosity, tortuosity, binder content, and electrode thickness. Optimizing graphite is essential for SSLBs for two primary reasons. Firstly, it continues to be the benchmark material in LIBs, whereas new high-energy anodes such as silicon and lithium metal encounter difficulties related to volume expansion and cycle stability. Secondly, the dependability and consistency of graphite position it as a suitable option for process development during the initial phases of SSLB commercialization. Through the refinement of graphite anodes for solid-state systems, it is possible to connect conventional lithium-ion batteries with next-generation all-solid-state batteries, facilitating a more seamless transition to high-performance, scalable energy storage solutions.

To mitigate the low energy density of SSLBs utilizing graphite anodes, it is essential to implement strategies that involve reducing the negative/positive (N/P) capacity ratio and enhancing the charging voltage limit. Nonetheless, these methodologies present a significant challenge: the occurrence of lithium metal plating and dendrite formation on the graphite electrode, which can result in short circuits and ultimately compromise battery integrity. Moreover, graphite anodes encounter slow intercalation kinetics, inferior interfacial contact, and insufficient carrier transport. This situation necessitates charging at low current densities in SSLBs.

In SSBs that utilize graphite anodes, the phenomenon of lithium plating frequently arises when subjected to charging at higher current densities. During the charging process, besides intercalation, Li^+ may also get reduced as metallic lithium as given in the below reactions [86]:

$$\text{Lithium intercalation reaction (desired):}\quad C_6 + yLi^+ + ye^- \rightarrow Li_yC_6$$

$$\text{Lithium plating reaction (undesired):}\quad (1-y)Li^+ + (1-y)e^- \rightarrow (1-y)Li$$

$$\text{Overall charging reaction at graphite anode:}\quad C_6 + Li^+ + e^- \rightarrow Li_yC_6 + (1-y)Li$$

The process of lithium intercalation in graphite is thermodynamically favored initially, as indicated by the intercalation potential of approximately 0.1 V versus Li/Li^+, which exceeds the lithium plating potential of <0 V versus Li/Li^+. At elevated charging rates, the potential of the anode decreases, leading to intercalation potential approaching the threshold for lithium plating. The accumulation of lithium on the anode surface leads to significant volume changes during the charging process, which can result in contact loss and subsequently exacerbate current inhomogeneity. Lithium dendrites form at defects within the solid electrolytes and the electrode interface, passing through the solid electrolyte and reaching the cathode, which results in short circuits. The significant amount of current going through the dendrites induces thermal effects, resulting in melting due to Joule heating, which subsequently mitigates the short circuit. However, this phenomenon recurs, leading to voltage drops and variations in the charging profiles.

A promising approach involves the direct coating of graphite-based active materials to improve their compatibility with different SEs. A LiI-coated graphite has been synthesized using a wet chemical method and subsequently evaluated with sulfide-based solid electrolytes (Fig. 24) [87]. The intercalation potential of lithium ions in uncoated graphite is lower than the reduction limit of Li_3PS_4. This results in the reductive decomposition of Li_3PS_4 solid electrolyte into Li_2S and Li_3P, consequently increasing interfacial resistance. The LiI-coated graphite composite inhibited this decomposition and demonstrated enhanced electrochemical performance within the Li_6PS_5Cl SE system in comparison with uncoated graphite. This improvement can be attributed to the ability of LiI to suppress reductive decomposition and its superior ionic conductivity (1.2×10^{-3} S/cm) in comparison to Li_3PS_4 (5.0×10^{-4} S/cm), leading to a notable enhancement in rate performance.

Alloy Anodes: Despite extensive research on lithium-metal and graphite-based SSLBs, challenges such as dendrite formation at elevated currents and the instability

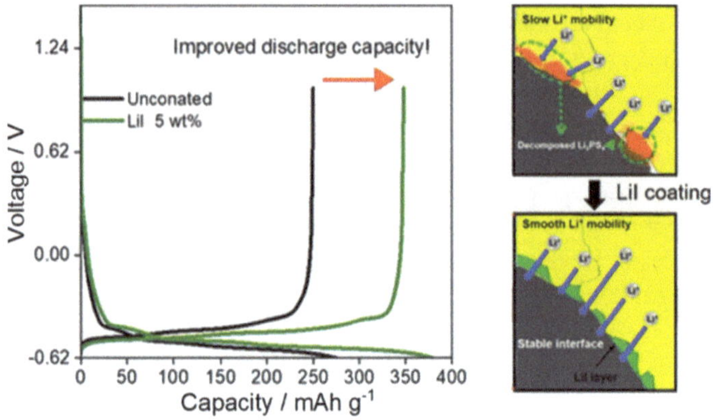

Fig. 24 LiI-coated graphite anode: electrochemical performance (left), and schematic of interface reactions (right). Adopted with permission from Ref. [87]

of lithium in atmospheric conditions continue to be significant concerns. Consequently, there has been an exploration of alternative high-energy anode materials, particularly alloy-type electrode materials such as Si, Sn, Sb, and Pb. Silicon has attracted significant interest owing to its impressive capacity (4200 mAh/g, 9786 mAh/cm^3), widespread availability, cost-effectiveness, environmentally benign characteristics, low operating potential (~0.4 V), and its capability to circumvent lithium plating processes at low voltage. Nonetheless, silicon experiences significant volume fluctuations (approximately 360% for $Li_{4.4}Si$) during the processes of lithiation and de-lithiation, which results in the growth of the SEI and contributes to irreversible capacity loss. It leads to pulverization, cracking, and the generation of stress during the lithiation and de-lithiation processes. The observed effects result in structural damage, disconnection from the current collector, and ongoing consumption of lithium ions due to the formation and disruption of the SEI layer. The volume expansion of the various alloy anodes is tabulated in Table 1. The advantages and disadvantages of the use of alloy anodes in comparison to lithium anodes are shown [10] in Fig. 25. Strategies to improve silicon anodes for sulfide all-solid-state batteries require material engineering techniques such as size control, surface modification, thin films, and alloying, alongside system optimization approaches that involve applied pressure, composite anodes with advanced binders, and conductive materials.

A recent study conducted by Zhang et al. [88] examined a composite electrode composed of Si and $Li_{21}Si_5$, focusing on an optimization of the ratio of the two components (0–100%). The $Li_{21}Si_5$ alloy was produced via a quick Li–Si alloying reaction and subsequently combined with Si particles (Fig. 26). In this composite, $Li_{21}Si_5$ grains function as a soft buffer, providing stabilization to the Si structure and mitigating the expansion stress experienced during cycling. The findings indicated that the incorporation of $Li_{21}Si_5$ resulted in a decrease of the charging plateau by 0.1–0.2 V in comparison to pure Si, minimized polarization, and inhibited the deposition of Li metal. During discharge, the composites containing varying amounts of $Li_{21}Si_5$

Table 1 Specific capacities and volume changes of various alloy anodes [10]

Alloy	Specific capacity (mAh/g)	Volume change (%)
Si	3579	320
Sn	993	260
Al	990	96
Sb	660	200
In	1012	105
Mg	33,350	100
Zn	410	98

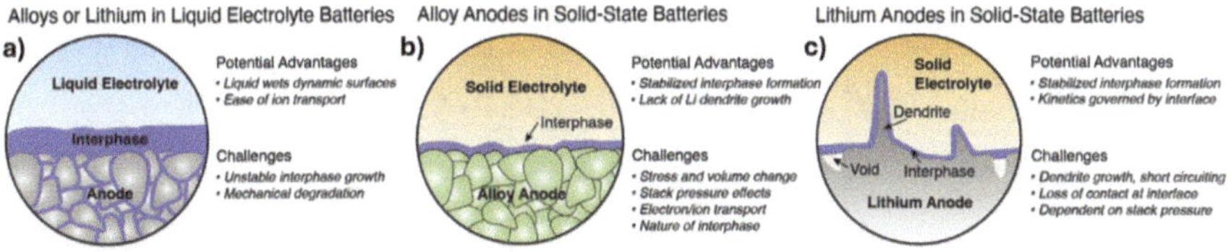

Fig. 25 Advantages and challenges of alloy anodes: **a** alloy anodes in liquid electrolytes, **b** alloy anodes in SEs, and **c** Li anodes in SEs. Adopted with permission from Ref. [10]

demonstrated significant discharge capacities ranging from 72.1 to 139.1 mAh/g, along with impressive coulombic efficiency between 96.7 and 97.8%. This indicates that $Li_{21}Si_5$ effectively enhances conductivity and serves as a lithium source, thereby improving cycle stability.

6 Exploration of Other Alkali-Metal Ion Batteries

Investigating alternative battery technologies is essential for attaining long-term sustainability and creating avenues beyond LIBs. Sodium (Na), potassium (K), calcium (Ca), magnesium (Mg), zinc (Zn), and aluminum (Al) are readily accessible elements that exhibit significant potential to foster the technology alternative to the lithium-ion battery technology in applications such as electric vehicles, electronics, and grid storage. Sodium-ion batteries (SIBs) have garnered considerable interest as viable and economically advantageous alternatives due to their abundant resources. Nonetheless, the significant reactivity of sodium metal, coupled with the larger Na^+ ionic radius, presents substantial interfacial challenges for all-solid-state sodium batteries (SSNBs), akin to those faced by lithium-based solid-state batteries.

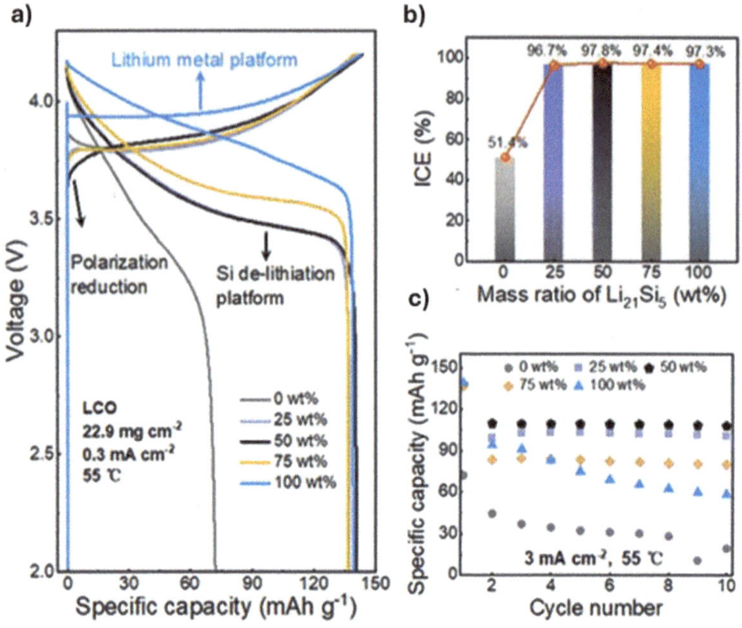

Fig. 26 **a** Electrochemical performance comparison, **b** initial coulombic efficiency (ICE), and **c** specific capacity, with and without $Li_{21}Si_5$. Adopted with permission from Ref. [88]

6.1 Electrode Materials for Solid-State Sodium Batteries (SSNBs)

6.1.1 Cathode Materials

The advancement of solid-state sodium batteries (SSNBs) is enhanced by cathode materials similar to those utilized in liquid-electrolyte sodium-ion batteries, such as layered metal oxides, sodium polyanionic compounds, and Prussian blue analogs. Nonetheless, sodium's larger ionic radius and mass result in reduced practical capacities compared to lithium-based materials. Na_xCoO_2 exhibits a capacity of approximately 70.4 mAh/g, which is significantly lower than that of Li_xCoO_2, which stands at around 175 mAh/g. Metal oxides characterized by the composition Na_xMO_2 (where M = Co, Mn, V, Cr or Ni) demonstrate significant electrochemical properties, attaining capacities exceeding 200 mAh/g at approximately 3 V versus Na/Na^+. While sodium-based cathodes offer certain benefits, they encounter significant challenges, including inadequate electronic conductivity, inflexible interfacial contact, and instability at elevated voltages. These issues impede ion transport and adversely affect cycling performance.

To address these challenges, researchers have implemented sophisticated coating methodologies. Banerjee et al. demonstrated that a Na_3SbS_4 coating (13 wt%) on

$NaCrO_2$ significantly enhanced ionic connectivity and reduced interfacial resistance (Fig. 27). In contrast to traditional mixed electrodes characterized by the uneven distribution of solid electrolytes, the coated electrode demonstrated a well-percolated structure that facilitated intimate ionic contact [89]. The structural refinement resulted in enhanced electrochemical performance through improved ionic conduction and minimized interfacial voids, thereby contributing to superior cycling stability and efficiency.

In addition, sodium manganese layered oxides (Na_xMnO_{2+y}) have attracted significant attention due to their high theoretical capacity of ~240 mAh/g and low cost. However, they suffer from low ionic conductive, inferior cyclic stability, and undesired side reactions. To solve these problems, Li et al. [90] developed the $NaPO_3$ coating on the P2-$Na_{0.7}MnO_{2.05}$ by reacting $(NH_4)_3PO_4$ and $NaMnO_2$ under high temperatures (Fig. 28a). The resulting structure $NaPO_3$@$Na_{0.7}MnO_{2.05}$ inhibits the phase transition and side reactions, thereby enhancing the rate capability and cyclic performance (Fig. 28b and c).

Compared to P2-layered oxides, the O3-layered oxides contain more sodium content in their skeleton, resulting in higher capacity. However, O3-type oxides suffer from significant phase transition and low diffusion kinetics. Cai et al. [91] introduced a high-entropy concept in an O3-layered structure and developed O3-$Na_{0.95}Li_{0.06}Ni_{0.25}Cu_{0.05}Fe_{0.15}Mn_{0.49}O_2$, which, owing to high entropy stabilization, showed good structural stability, improved ionic conductivity, and reduced JT effect (Fig. 29). Consequently, the resultant cathode exhibits a high cyclic stability of 85.1% for 1000 cycles at 4 A/g current density, which is one of the best cathode performances in the O3-type structure so far.

In addition to the layered oxides, NASICON-based polyanionic cathodes with the general formula of $A_xM_2(XO_4)_3$ (A = Li, Na, K; M = V, Ti, etc.; and X = P or Si) have also shown promise as effective cathode materials for SSNBs due to high ionic conductivity, good thermal and chemical stability, and less thermal expansion. The typical structure of $Na_xM_2(PO_4)_3$ is shown in Fig. 30 [92]. However, they suffer from poor electronic conductivity and rate performance. Recently, Pandit et al. developed the carbon-coated $NaTi_2(PO_4)_3$ (C-NTO) using ball milling followed by heat treatment of the mixture of NTO and sugar to improve the inferior electronic conductivity of the pristine NTO cathodes. The introduced carbon coating enhanced the overall conductivity and rate performance. The resultant SSNB maintained nearly 100% coulombic efficiency over 950 cycles [93].

Besides insertion-type electrodes, conversion-type sulfur cathodes have also been explored for SSNB [94]. However, the huge volume expansion of the active materials still needs more investigation.

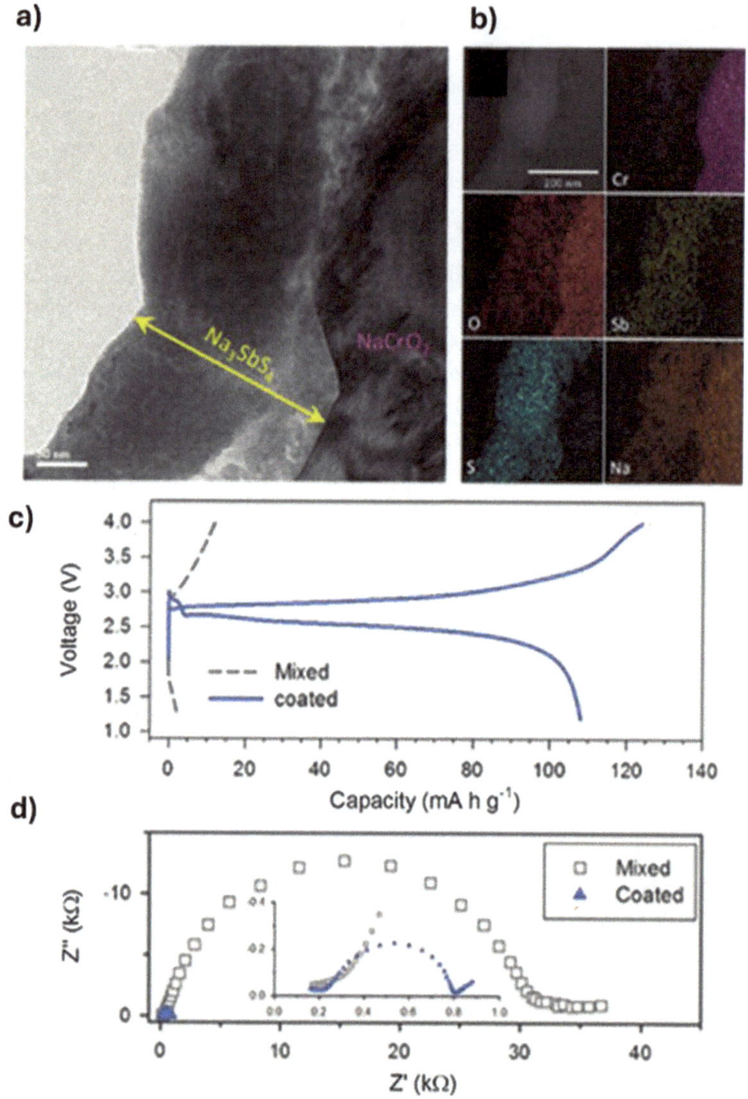

Fig. 27 **a** TEM images of Na_3SbS_4 coated $NaCrO_2$, and **b** and **c** first charge/discharge curves and EIS of coated and mixed composite, respectively. Adopted with permission from Ref. [89]

6.1.2 Anode Active Materials

Sodium metal has a low melting point and mechanical modulus, which contributes to its high wettability at interfaces. However, its increased chemical reactivity exacerbates interfacial issues in SSNBs, including side reactions and dendrite formation, which are more pronounced than in lithium-based systems. To address these

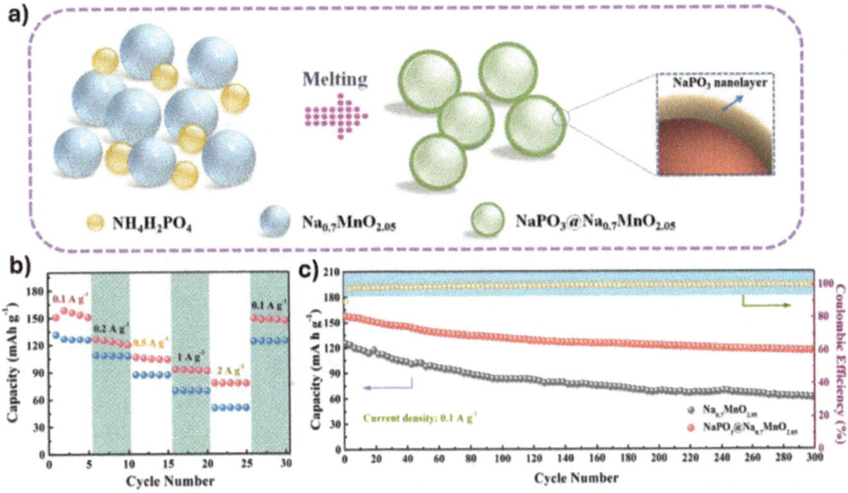

Fig. 28 **a** Schematic illustration of surface coating on the layered oxide, **b** and **c** rate performance and cyclic stability with and without coating, respectively. Adopted with permission from Ref. [90]

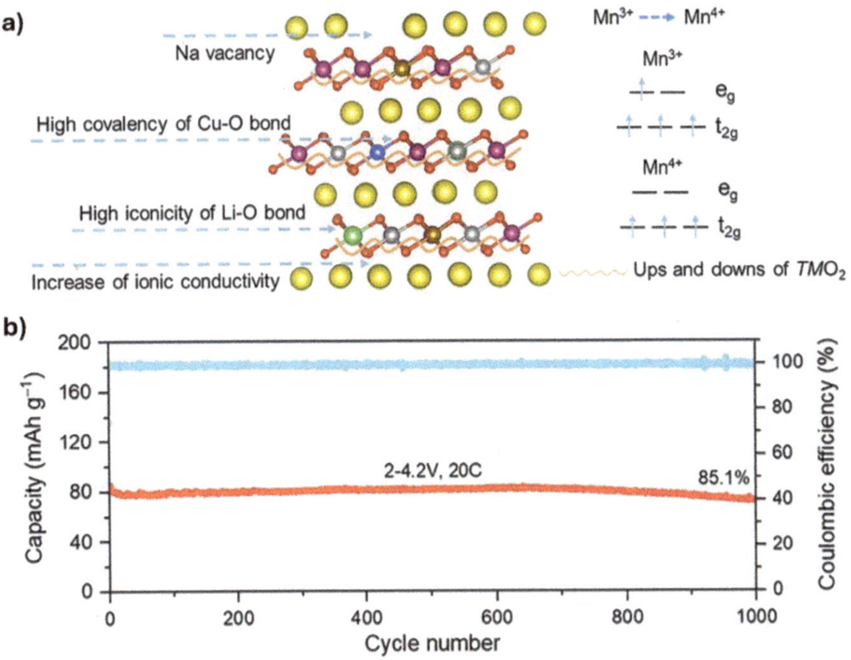

Fig. 29 **a** Schematic illustration of O3-type high-entropy oxides, and **b** long-term cyclic performance of cathode material @ 20C (4000 mA/g). Adopted under Creative Commons CC BY 4.0 license from Ref. [91]

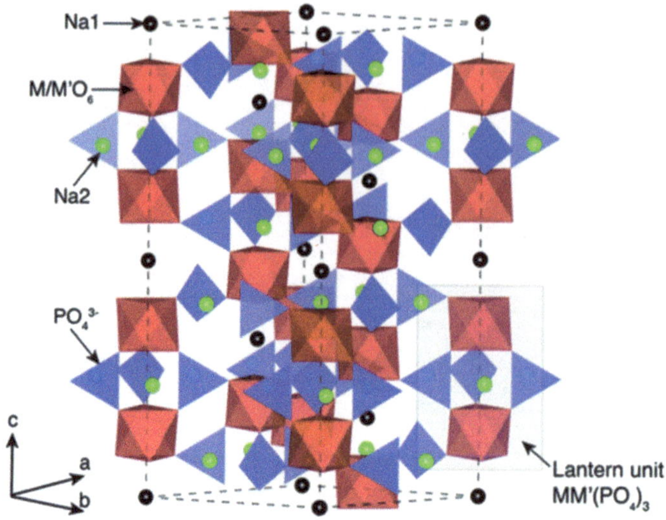

Fig. 30 Structure of $Na_xM_2(PO_4)_3$ where M_2 is denoted as M and M' in the structure. Adopted with permission from Ref. [92]

challenges, various strategies for modifying interfaces have been explored. These include the development of sodiophilic interfaces, the incorporation of polymer or plastic-crystal interlayers, and the use of three-dimensional sodium hosts as current collectors. The implementation of these methods enhances interfacial wettability and electrochemical stability, promoting uniform Na^+ transport and improving anode performance.

Yang et al. [95] developed an ultrastable composite sodium anode featuring gradient ion–electron conductive phases through an in situ reaction between SbF_3 and sodium (Fig. 31). This approach creates a robust solid–solid interface with a surface enriched in NaF and Na_3Sb within the anode, which effectively reduces dendrite formation and improves sodium transport efficiency. Electrochemical impedance spectroscopy (EIS) and area-specific resistance (ASR) measurements indicate that the interface between the sodium anode and the $Na_3Zr_2Si_2PO_{12}$ (NZSP) SE suffers from poor contact, resulting in a significant ASR of 429.3 Ω cm^2. However, the introduction of SbF_3 notably enhances interfacial contact, leading to improved electrochemical performance.

Various alternative anode materials have also been studied for SSNBs, improving their stability and efficiency. Yoshida and colleagues investigated the compatibility of hard carbon with the sulfide electrolyte Na_3PS_4 and found that the electrolyte's reductive decomposition at low voltage leads to significant irreversible capacity. They demonstrated that a composite anode of Na_3BS_3 and hard carbon effectively reduced these irreversible capacity issues, with Na_3BS_3 serving as a buffering agent [96]. Figure 32 shows the SEM cross-sectional image of SSNBs with and without buffer layers.

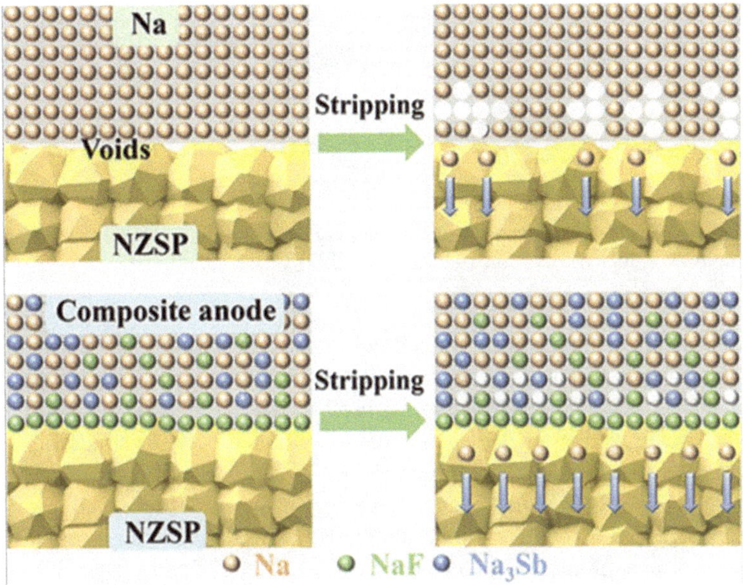

Fig. 31 Cartoon picture of the interfaces such as Na/NZSP (top), and NSF5(Composite anode)/NZSP (bottom). Adopted with permission from Ref. [95]

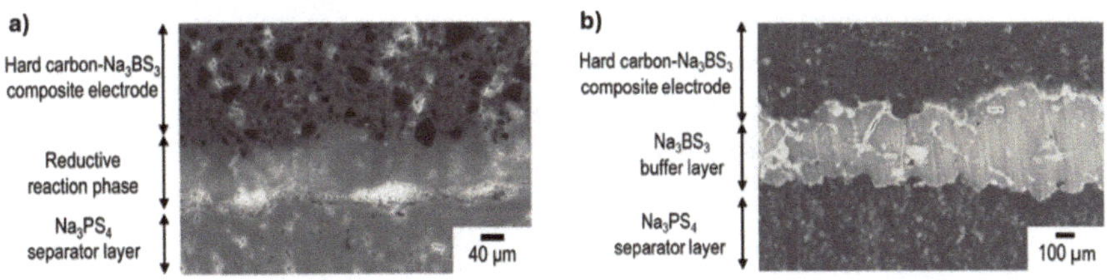

Fig. 32 Cross-sectional SEM image of SSNB: **a** without buffer layer and with buffer layer. Adopted with permission from Ref. [96]

6.2 Electrode Materials for Solid-State Potassium Batteries (SSKBs)

SSKBs primarily use potassium metal anodes. However, the high reactivity of potassium leads to significant interfacial parasitic reactions, an unstable solid electrolyte interface (SEI), and the formation of dendrites. These challenges are compounded by considerable volume fluctuations during cycling, inadequate stability of the SEI, and insufficient contact between the potassium and the solid electrolyte. These issues not only pose safety concerns but also hinder battery performance. Recent research by Wu et al. introduced a fused modeling methodology that employed potassiophilic carbon alongside molten potassium. This innovative approach significantly

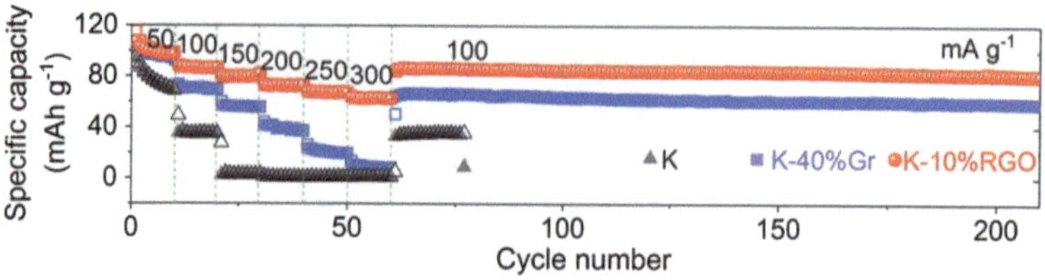

Fig. 33 Specific capacity versus cycle number graph of the pristine K anode, K-40%Graphene, and K-10%rGO. Adopted with permission from Ref. [97]

improves ion diffusion, reduces interfacial resistance, and effectively suppresses dendrite formation. Additionally, it enhances cycling stability and thermal tolerance, resulting in outstanding electrochemical performance (Fig. 33), increased energy density, and extended cycle life, even under extreme temperature conditions [97].

Prussian blue analogs (PBAs) ($A_xM'[M(CN)_6]_y\square_{1-y}\cdot nH_2O$: A = Li, Na, or K; M and M′ = Fe, Co, Mn, Ni, or Cu; □ = vacancy in the crystal lattice) are also considered as potential cathode materials, attributed to their open structures and the presence of multiple redox-active sites. Khudyshkina et al. [98] established the first SSKB utilizing a potassium metal anode, a PBA cathode, along with poly(ethylene oxide) and potassium bis(trifluoromethanesulfonyl)imide (PEO-KTFSI) polymer electrolyte. This combination demonstrated enhanced cycling stability and Coulombic efficiency, surpassing the performance of liquid-electrolyte counterparts, even with a marginally lower initial capacity. The development of SSKB is in the early stages. Therefore, an extensive investigation needs to be done to establish SSKB for commercial applications.

7 Conclusions and Future Directions

The rapid expansion of SSBs research has greatly enhanced our comprehension of the essential challenges that restrict large-scale production and commercialization. In light of these advancements, numerous critical issues continue to exist, necessitating the development of innovative strategies to improve performance, stability, and manufacturing capacity. The following sections mention some critical domains for future research directions:

(a) **Engineering Cathode-Solid Electrolyte Interface**

The interface between the cathode and solid electrolyte is crucial in influencing the performance of SSBs; however, it continues to be one of the most intricate and insufficiently studied domains due to its heterogeneous and dynamic characteristics. Future work should concentrate on:

- Analyzing the formation mechanisms and electrochemical behavior of cathode/electrolyte interphases, especially under prolonged cycling conditions.
- Creating cathodes with high energy density and minimal SE content while maintaining effective ionic conductivity greater than 10 mS/cm in composites that include binders, conductive agents, and pores.
- Advanced 3D electrode configurations have the potential to enhance interfacial contact and minimize charge accumulation; however, the establishment of a stable CEI is essential for maintaining high coulombic efficiency throughout extended cycling periods.

(b) **Development of Anodes and Stabilization of Anode-SE Interfaces**

The performance and safety of SSBs are significantly affected by the anode-SE interface, particularly with lithium-metal anodes presenting distinct challenges.

- Dendritic growth, interfacial decomposition, and pore formation at the Li/SE boundary significantly restrict fast charging capabilities. In addition, achieving long-term stability at elevated current densities necessitates additional research.
- Silicon-based anodes are emerging as promising alternatives to lithium metal, providing high capacity while presenting various challenges, especially solid electrolyte compatibility, interface stability, and volume expansion.
- A comprehensive analysis of morphology, formation mechanisms, and properties of the SEI layer can facilitate precise control over interfacial characteristics, thereby minimizing dendrite formation, decreasing interfacial resistance, and enhancing compatibility.

(c) **Sustainability of Materials and Optimization of Composition**

To enable SSBs to effectively replace conventional LIBs on a large scale, it is essential to focus on sustainable material selection and the optimization of battery architecture.

- Minimizing lithium dependency is one of the critical factors. Investigating SE compositions with decreased lithium content and less dependence on critical materials will facilitate cost-effective and sustainable manufacturing practices.
- SSBs present potential for more compact and thermally stable battery configurations. Optimized battery pack design has the potential to achieve enhanced packing densities and wider operational temperature ranges, which may exceed the capabilities of current LIB technologies, particularly in applications utilizing LFP or NCM-based materials.

(d) **State-of-the-Art Characterization**

- Investigating the ionic transport at electrode/SE interfaces is essential for the progression of SSBs.

- A critical examination of the physical, electrochemical, and mechanical properties, along with the thermodynamic and kinetic stability of lithium and other cathodes, is essential. By gaining a comprehensive understanding of these mechanisms, we can identify viable solutions to current challenges, which will help accelerate the commercialization of solid-state batteries (SSBs).
- Advanced methodologies such as synchrotron X-ray techniques, specifically XAS, XPS, etc., provide critical insights into interfacial chemistry. Theoretical calculations, when integrated with high-throughput methods (artificial intelligence/machine learning methods), can effectively predict the behavior of SEs, thereby enhancing the insights gained from experimental studies. The integration of these approaches is essential for the advancement of high-performance SSBs.

(e) **Safety of SSBs**

- Although solid-state batteries are regarded as safer alternatives to batteries having liquid electrolytes, the definitive proof of their safety advantages remains to be established. Concerns, including short circuits due to dendrite growth, mechanical failure due to gas formation, and the hazardousness of SEs, continue to present safety challenges.
- Additionally, the significant interfacial resistance present at the electrode/solid electrolyte interface may result in heat generation via the Joule effect, which could potentially initiate thermal runaway and pose safety risks. Thorough safety investigations and the establishment of targeted safety standards for SSBs are essential.

(f) **Standardization**

- The advancement of SSBs presents various challenges, including microstructural, chemical, electrochemical, and processing complexities in the electrodes, alongside ambiguities in ion transport behavior. The standardization of experimental setups and procedures that accurately replicate real-world conditions is essential for the progression of SSB technology.

(g) **Beyond Lithium Technology**

So far, significant studies have been conducted in the area beyond lithium chemistry (sodium, potassium, and other alkali/alkali-earth metals) in liquid batteries. However, the development of solid-state alkali-metal and alkaline-earth metal batteries remains in their early stages and needs more attention to achieve greater sustainability.

References

1. Bates AM, Preger Y, Torres-Castro L, Harrison KL, Harris SJ, Hewson J (2022) Are solid-state batteries safer than lithium-ion batteries? Joule 6:742–755
2. Xiao Y, Wang Y, Bo SH, Kim JC, Miara LJ, Ceder G (2020) Understanding interface stability in solid-state batteries. Nat Rev Mater 5:105–126
3. Wang S, Liu S, Chen W, Hu Y, Chen D, He M, Zhou M, Lei T, Zhang Y, Xiong J (2024) Designing reliable cathode system for high-performance inorganic solid-state pouch cells. Adv Sci 11:2401889
4. Jetybayeva A, Aaron DS, Belharouak I, Mench MM (2023) Critical review on recently developed lithium and non-lithium anode-based solid-state lithium-ion batteries. J Power Sour 566:232914
5. Pourzolfaghar H, Wang PY, Jiang XY, Kositsarakhom S, Jirasupcharoen W, Suwantri C, Jyothi D, Prabhakaran K, Li YY (2024) Emerging trends and innovations in all-solid-state lithium batteries: a comprehensive review. Chem Eng J 500:157394
6. Shalaby MS, Alziyadi MO, Gamal H, Hamdy S (2023) Solid-state lithium-ion battery: the key components enhance the performance and efficiency of anode, cathode, and solid electrolytes. J Alloys Compd 969:172318
7. Zhang XD, Yue FS, Liang JY, Shi JL, Li H, Guo YG (2020) Structure design of cathode electrodes for solid-state batteries: challenges and progress. Small Struct 1:2000042
8. Li L, Duan H, Li J, Zhang L, Deng Y, Chen G (2021) Toward high performance all-solid-state lithium batteries with high-voltage cathode materials: design strategies for solid electrolytes, cathode interfaces, and composite electrodes. Adv Energy Mater 11:2003154
9. Zhong Y, Yang X, Guo R, Zhai L, Wang X, Wu F, Wu C, Bai Y (2024) Protecting lithium metal anodes in solid-state batteries. Electrochem Energy Rev 7:1–53
10. Lewis JA, Cavallaro KA, Liu Y, McDowell MT (2022) The promise of alloy anodes for solid-state batteries. Joule 6:1418–1430
11. York M, Larson K, Harris KC, Carmona E, Albertus P, Sharma R, Noked M, Strauss E, Ragones H, Golodnitsky D (2023) Solid batteries chemistries beyond lithium, sustainable energy storage in the scope of circular economy. Adv Mater Dev Des 12:69–122
12. York M, Larson K, Harris KC, Carmona E, Albertus P, Sharma R, Noked M, Strauss E, Ragones H, Golodnitsky D (2022) Recent advances in solid-state beyond lithium batteries. J Solid State Electrochem 26:1851–1869
13. Huang Y, Shao B, Han F (2022) Solid-state batteries: an introduction. In: ACS symposium series, American chemical society, pp 1–20
14. Sarfraz N, Kanwal N, Ali M, Ali K, Hasnain A, Ashraf M, Ayaz M, Ifthikar J, Ali S, Hendi A, Baig N, Ehsan MF, Shah SS, Khan R, Khan I (2024) Materials advancements in solid-state inorganic electrolytes for highly anticipated all solid Li-ion batteries. Energy Storage Mater 71:103619
15. Famprikis T, Canepa P, Dawson JA, Islam MS, Masquelier C (2019) Fundamentals of inorganic solid-state electrolytes for batteries. Nat Mater 18:1278–1291
16. Weiss M, Simon FJ, Busche MR, Nakamura T, Schröder D, Richter FH, Janek J (2020) From liquid- to solid-state batteries: ion transfer kinetics of heteroionic interfaces. Electrochem Energy Rev 3:221–238
17. Grady ZA, Wilkinson CJ, Randall CA, Mauro JC (2020) Emerging role of non-crystalline electrolytes in solid-state battery research. Front Energy Res 8:557870
18. Moon H, Mandai T, Tatara R, Ueno K, Yamazaki A, Yoshida K, Seki S, Dokko K, Watanabe M (2015) Solvent activity in electrolyte solutions controls electrochemical reactions in Li-ion and Li-sulfur batteries. J Phys Chem C 119:3957–3970
19. Aurbach D (2000) Review of selected electrode–solution interactions which determine the performance of Li and Li ion batteries. J Power Sources 89:206–218

20. Aurbach D, Markovsky B, Weissman I, Levi E, Ein-Eli Y (1999) On the correlation between surface chemistry and performance of graphite negative electrodes for Li ion batteries. Electrochim Acta 45:67–86
21. Lee HJ, Liu X, Chart Y, Tang P, Bae JG, Narayanan S, Lee JH, Potter RJ, Sun Y, Pasta M (2022) $LiNi_{0.5}Mn_{1.5}O_4$ cathode microstructure for all-solid-state batteries. Nano Lett 22:7477–7483
22. Jiang W, Zhu X, Liu Y, Zhao S, Huang R, Ling M, Wang L, Liang C (2023) Design of composite cathodes for sulfide-based all-solid-state batteries. ETransportation 17:100246
23. Jacob M, Wissel K, Clemens O, Wang Y, Li G, Shen K, Mishra AK, Chaliyawala HA, Patel R, Milan E, Pasta M (2022) The role of grain boundaries in solid-state Li-metal batteries. Mater Fut 2:013501
24. Liu QS, An HW, Wang XF, Kong FP, Sun YC, Gong YX, Lou SF, Shi YF, Sun N, Deng B, Wang J, Wang JJ (2023) Effective transport network driven by tortuosity gradient enables high-electrochem-active solid-state batteries. Natl Sci Rev 10:NWAC272
25. Zhang W, Weber DA, Weigand H, Arlt T, Manke I, Schröder D, Koerver R, Leichtweiss T, Hartmann P, Zeier WG, Janek J (2017) Interfacial processes and influence of composite cathode microstructure controlling the performance of all-solid-state lithium batteries. ACS Appl Mater Interfaces 9:17835–17845
26. Liu Y, Liu J, Sun Q, Wang D, Adair KR, Liang J, Zhang C, Zhang L, Lu S, Huang H, Song X, Sun X (2019) Insight into the microstructure and ionic conductivity of cold sintered NASICON solid electrolyte for solid-state batteries. ACS Appl Mater Interfaces 11:27890–27896
27. Choi S, Jeon M, Ahn J, Jung WD, Choi SM, Kim JS, Lim J, Jang YJ, Jung HG, Lee JH, Sang BI, Kim H (2018) Quantitative analysis of microstructures and reaction interfaces on composite cathodes in all-solid-state batteries using a three-dimensional reconstruction technique. ACS Appl Mater Interf 10:23740–23747
28. Strauss F, Bartsch T, De Biasi L, Kim AY, Janek J, Hartmann P, Brezesinski T (2018) Impact of cathode material particle size on the capacity of bulk-type all-solid-state batteries. ACS Energy Lett 3:992–996
29. Koerver R, Aygün I, Leichtweiß T, Dietrich C, Zhang W, Binder JO, Hartmann P, Zeier WG, Janek J (2017) Capacity fade in solid-state batteries: interphase formation and chemomechanical processes in nickel-rich layered oxide cathodes and lithium thiophosphate solid electrolytes. Chem Mater 29:5574–5582
30. Molel EL, Lewis JA, Sandoval SE, Nelson DL, McDowell MT, Fuller TF (2024) Effects of contact loss at electrolyte/negative electrode interface on current density distribution in solid-state batteries. J Electrochem Soc 171:103504
31. Sazvar A, Hajibandeh M, Vafaei P, Hosseinzadeh E, Jabbari M (2024) Review of recent progress in sintering of solid-state batteries: application and modelling. J Energy Storage 101:113863
32. Rathore D, Garayt M, Liu Y, Geng C, Johnson M, Dahn JR, Yang C (2022) Preventing interdiffusion during synthesis of Ni-rich core-shell cathode materials. ACS Energy Lett 7:2189–2195
33. Zhao Y, Li J, Dahn JR (2017) Interdiffusion of cations from metal oxide surface coatings into $LiCoO_2$ during sintering. Chem Mater 29:5239–5248
34. Kim KH, Iriyama Y, Yamamoto K, Kumazaki S, Asaka T, Tanabe K, Fisher CAJ, Hirayama T, Murugan R, Ogumi Z (2011) Characterization of the interface between $LiCoO_2$ and $Li_7La_3Zr_2O_{12}$ in an all-solid-state rechargeable lithium battery. J Power Sources 196:764–767
35. Lubben D, Modine FA (1996) Enhanced ionic conduction mechanisms at LiI/Al_2O_3 interfaces. J Appl Phys 80:5150–5157
36. Chen C, Guo X (2016) Space charge layer effect in solid state ion conductors and lithium batteries: principle and perspective. Acta Chim Slov 63:489–495
37. Takada K, Ohno T, Ohta N, Ohnishi T, Tanaka Y (2018) Positive and negative aspects of interfaces in solid-state batteries. ACS Energy Lett 3:98–103
38. Haruyama J, Sodeyama K, Han L, Takada K, Tateyama Y (2014) Space–charge layer effect at interface between oxide cathode and sulfide electrolyte in all-solid-state lithium-ion battery. Chem Mater 26:4248–4255

39. Yada C, Ohmori A, Ide K, Yamasaki H, Kato T, Saito T, Sagane F, Iriyama Y (2014) Dielectric modification of 5V-class cathodes for high-voltage all-solid-state lithium batteries. Adv Energy Mater 4:1301416
40. Gittleson FS, El-Gabaly F (2017) Non-faradaic Li^+ migration and chemical coordination across solid-state battery interfaces. Nano Lett 17:6974–6982
41. Takada K, Ohta N, Zhang L, Xu X, Hang BT, Ohnishi T, Osada M, Sasaki T (2012) Interfacial phenomena in solid-state lithium battery with sulfide solid electrolyte. Solid State Ion 225:594–597
42. Li P, Zhao Y, Shen Y, Bo S-H (2020) Fracture behavior in battery materials. J Phys Energy 2:022002
43. Padhi AK, Nanjundaswamy KS, Goodenough JB (1997) Phospho-olivines as positive-electrode materials for rechargeable lithium batteries. J Electrochem Soc 144:1188–1194
44. Lv Z, Li M, Yang H, Lin J, Luo J, Hong R, Wu B, Cao SC (2024) The first-principles study on electrochemical performance, mechanical properties, and lithium-ion migration of LiFePO4 modified by doping with Co and Nb. J Solid State Electrochem 28:2873–2883
45. Çapraz ÖÖ, Bassett KL, Gewirth AA, Sottos NR (2017) Electrochemical stiffness changes in lithium manganese oxide electrodes. Adv Energy Mater 7:1601778
46. Wang L, Wu Z, Zou J, Gao P, Niu X, Li H, Chen L (2019) Li-free cathode materials for high energy density lithium batteries. Joule 3:2086–2102
47. Zhu Y, He X, Mo Y (2016) First principles study on electrochemical and chemical stability of solid electrolyte–electrode interfaces in all-solid-state Li-ion batteries. J Mater Chem A Mater 4:3253–3266
48. Nolan AM, Zhu Y, He X, Bai Q, Mo Y (2018) Computation-accelerated design of materials and interfaces for all-solid-state lithium-ion batteries. Joule 2:2016–2046
49. Han F, Zhu Y, He X, Mo Y, Wang C (2016) Electrochemical stability of $Li_{10}GeP_2S_{12}$ and $Li_7La_3Zr_2O_{12}$ Solid Electrolytes. Adv Energy Mater 6:1501590
50. Zhu Y, He X, Mo Y (2015) Origin of outstanding stability in the lithium solid electrolyte materials: insights from thermodynamic analyses based on first-principles calculations. ACS Appl Mater Interf 7:23685–23693
51. Zhang W, Leichtweiß T, Culver SP, Koerver R, Das D, Weber DA, Zeier WG, Janek J (2017) The detrimental effects of carbon additives in $Li_{10}GeP_2S_{12}$-based solid-state batteries. ACS Appl Mater Interf 9:35888–35896
52. Huo S, Sheng L, Xue W, Wang L, Xu H, Zhang H, Su B, Lyu M, He X (2023) Challenges of stable ion pathways in cathode electrode for all-solid-state lithium batteries: a review. Adv Energy Mater 13:2204343
53. Alsaç EP, Nelson DL, Yoon SG, Cavallaro KA, Wang C, Sandoval SE, Eze UD, Jeong WJ, McDowell MT (2025) Characterizing electrode materials and interfaces in solid-state batteries. Chem Rev 125:2009–2119
54. Yang G, Abraham C, Ma Y, Lee M, Helfrick E, Oh D, Lee D (2020) Advances in materials design for all-solid-state batteries: from bulk to thin films. Appl Sci 10:4727
55. Liu D, Wu B, Xu Y, Ellis J, Baranovskiy A, Lu D, Lochala J, Anderson C, Baar K, Qu D, Yang J, Galvez-Aranda D, Lopez KJ, Balbuena PB, Seminario JM, Liu J, Xiao J (2024) Controlled large-area lithium deposition to reduce swelling of high-energy lithium metal pouch cells in liquid electrolytes. Nat Energy 9:559–569
56. Cronk A, Chen YT, Deysher G, Ham SY, Yang H, Ridley P, Sayahpour B, Nguyen LHB, Oh JAS, Jang J, Tan DHS, Meng YS (2023) Overcoming the interfacial challenges of $LiFePO_4$ in inorganic all-solid-state batteries. ACS Energy Lett 8:827–835
57. Zaghib K, Charest P, Guerfi A, Shim J, Perrier M, Striebel K (2005) $LiFePO_4$ safe Li-ion polymer batteries for clean environment. J Power Sour 146:380–385
58. Julien CM, Mauger A, Zaghib K, Groult H (2014) Comparative issues of cathode materials for Li-ion batteries. Inorganics 2:132–154
59. Akhmetova K, Sultanov F, Mentbayeva A, Umirov N, Bakenov Z, Tatykayev B (2024) Advances in multi-element doping of $LiFePO_4$ cathode material for capacity enhancement in Li-ion batteries. J Power Sour 624:235531

60. Cherkashinin G, Yu Z, Eilhardt R, Alff L, Jaegermann W, Cherkashinin G, Yu Z, Eilhardt R, Alff L, Jaegermann W (2020) The effect of interfacial charge distribution on chemical compatibility and stability of the high voltage electrodes ($LiCoPO_4$, $LiNiPO_4$)/solid electrolyte (LiPON) interface. Adv Mater Interf 7:2000276
61. Bruce PG, Armstrong AR, Gitzendanner RL (1999) New intercalation compounds for lithium batteries: layered $LiMnO_2$. J Mater Chem 9:193–198
62. Liu W, Oh P, Liu X, Lee MJ, Cho W, Chae S, Kim Y, Cho J (2015) Nickel-rich layered lithium transition-metal oxide for high-energy lithium-ion batteries. Angew Chem Int Ed 54:4440–4457
63. Liu T, Yu L, Liu J, Lu J, Bi X, Dai A, Li M, Li M, Hu Z, Ma L, Luo D, Zheng J, Wu T, Ren Y, Wen J, Pan F, Amine K (2021) Understanding Co roles towards developing Co-free Ni-rich cathodes for rechargeable batteries. Nat Energy 6:277–286
64. Kim S, Harada K, Toyama N, Oguchi H, Kisu K, Orimo SI (2020) Room temperature operation of all-solid-state battery using a closo-type complex hydride solid electrolyte and a $LiCoO_2$ cathode by interfacial modification. J Energy Chem 43:47–51
65. Liang J, Zhu Y, Li X, Luo J, Deng S, Zhao Y, Sun Y, Wu D, Hu Y, Li W, Sham TK, Li R, Gu M, Sun X (2023) A gradient oxy-thiophosphate-coated Ni-rich layered oxide cathode for stable all-solid-state Li-ion batteries. Nat Commun 14:1–12
66. Walther F, Strauss F, Wu X, Mogwitz B, Hertle J, Sann J, Rohnke M, Brezesinski T, Janek J (2021) The working principle of a Li_2CO_3/$LiNbO_3$ coating on NCM for thiophosphate-based all-solid-state batteries. Chem Mater 33:2110–2125
67. Woo SW, Myung ST, Bang H, Kim DW, Sun YK (2009) Improvement of electrochemical and thermal properties of $Li[Ni_{0.8}Co_{0.1}Mn_{0.1}]O_2$ positive electrode materials by multiple metal (Al, Mg) substitution. Electrochim Acta 54:3851–3856
68. Zhao W, Zou L, Jia H, Zheng J, Wang D, Song J, Hong C, Liu R, Xu W, Yang Y, Xiao J, Wang C, Zhang JG (2020) Optimized Al doping improves both interphase stability and bulk structural integrity of Ni-rich NMC cathode materials. ACS Appl Energy Mater 3:3369–3377
69. Qian H, Ren H, Zhang Y, He X, Li W, Wang J, Hu J, Yang H, Sari HMK, Chen Y, Li X (2022) Surface doping versus bulk doping of cathode materials for lithium-ion batteries: a review. Electrochem Energy Rev 5:1–32
70. Xu X, Huo H, Jian J, Wang L, Zhu H, Xu S, He X, Yin G, Du C, Sun X, Zhu H, Sun X, Xu X, Huo H, Jian J, Wang L, Xu S, He X, Yin G, Du C (2019) Radially oriented single-crystal primary nanosheets enable ultrahigh rate and cycling properties of $LiNi_{0.8}Co_{0.1}Mn_{0.1}O_2$ cathode material for lithium-ion batteries. Adv Energy Mater 9:1803963
71. Strauss F, De Biasi L, Kim AY, Hertle J, Schweidler S, Janek J, Hartmann P, Brezesinski T (2020) Rational design of quasi-zero-strain NCM cathode materials for minimizing volume change effects in all-solid-state batteries. ACS Mater Lett 2:84–88
72. Yamada I, Miyazaki K, Fukutsuka T, Iriyama Y, Abe T, Ogumi Z (2015) Lithium-ion transfer at the interfaces between $LiCoO_2$ and $LiMn_2O_4$ thin film electrodes and organic electrolytes. J Power Sour 294:460–464
73. Zuo C, Hu Z, Qi R, Liu J, Li Z, Lu J, Dong C, Yang K, Huang W et al (2020) Double the capacity of manganese spinel for lithium-ion storage by suppression of cooperative Jahn-Teller distortion. Adv Energy Mater 10:2000363
74. Hunter JC (1981) Preparation of a new crystal form of manganese dioxide: λ-MnO_2. J Solid State Chem 39:142–147
75. Whang G, Zeier WG (2023) Transition metal sulfide conversion: a promising approach to solid-state batteries. ACS Energy Lett 8:5264–5274
76. Ma Y, Qing S, Liu H, Ma C, Yu Y, Yu C, Wang L (2025) Conversion-type cathode materials for high energy density solid-state lithium batteries. J Energy Chem 100:409–425
77. Hua X, Eggeman AS, Castillo-Martínez E, Robert R, Geddes HS, Lu Z, Pickard CJ, Meng W, Wiaderek KM, Pereira N, Amatucci GG, Midgley PA, Chapman KW, Steiner U, Goodwin AL, Grey CP (2021) Revisiting metal fluorides as lithium-ion battery cathodes. Nat Mater 20:841–850
78. Morita Y, Tanaka A, Nakanishi K, Kiuchi H, Haruyama J, Takami T, Uchimoto Y, Abe T, Ogumi Z (2020) Redox behavior of a CuF_2/Cu electrode for an all-solid-state fluoride shuttle battery. In: ECS meeting abstracts MA2020-01, p 71

79. Zheng Y, Zhang P, Wu SQ, Wen YH, Zhu ZZ, Yang Y (2012) First-principles studies on the structural and electronic properties of Li-ion battery cathode material CuF_2. Solid State Commun 152:1703–1706
80. Xing C, Chen H, Qian S, Wu Z, Nizami A, Li X, Zhang S, Lai C (2022) Regulating liquid and solid-state electrolytes for solid-phase conversion in Li–S batteries. Chemistry 8:1201–1230
81. Oh P, Yun J, Choi JH, Saqib KS, Embleton TJ, Park S, Lee C, Ali J, Ko K, Cho J (2022) Development of high-energy anodes for all-solid-state lithium batteries based on sulfide electrolytes. Angew Chem Int Ed 61:e202201249
82. Zhang R, Chen B, Ma Y, Li Y, Sha J, Ma L, Shi C, Zhao N (2024) Armoring lithium metal anode with soft–rigid gradient interphase toward high-capacity and long-life all-solid-state battery. Green Energy Environ 9:1279–1289
83. Zhou S, Li M, Wang P, Cheng L, Chen L, Huang Y, Cao B, Yu S, Liu Q, Wei J (2024) Liquid metal as an efficient protective layer for lithium metal anodes in all-solid-state batteries. Carbon Energy 6:e462
84. Deng T, Wang C, Wan H, Li AM, He X, Wang Z, Cao L, Fan X, Wang C (2025) A universal design of lithium anode via dynamic stability strategy for practical all-solid-state batteries. Angew Chem Int Ed 64:e202418811
85. Kim J, Jeong J, Lee D, Kim J, Naito M, Park H, Song T, Paik U (2025) Rational material design on high capacity and long-term-cyclability of graphite/Si-based anodes for room temperature all-solid-state batteries. Chem Eng J 505:159328
86. Zhang Z, Wang J, Jin Y, Liu G, Yang S, Yao X (2023) Insights on lithium plating behavior in graphite-based all-solid-state lithium-ion batteries. Energy Storage Mater 54:845–853
87. Yang S, Yamamoto K, Mei X, Sakuda A, Uchiyama T, Watanabe T, Takami T, Hayashi A, Tatsumisago M, Uchimoto Y (2022) High rate capability from a graphite anode through surface modification with lithium iodide for all-solid-state batteries. ACS Appl Energy Mater 5:667–673
88. Zhang Z, Sun Z, Han X, Liu Y, Pei S, Li Y, Luo L, Su P, Lan C, Zhang Z, Xu S, Guo S, Huang W, Chen S, Wang MS (2024) An all-electrochem-active silicon anode enabled by spontaneous Li–Si alloying for ultra-high performance solid-state batteries, Energy. Environ Sci 17:1061–1072
89. Banerjee A, Park KH, Heo JW, Nam YJ, Moon CK, Oh SM, Hong ST, Jung YS (2016) Na_3SbS_4: a solution processable sodium superionic conductor for all-solid-state sodium-ion batteries. Angew Chem Int Ed 55:9634–9638
90. Li W, Yao Z, Zhang S, Wang X, Xia X, Gu C, Tu J (2021) Building superior layered oxide cathode via rational surface engineering for both liquid and solid-state sodium ion batteries. Chem Eng J 421:127788
91. Cai T, Cai M, Mu J, Zhao S, Bi H, Zhao W, Dong W, Huang F (2024) High-entropy layered oxide cathode enabling high-rate for solid-state sodium-ion batteries. Nanomicro Lett 16:10
92. Singh B, Wang Z, Park S, Gautam GS, Chotard JN, Croguennec L, Carlier D, Cheetham AK, Masquelier C, Canepa P (2021) A chemical map of NaSICON electrode materials for sodium-ion batteries. J Mater Chem A Mater 9:281–292
93. Pandit B, Johansen M, Andersen BP, Martínez-Cisneros CS, Levenfeld B, Ravnsbæk DB, Varez A (2023) All-solid-state sodium-ion batteries operating at room temperature based on NASICON-type $NaTi_2(PO_4)_3$ cathode and ceramic NASICON solid electrolyte: a complete in situ synchrotron X-ray study. Chem Eng J 472:144509
94. Jhang L-J, Wang D, Silver A, Li X, Reed D, Wang D (2023) Stable all-solid-state sodium-sulfur batteries for low-temperature operation enabled by sodium alloy anode and confined sulfur cathode. Nano Energy 105:107995
95. Yang Y, Chang G, Liu Z, Chen X, Huang C, Qian Y, Liu Z, Tang Q, Hu A, Chen X (2024) Stabilized solid-solid interface for solid-state sodium batteries using gradient ion-electron conductive phases modified sodium metal anode. ACS Mater Lett 6:3564–3571
96. Yoshida W, Nasu A, Motohashi K, Tatsumisago M, Sakuda A, Hayashi A (2023) Hard carbon–sulfide solid electrolyte interface in all-solid-state sodium batteries. Electrochemistry 91:037009

97. Wu JF, Zhou W, Wang Z, Wang WW, Lan X, Yan H, Shi T, Hu R, Cui X, Xu C, He X, Mao BW, Zhang T, Liu J (2023) Building K-C anode with ultrahigh self-diffusion coefficient for solid state potassium metal batteries operating at −20 to 120 °C. Adv Mater 35:2209833
98. Khudyshkina AD, Morozova PA, Butzelaar AJ, Hoffmann M, Wilhelm M, Theato P, Fedotov SS, Jeschull F (2022) Poly(ethylene oxide)-based electrolytes for solid-state potassium metal batteries with a Prussian blue positive electrode. ACS Appl Polym Mater 4:2734–2746

Next-Generation Electrolytes for Solid-State Batteries

Komal Yadav, Banti, and Sumanta Sarkar

Abstract Since the first commercialization in the early 1990s, lithium-ion batteries have now become indispensable in portable electronic devices such as mobile phones, laptops, etc., to much larger transportation mediums such as electric bikes and cars. However, two major roadblocks in the current Li-ion batteries are—low energy density primarily originating from the cathode part and the fire safety concerns owing to the low flash point of the organic additives used in the liquid electrolytes. All-solid-state batteries are promising candidates for next-generation batteries, where sulfide-based electrolytes, the current state-of-the-art materials among solid-state electrolytes, possess respectable mechanical strength and ionic conductivity (~10^{-3} S cm^{-1} at room temperature). Solid-state electrolytes are ionic conductors that operate in a solid phase, in contrast to the conventional liquid electrolytes commonly used in lithium ion and other batteries. They enable the transfer of ions between the electrodes, which is crucial for energy storage and conversion processes. Key advantages of solid-state electrolytes include improved safety by eliminating the risk of leakage and flammability associated with liquid electrolytes, as well as the ability to operate at higher voltages and temperatures. Additionally, they allow for the use of metallic lithium anodes, which can significantly increase the energy density of batteries. However, challenges such as low ionic conductivity at ambient temperatures, poor mechanical stability, and incompatibility with the existing electrode materials have hindered their widespread adoption in commercial energy storage systems. In this book chapter, we will discuss the recent progress in the research and development of different classes of next-generation solid-state electrolytes in detail encompassing the synthesis strategies, advanced characterization techniques used to understand their working as well as degradation mechanisms, and the recent trend of predicting new solid electrolytes with high ionic conductivity using machine learning (ML) technique. Some of the electrolyte classes that will be discussed in detail are: anti-perovskite-type oxyhalides and oxysulfides, lithium and sodium-based metal halides, multivalent ion-based electrolytes such as Ca^{2+}, Mg^{2+}, Al^{3+}, etc., solid-state thin-film electrolytes, and solid composite electrolytes.

K. Yadav · Banti · S. Sarkar (✉)
Department of Chemistry, University of Delhi, Delhi 110007, India
e-mail: ssarkar@chemistry.du.ac.in

A. K. Rai et al. (eds.), *Electrochemistry of Next-Generation Solid-State Batteries*,
Green Energy and Technology, https://doi.org/10.1007/978-981-95-7603-6_4

Keywords Ionic conductors · Solid-state electrolytes · Interface · Argyrodites · NASICON electrolyte

1 Introduction

Solid-state electrolytes (SSEs) are the foundational materials in all-solid-state lithium batteries (SSLBs), playing a decisive role in determining key performance characteristics such as energy and power density, cycling stability, safety, operational temperature range, and overall service life [1]. The ideal solid-state electrolyte must meet a demanding set of requirements to be viable for next-generation battery systems. These include: high lithium-ion conductivity at room temperature ($\geq 10^{-4}$ S cm^{-1}), excellent electronic insulation (with a lithium-ion transference number close to 1), a wide electrochemical stability window (>5.5 V vs. Li/Li$^+$), chemical and electrochemical compatibility with both anode and cathode materials, strong thermal stability, resistance to ambient moisture, favorable mechanical properties, cost-effective and abundant raw materials, and scalable and straightforward synthesis methods [2].

Significant progress has been made over the past decades toward developing solid electrolytes that satisfy these requirements. A range of materials with high lithium-ion conductivity (σLi$^+$ > 1 mS cm^{-1}) have been successfully synthesized—some even outperforming traditional organic liquid electrolytes (OLEs) in terms of ionic transport efficiency. Despite these achievements, considerable challenges remain in translating these advances into large-scale commercial SSLB applications.

This chapter provides a comprehensive overview of current developments in the field of solid-state electrolytes. It explores the synthesis methods, chemical compositions, physicochemical properties, and metal-ion transport mechanisms of major SSE types—including solid polymer electrolytes (SPEs), inorganic electrolytes (such as oxides, sulfides, and halides), and organic–inorganic composite electrolytes. Each category is discussed in terms of its advantages and limitations, with particular attention given to factors affecting performance, stability, and processability.

To support future advancements, this chapter proposes a rational design approach to SSEs that integrates insights from experimental research, theoretical modeling, and machine learning. Advanced simulation tools and high-throughput computational screening are emphasized as essential for accelerating the discovery of new electrolyte chemistries and optimizing electrode–electrolyte interfaces. This multipronged approach is vital for overcoming the remaining barriers to industrial-scale deployment and enabling the reliable mass production of solid-state lithium batteries.

Before we delve into the details of different classes of next-generation solid-state ionic conductors, it is important to briefly discuss the key structural features that most of these conductors share [3]. Understanding these commonalities in different classes of solid electrolyte structures will be helpful in rationally designing new materials with high ionic conductivities and wider electrochemical voltage window.

To enable fast ion transport in solid-state materials, several structural characteristics must be carefully considered and optimized:

I. **Continuous Ion Conduction Pathways**: Efficient ion migration in solids relies on thermally activated jumps between neighboring energy minima, often described as hopping between adjacent lattice sites. For long-range conduction, these hops must occur along well-connected channels that provide an energetically favorable route. The geometry of these channels must strike a balance—too narrow, and they restrict ion flow; too wide, and ion movement becomes inefficient due to weakened interactions.

II. **Polyhedral Connectivity**: Structures composed of face-sharing polyhedra are generally more conducive to ion mobility than those with edge-sharing units. This is because face-sharing geometries create larger bottlenecks through which ions can more easily pass. A continuous network of such polyhedra within the electrolyte structure facilitates the formation of high-mobility ion channels [4].

III. **Uniform Site Energetics**: For smooth ion migration, the energy landscape along the conduction path should be relatively flat. This means that adjacent alkali ion sites should have similar potential energies, and the migration barriers between them should be low. Significant energy differences between sites can act as traps, hindering ionic movement.

IV. **Defects and Structural Disorder**: Partial occupation of ion sites through vacancies or interstitials—is essential for efficient ionic conduction. Materials that rely solely on a low concentration of intrinsic point defects, such as those formed by Frenkel or Schottky reactions, typically exhibit poor ionic conductivity at room temperature. This is because the energy required to generate these defects contributes to the overall activation energy of ion transport.

V. **Anion Framework Configuration**: The spatial arrangement of the anions plays a critical role in shaping the ion migration pathways. Anions define the electrostatic environment and steric constraints through which cations must move. Optimizing the configuration of the anion lattice can significantly enhance ion diffusion. The argyrodite-type thiophosphates with *fcc*-type anionic framework is a notable exception to this general rule [4] (Fig. 1).

VI. **Anion Polarizability**: A highly polarizable anion sub-lattice tends to facilitate cation movement by reducing the electrostatic barriers along the conduction path. Sulfides, thiophosphates, and iodides, for example, typically exhibit higher ionic conductivities than their oxide counterparts due to their softer, more polarizable lattices. Studies of lithium superionic argyrodites (e.g., Li_6PS_5X, where X = Cl, Br, I) and selenium-substituted sodium analogs (e.g., $Na_3PS_{4-n}Se_n$) highlight how increased polarizability can lower activation energies [5]. However, this softening also reduces the attempt frequency of ion hopping, which must be accounted for when designing new materials. Thus, both the energy barrier and the vibrational dynamics of the lattice (phonon spectrum) must be considered in electrolyte design.

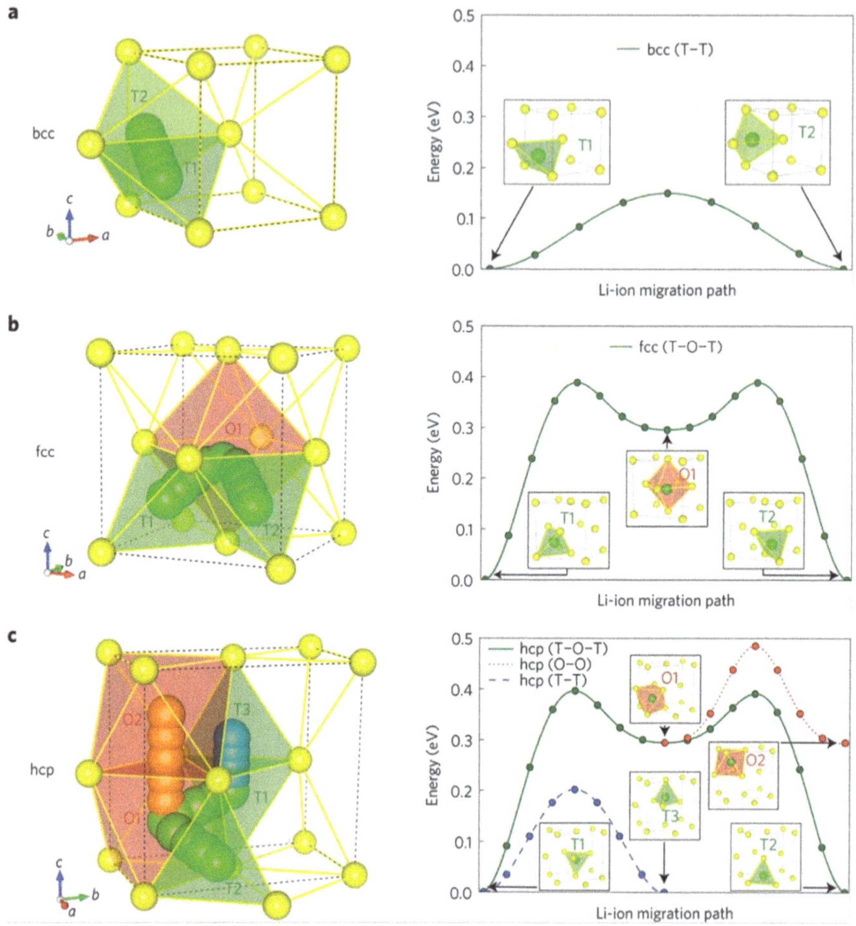

Fig. 1 **a–c**, Left panel shows the Li-ion migration path and the right panel shows the calculated energy path in bcc (**a**), fcc (**b**), and hcp (**c**) sulfur lattices. Legend: the sulfur anions are in yellow, and the Li ions are in green, blue and red for different paths. LiS_4 tetrahedra and LiS_6 octahedra are ingreen and red, respectively. Reproduced with permission from Wang et al., Nat. Mater. 2015, Copy right@ springer Nature [4]

2 General Synthesis and Fabrication Techniques

The synthesis and fabrication approach used for solid-state electrolytes (SSEs) profoundly impacts critical properties such as ionic conductivity, microstructure, interfacial contact, and production scalability [6]. Key techniques are outlined and discussed below:

2.1 Solid-State Heating

This conventional approach involves high-temperature reaction of precursor powders—typically oxides or phosphates at temperatures ranging from 700–1200 °C for extended durations. The goal is to achieve dense, crystalline pellets with large, well-connected grains. The lithium-containing garnets such as $Li_3Ln_3Te_2O_{12}$ (Ln = Y, Pr, Nd, Sm – Lu) have been synthesized by the classical solid-state heating [7, 8]. While this method has proven effective for a vast number of ceramic as well as sulfide-based electrolytes, drawbacks include high energy use, long processing times, potential lithium loss, and formation of unwanted secondary phases.

2.2 Ultrafast High-Temperature Sintering (UHS)

UHS is a rapid alternative, employing extreme heating rates to sinter and crystallize materials in mere minutes rather than hours. A notable example is ultrafast sintering of $Li_{1.5}Al_{0.5}Ge_{1.5}(PO_4)_3$ (LAGP) in just 180 s under ambient pressure—a process yielding ~1.1×10^{-4} S/cm ionic conductivity, close to conventional sintering value [9]. Tailoring additives like B_2O_3 or Li_3BO_3 during UHS can further enhance grain-boundary conductivity by manifolds. This approach significantly cuts energy use and time, offering potential for continuous industrial-scale production [10].

2.3 Mechanochemical Synthesis

Mechanochemistry leverages mechanical milling to initiate or accelerate solid-state reactions under solvent-free conditions, thus offering an eco-friendlier, energy-efficient synthesis pathway. It has successfully produced various sulfide and oxide-based SSEs. Mechanochemical synthesis is often followed by heat treatment for final phase formation. Notably, most pure phases of lithium metal halides are synthesized by ball-milling a mixture of lithium and metal halides [11]. However, industrial scaling faces challenges, including controlling milling-induced contamination and achieving consistent batch-to-batch quality.

2.4 Wet Chemical Synthesis

In recent years, many thiophosphate-based lithium-ionic conductors have been successfully synthesized at low or room temperature using a single or a combination of solvents [12]. One of the first reports of solution-phase synthesis of a solid electrolyte—β-Li_3PS_4 was by Liang et al. using THF as a solvent [13]. Later on, many other phases such as $Li_7P_3S_{11}$, Li_6PS_5Br as well as multianionic compounds have also been reported [14–16]. Understandably, the chemistry and formation mechanism of the solution-assisted techniques are different than that of solid-state synthesis. As a result, there are both advantages and challenges associated with such a technique, such as: no need for high temperature, surface passivation leading to higher resistance to air and moisture sensitivity; limited understanding of choice of solvent, formation of impurities requiring often tedious separation processes [17–19].

3 Lithium-Based Solid Electrolytes

Lithium-based solid electrolytes are at the forefront of solid-state battery research due to their potential to enhance safety, energy density, and cycle life compared to conventional liquid electrolytes. These materials serve as the medium through which lithium ions move between the anode and cathode in all-solid-state lithium batteries (ASSLBs) [20] (Fig. 2).

They can be broadly categorized into three major classes:

3.1 Oxide-Based Electrolytes

3.1.1 Garnet-Type Conductors

Garnet-type oxides, particularly those with the formula $Li_3Ln_3Te_2O_{12}$ (where Ln = Y^{3+}, Pr^{3+}, Nd^{3+}, or any lanthanide from Sm^{3+} to Lu^{3+}), represent one of the most extensively studied families of solid-state lithium-ion conductors. In the case of lithium garnets like $Li_3Ln_3Te_2O_{12}$, the Ln^{3+} and Te^{6+} ions occupy the eight- and six-coordinated sites, respectively, while lithium ions fully occupy the tetrahedral 24d sites [7]. Another class of the garnet-type lithium conductors, cubic $Li_7La_3Zr_2O_{12}$ (LLZO), is considered to be one of the most promising candidates for solid-state electrolytes. It offers high ionic conductivity at room temperature (greater than 10^{-4} S/cm), excellent chemical stability against lithium metal, and a wide electrochemical stability window [22]. However, LLZO is sensitive to moisture, which can affect its performance. In cubic LLZO (space group: *Ia*-3*d*), lithium ions are distributed in a disordered manner across tetrahedral and octahedral sites. A more thermodynamically stable tetragonal phase of LLZO (space group: $I4_1/acd$) also exists, but it exhibits much lower conductivity. As a result, considerable effort has been dedicated to stabilizing the highly conductive cubic phase, primarily through substitution strategies that reduce lithium content or increase lithium vacancy concentrations. One of the earliest and most effective approaches involved the use of Al^{3+} doping—either intentionally added or inadvertently introduced from alumina crucibles. The aliovalent substitution of Li^+ with Al^{3+} creates lithium vacancies, which disrupt the ordered Li sub-lattice and help stabilize the cubic structure [23].

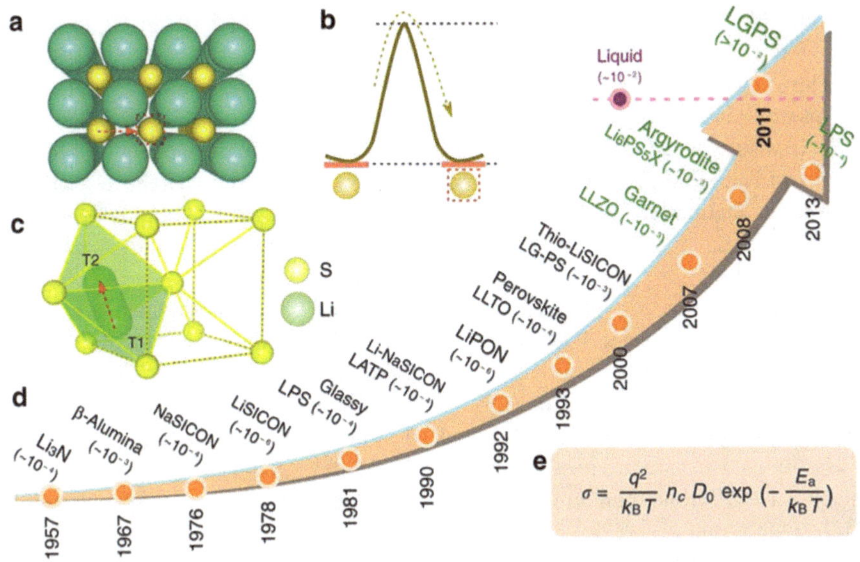

Fig. 2 Schematic of the activation energy barrier corresponding to the hopping of ions from site to site within the crystal structure (**a, b**). The overall equation defining the ionic conductivity σ is also shown (**c**). Development of solid electrolytes over times together with their superionic conductivities (in $S{\cdot}cm^{-1}$) indicated in the parentheses (**d**). The most favorable migration path between two tetrahedral sites T_1 and T_2 in the *bcc* anionic structure (**e**) The legend: NaSICON: $Na_3Zr_2PSi_2O_{12}$; LiSICON: $Li_{14}ZnGe_4O_{16}$; Glassy LPS: Li_2S–P_2S_5; LATP: $Li_{1.3}Al_{0.3}Ti_{1.7}(PO_4)_3$; LiPON: $Li_{2.98}PO_{3.30}N_{0.46}$; LLTO: $Li_xLa_{2/3-x/3}TiO_3$; LGPS: $Li_{4-x}Ge_{1-x}P_xS_4$; LLZO: $Li_7La_3Zr_2O_{12}$; LGPS: $Li_{10}GeP_2S_{12}$; LPS: Li_3PS_4. Reproduced from Abakumov et al., Nat. Commun. 2020, Copy right@ The Author(s) under CC BY 4.0 license [21]

3.1.2 Anti-Perovskite-Type Conductors

Driven by the excellent high-temperature ion transport characteristics observed in fluoride-based perovskites, a new family of structurally distinct materials-anti-perovskites with the general formula Li_3OX (where $X = Cl^-$, Br^-, I^-, or halide mixtures) has emerged as a promising platform for solid-state lithium-ion conductors [24]. These materials adopt a cubic structure analogous to traditional perovskites but feature an inverted arrangement of their ionic sub-lattices, effectively reversing the typical charge distribution. In their idealized structure, halide ions (X^-) occupy the corners of the cube, the oxide ion (O^{2-}) sits at the center, and lithium ions (Li^+) form an octahedral cage surrounding the central oxygen, creating a body-centered cubic (bcc) network. This Li^+-rich bcc framework facilitates efficient ion migration via interconnected diffusion pathways. At room temperature, Li_3OCl exhibits an ionic conductivity of 8.5×10^{-5} $S{\cdot}cm^{-1}$. Notably, this conductivity can be markedly improved through strategic chemical modifications. For example, replacing Cl^- with the larger Br^- or I^- ions, or introducing aliovalent dopants by partially substituting

Li^+ with divalent cations, has been shown to raise conductivity levels to above 10^{-3} S·cm^{-1}.

The mixed-halide anti-perovskite compound $Li_3OCl_{0.5}Br_{0.5}$ has demonstrated enhanced lithium-ion conductivity, achieving values as high as 1.94×10^{-3} S·cm^{-1} [47]. In a separate study, glassy, cation-doped anti-perovskite electrolytes with the formula $Li_{3-2n}M_nOX$ (where M = Ca, Mg, or Ba) were also shown to have relatively high ionic conductivity [25]. Among these, $Li_{3-2n}Ba_nClO$ at a doping level of x = 0.005 exhibited an exceptionally high room-temperature ionic conductivity of 2.5×10^{-2} S·cm^{-1}.

These anti-perovskite materials are characterized by low electronic conductivity, typically ranging between 10^{-9} and 10^{-7} S·cm^{-1}, and exhibit broad electrochemical stability windows surpassing the oxidation potential of chloride ions [26]. Despite these promising attributes, a major limitation is their high reactivity with ambient moisture, necessitating careful handling under controlled environments.

Computational studies based on density functional theory (DFT) have shown that Li_3OCl, Li_3OBr, and their halide-mixed variants are thermodynamically metastable. Molecular dynamics (MD) simulations further revealed that perfectly ordered, defect-free anti-perovskite crystals exhibit relatively poor Li-ion conductivity [27]. The incorporation of lithium vacancies and structural disorder significantly improves ion mobility by reducing the energy barriers for diffusion. Additionally, recent theoretical analyses have pointed to grain boundaries in polycrystalline Li_3OCl as major bottlenecks, with local ionic conductivities reduced by up to an order of magnitude compared to the bulk. Adding complexity to the discussion, Li et al. have challenged the assumed composition of "Li_3OX," proposing that the material might instead be Li_2OHX. They reported that substituting a small fraction of hydroxide (OH^-) with fluoride (F^-) to form $Li_2(OH)_{0.9}F_{0.1}Cl$ resulted in a compound that is not only stable in contact with lithium metal but also exhibits an extended electrochemical stability window up to 9 V versus Li^+/Li [28].

3.2 *Sulfide-Based Electrolytes*

Sulfide-based solid electrolytes with the argyrodite structure are among the most prominent materials in this category and currently represent the leading choice for fabricating all-solid-state batteries. These materials follow the general formula $Li_{7+n-Y}M^4{}_nM^5{}_{1-n}Ch_{6-Y}X_Y$ (where M^4 = Si, Ge, Sn; M^5 = P, Sb; Ch = O, S, Se; X = Cl, Br, I, BH_4; $0 \leq x \leq 1$; $0 \leq y \leq 2$). Their popularity stems from their exceptionally high lithium-ion conductivities—reaching up to 24 mS·cm^{-1} as well as the availability of scalable, solution-based synthesis routes, both of which are essential for commercial viability [29].

The argyrodite crystal structure was originally discovered in the mineral Ag_8GeS_6 [30]. This was followed by several compounds with related crystal structure that possessed high mobility of Cu^+ and Ag^+ ions. Lithium argyrodites exhibit two structural polymorphs, and the temperature at which the phase transition occurs depends on the specific chemical composition. For instance, compounds like Li_7PS_6, which adopt the low-temperature orthorhombic structure, show very low ionic conductivity at room temperature. High conductivity is only achieved in the high-temperature cubic phase, which can be stabilized by substituting one or more sulfide ions (S^{2-}) with halide ions (X^-). This leads to compounds such as Li_6PS_5X (where X = Cl, Br, I) [31].

In these cubic Li_6PS_5X structures, the halide ions (X^-) form a face-centered cubic (*fcc*) lattice, while non-bonded ("free") sulfide ions (S^{2-}) and PS_4^{3-} tetrahedral units fill the tetrahedral and octahedral voids within the lattice, respectively. Depending on the halide component, varying degrees of site disorder between S^{2-} and X^- ions can occur. For example, in Li_6PS_5I, due to the large ionic size difference, the iodide and sulfide ions maintain distinct crystallographic positions. However, for Li_6PS_5Br and Li_6PS_5Cl, significant site disorder is observed due to the closer match in ionic radii. Aside from anion site disorder, high cation site disorder at the lithium sites may significantly modify the activation barrier for ionic movement. Such a scenario arises in the cases of Li-rich $Li_{6+x}M_xSb_{1-x}S_5I$ (M = Si, Ge, Sn) group of compounds wherein the highly disordered lithium site leads to a drop in the activation energy leading to an enhanced ionic mobility [32].

3.3 *Lithium Metal Halide-Based Electrolytes*

Prior to 2018, halide SSEs received limited research attention due to their relatively low room-temperature ionic conductivity ($\sigma < 10^{-3}$ S/cm), which made them less competitive compared to oxide and sulfide electrolytes. This perception changed dramatically in 2018, when Asano et al. synthesized Li_3YCl_6 and Li_3YBr_6, achieving conductivities above 10^{-3} S/cm at room temperature using high-energy ball milling followed by high-temperature annealing. This breakthrough triggered a surge of interest in halide-based SSEs for lithium metal batteries (LMBs). A key focus of recent research has been on ternary halide electrolytes with the general formula Li_3MX_6, where M is a metal (such as Y, Sc, In, or Er) and X is a halide (Cl, Br, or I). These materials have consistently demonstrated room-temperature conductivities in the 10^{-3} S/cm range, making them strong candidates for solid-state battery applications.

In 2020, Sun and co-workers developed monoclinic Li_3ScCl_6 with enhanced Li^+ conductivity by optimizing the occupancy of Li and Sc within the structure. Prepared by cold pressing, this SSE delivered a conductivity of 3.0×10^{-3} S/cm at room temperature. Additionally, it showed a broad electrochemical stability window (ESW) from 0.9 to 4.3 V versus Li^+/Li, and demonstrated excellent interfacial compatibility with both lithium metal and LCO cathodes.

Building on this progress, Liu et al. reported a mixed-halide composition, $Li_3Y(Br_3Cl_3)$, which reached an impressive σ of 7.2×10^{-3} S/cm at room temperature. This was accomplished via hot pressing at 170 °C under 294 MPa, setting a new benchmark in halide electrolyte performance. The lanthanide-based chloride

SSE with the composition $Li_{0.388}Ta_{0.238}La_{0.475}Cl_3$ achieved a conductivity of 3.0×10^{-3} S/cm at 30 °C and excellent compatibility with lithium metal. An all-solid-state lithium metal battery using this electrolyte and an NMC523 cathode delivered an initial discharge capacity of 163.96 mAh/g and maintained 81.6% capacity after 100 cycles at 30 °C.

Although many halide SSEs now achieve room-temperature conductivities above 10^{-3} S/cm, achieving electrochemical stability windows >4.5 V versus Li^+/Li remains challenging. To address this, Ye et al. introduced Ta^{5+} doping into Li_3InCl_6, producing an optimized composition $Li_{2.6}In_{0.8}Ta_{0.2}Cl_6$. This SSE exhibited both high conductivity (4.5×10^{-3} S/cm at 30 °C) and an extended ESW of 5.1 V, one of the widest reported among halide electrolytes. Halide SSEs are now gaining momentum thanks to several key advantages: (i) high Li^+ conductivity at room temperature without requiring high-temperature sintering, (ii) good chemical stability against lithium metal, (iii) mechanical flexibility, allowing cold pressing for easy processing, (iv) wider ESWs compared to sulfide electrolytes, (v) environmental safety, as they do not release toxic gases upon moisture exposure. Despite their high moisture sensitivity, halide SSEs are still more environmentally benign than sulfides. Their unique balance of conductivity, interfacial compatibility, and processability makes them highly promising candidates for use in next-generation all-solid-state lithium batteries.

4 Sodium Ion Solid Electrolytes

4.1 NaSICON-Type Sodium Ion Conductors

NaSICON-type sodium-ion conductors, structurally derived from the base compound $NaZr_2(PO_4)_3$, have garnered considerable interest due to their adaptable crystal framework and favorable sodium-ion transport characteristics [33]. When silicon partially replaces phosphorus in the structure, a compositional series with the formula $Na_{1+n}Zr_2Si_xP_{3-n}O_{12}$ (with $0 \leq x \leq 3$) is formed. The composition with x approximately equal to 2, corresponding to $Na_3Zr_2Si_2PO_{12}$, demonstrates the highest ionic conductivity—measured at 6.7×10^{-4} S/cm at room temperature (25 °C) and reaching 0.2 S/cm at 300 °C. This enhanced conductivity is attributed to the material's three-dimensional framework made up of interconnected SiO_4 and PO_4 tetrahedra linked to ZrO_6 octahedra, creating channels that facilitate rapid Na^+ ion migration [34].

Beyond their high ionic mobility, NaSICON ceramics also show low thermal expansion, which supports their stability and functionality at elevated temperatures. Their structure allows a broad range of chemical substitutions. The Zr^{4+} site can accommodate cations with varying valence states—divalent, trivalent, tetravalent, and even pentavalent—while the Si/P positions can be replaced by elements like Ge^{4+} or As^{4+} to tailor the material's properties. One notable example is the replacement of Zr with hafnium, as in $Na_{3.2}Hf_2(SiO_4)_{2.2}(PO_4)_{0.8}$, which exhibits a significantly improved ionic conductivity of 2.3×10^{-3} S/cm at ambient temperature compared to its Zr-based counterpart.

NaSICON materials are frequently not phase-pure and often contain a secondary amorphous or glassy phase composed primarily of sodium, phosphate or silicate species, and the introduced dopants. Such multiphase behavior has been reported in materials doped with various cations, including Mg^{2+}, Co^{2+}, Ce^{4+}, Yb^{3+}, and Gd^{3+}. Moreover, during the sintering process, monoclinic zirconia often crystallizes from the melt. Both dopant selection and the processing parameters, such as sintering temperature and time, play critical roles in determining the microstructure, especially the grain boundary characteristics, which in turn affect the overall ionic transport.

A significant advancement in this area involves scandium doping. Sc-substituted NaSICON materials have achieved record-setting ionic conductivities, with values as high as 4.0×10^{-3} S/cm at room temperature—currently the highest reported among all NaSICON-type compounds. Through detailed analysis of both structural and electrochemical data, Guin et al. identified that the most favorable ionic conductivity occurs when the sodium content nears 3.3 per formula unit and the average ionic radius of the cations occupying the M-site is about 0.72 Å.

Furthermore, doping has a pronounced effect on the crystallographic phase behavior. Specifically, the temperature at which the material transitions from the monoclinic to the rhombohedral phase is highly sensitive to the nature of the dopants. Research by Jolley et al. demonstrated that substituting Zr^{4+} with cations of different valencies can stabilize the high-symmetry rhombohedral phase. Among the various dopants tested, Y^{3+} was found to be the most effective in lowering the phase transition temperature while maintaining minimal distortion of the crystal lattice.

4.2 *Sulfide-Based Sodium Ion Conductors*

Like their lithium counterparts, sodium thiophosphate materials have recently gained considerable attention due to their relatively high ionic conductivity, minimal grain boundary resistance, and favorable mechanical properties such as ductility. However, the number of sodium superionic conductors remains limited compared to lithium-based sulfides, and their room-temperature ionic conductivities are generally lower. As a result, the search for high-performance sulfide-based sodium-ion conductors is actively ongoing.

A significant advancement in this field was the stabilization of the high-temperature cubic phase of Na_3PS_4 (c-Na_3PS_4, space group: $I - 43\ m$) through crystallization from the glassy state. This cubic phase demonstrated promising Na^+ conductivity (~2×10^{-4} S/cm). Subsequent improvements in synthetic purity—specifically, using high-purity (>99%) Na_2S as a precursor—further increased conductivity to 4.6×10^{-4} S/cm. This success revived interest in sodium thiophosphates, especially since the more common tetragonal phase of Na_3PS_4 (t-Na_3PS_4, space group: $P - 42_1c$) shows an order of magnitude lower conductivity [35].

Recent research efforts have focused on understanding the origin of the enhanced ionic conductivity in c-Na_3PS_4 and determining whether the differences in transport behavior between the cubic and tetragonal phases stem from structural, microstructural, or defect-related factors. Using Rietveld refinement and pair distribution function (PDF) analyses, researchers examined Na_3PS_4 samples synthesized via ball-milling and high-temperature methods. While Rietveld refinement identified the average structures as cubic and tetragonal, respectively, PDF analysis revealed that both samples share the same tetragonal-like local structure. Electrochemical impedance spectroscopy (EIS) further indicated that the high ionic conductivity observed in the "cubic" Na_3PS_4 is not primarily due to its long-range crystal structure, but rather the high density of defects introduced by the ball-milling process [36].

4.3 Sodium-Based LGPS-Type Sulfide Conductors

The sodium analog of the well-known LGPS ($Li_{10}GeP_2S_{12}$) system represents another promising class of sulfide-based solid electrolytes. The compound $Na_{10}GeP_2S_{12}$ (NGPS) was initially predicted through computational studies to have a high ionic conductivity [37]. Building on this, Richards et al. explored the $Na_{10}MP_2S_{12}$ (M = Si, Ge, Sn) series and predicted that Na^+ diffusivity would improve, and activation energy would decrease in the order Sn < Ge < Si [38]. Molecular dynamics (MD) simulations suggested that $Na_{10}SnP_2S_{12}$ behaves as a one-dimensional (1D) conductor, with Na^+ diffusion occurring along chains of NaS_4 tetrahedra aligned with the c-axis. A major advancement came with the successful synthesis of stoichiometric $Na_{11}Sn_2PS_{12}$ as a pure, polycrystalline phase. This material demonstrated high ionic conductivity (1.4×10^{-3} S/cm) and low-energy activation. The crystal structure was determined using single-crystal X-ray diffraction. Unlike the LGPS or NGPS structures, where Na^+ occupies a mixture of sites, in $Na_{11}Sn_2PS_{12}$, Na^+ ions occupy only octahedral sites, enabling fully three-dimensional (3D) ion conduction, as confirmed by ab initio molecular dynamics (AIMD) simulations [39].

5 Solid Electrolytes Containing Multivalent Mobile Ions

The quest for multivalent metal electrode-based solid electrolytes and batteries with high abundance and volumetric capacity that can one day potentially replace lithium-ion/metal batteries has been ongoing for more than a few decades [40]. Among these Magnesium (Mg)-based solid electrolytes are an emerging class of materials being investigated for next-generation multivalent batteries [41]. Magnesium metal offers several advantages over lithium, including higher volumetric capacity (3833 mAh/

cm^3 vs. 2061 mAh/cm^3), natural abundance, and improved safety due to dendrite-free deposition [42]. However, the development of viable magnesium solid-state electrolytes (Mg-SSEs) has been hindered by sluggish Mg^{2+} ion mobility and limited material options that support reversible Mg transport [43].

Despite their promise, magnesium (Mg) batteries face several critical challenges that must be addressed before they become commercially viable. One major issue is the formation of a passivation layer on the Mg anode surface when it contacts conventional carbonate-based electrolytes, which are commonly used in lithium-ion batteries. Unlike lithium, Mg^{2+} ions cannot penetrate this layer due to its insulating nature, rendering these electrolytes ineffective. To overcome this, novel electrolytes based on Grignard reagents in ethereal solvents like tetrahydrofuran (THF) have been developed. While they enable Mg plating/stripping, these electrolytes suffer from high flammability, high vapor pressure, corrosion due to Cl^- ions, and narrow electrochemical windows (<2 V vs. Mg/Mg^{2+}), limiting their practical application.

As a safer alternative, solid-state electrolytes (SSEs) are being explored for use in all-solid-state Mg batteries. They offer improved safety, mechanical stability, and resistance to thermal shock. However, designing solid Mg-ion conductors with sufficiently high ionic conductivity at room temperature remains difficult due to the strong Coulombic interactions and high charge density of Mg^{2+}. Current research efforts aim to enhance Mg-ion mobility and achieve target conductivities in the range of 10^{-3} to 10^{-4} S/cm, similar to those in lithium and sodium-based SSEs.

Recent efforts have explored borohydrides (e.g., $Mg(BH_4)_2$) and thiophosphates, which show promise due to their softer anionic substructures and increased ion mobility. Some complex hydrides, such as $MgSc_2Se_4$ and $MgZr_4(PO_4)_6$, have also shown moderate Mg^{2+} conductivity at elevated temperatures [44, 45]. Unlike Mg^{2+} electrolytes, the development of Ca^{2+} solid electrolytes and Ca-metal solid-state battery is hindered by poor interphase formation and sluggish Ca^{2+} kinetics [46]. Very recently, Skov et al. discovered a Ca-hydridoborate-based solid electrolyte $Ca(BH_4)_2 \cdot 2NH_2CH_3$, that was able to achieve an ionic conductivity of 10^{-5} S cm^{-1} at T = 70 °C [47]. A polymer hybrid solid-state electrolyte consisting of an ionic liquid, 1-ethyl-3-methylimidazolium chloride (EMImCl) and $AlCl_3$ was used to build an aluminum ion battery [48].

6 Computational and Machine Learning Models

Computational methods offer a powerful alternative for screening and predicting material properties. Density functional theory (DFT), molecular dynamics (MD), and nudged elastic band (NEB) methods are commonly employed to model ion diffusion pathways, calculate activation energies, and estimate stability windows.

6.1 Density Functional Theory (DFT)

DFT calculations also enable the prediction of Li^+ or Na^+ diffusion barriers by modeling the migration of ions through the lattice [49]. Lower diffusion barriers correlate with higher ionic conductivity, which is essential for effective SSE performance. DFT can help identify whether a compound is stable against decomposition when in contact with lithium metal or a cathode material [50].

6.2 Molecular Dynamics (MD)

Ab initio molecular dynamics (AIMD) simulations have become a powerful tool in the study and design of solid-state electrolytes (SSEs) for next-generation lithium and sodium batteries. AIMD enables accurate modeling of interatomic interactions and ion transport mechanisms without relying on empirical potentials [51]. This makes it especially well-suited for investigating the complex behavior of mobile ions within crystalline or disordered solid-state systems. A primary application of AIMD in SSE research is the evaluation of ionic conductivity and diffusion pathways. For example, AIMD simulations have elucidated lithium-ion migration mechanisms in sulfide-based conductors such as $Li_{10}GeP_2S_{12}$ (LGPS), revealing fast one-dimensional diffusion channels and identifying structural bottlenecks [52]. Similar techniques have been used to probe conduction in halide-based SSEs like Li_3YCl_6 and Li_3ScCl_6, as well as in oxide-based garnets such as $Li_7La_3Zr_2O_{12}$ (LLZO) [51, 53]. Galvez-Aranda et al. studied the interphase formed between the solid-state electrolyte and the lithium metal anode using the AIMD simulation. Their study revealed that the interface becomes unstable due to the formation of lithium and lanthanum oxides [54].

6.3 High-Throughput Computational Screening

High-throughput computational workflows have been developed to screen large materials databases such as the Materials Project and Open Quantum Materials Database (OQMD) [55, 56]. These workflows automate the calculation of properties like formation energy, band gaps, electrochemical stability windows, and Li^+ migration pathways. Notable examples include the discovery of lithium superionic conductors such as $Li_{10}GeP_2S_{12}$ (LGPS) [57]. Recently, Xiao et al. performed a high-throughput computational screening across a database of 6242 number of compounds based on known structural features Li-based garnets and NASICON-type superionic conductors (SICs). Their aim was to find structures with a network of Li sites showing fast Li-ion conductivity. They were able to identify seven promising candidates with high ionic mobility. Furthermore, the authors also cross-verified the pathway of lithium conduction in these materials using ab initio molecular dynamics (AIMD) simulations [58].

Despite their utility, computational models are often limited by their assumptions—such as perfect crystal structures and ideal stoichiometry—and computational expense. This is where machine learning offers significant advantages.

6.4 Machine Learning Models for Materials Discovery

Machine learning models, when trained on data generated from experiments or simulations, can learn complex structure-property relationships and make rapid predictions across large materials spaces [59]. This is why ML has been successfully used to predict viable element combinations in a quaternary chemical phases field such Li-M-X-X' (M is a metal, X and X' are two different anions) using certain experimentally known variables such as ionic radius, valence, atomic weight, etc. Such ML model significantly reduces the time and cost associated with the identification and optimization of promising SSE candidates [60, 61]. Interestingly, Xu et al. combined machine learning with molecular dynamics simulation to understand the correlation between the polyanion rotation and lithium mobility in two different classes of compounds such as thiophosphates, and metal chlorides [62]. Guo et al., went one step further and used artificial intelligence aided machine learning (AIML) to establish structure-composition correlation in a wide number of glass–ceramic lithium thiophosphate Electrolytes [63].

7 Conclusion and Future Outlook

In this chapter, we have examined the recent advancements across various classes of solid-state electrolytes, including sulfide-based, oxide-based, and polymer-based materials, as well as hybrid approaches. Each category offers distinct advantages—ranging from high ionic conductivity to enhanced mechanical strength and electrochemical stability—yet faces inherent challenges related to interfacial compatibility, dendrite suppression, and scalable processing.

The fundamental understanding of ion transport mechanisms, interfacial behavior, and mechanical properties has deepened significantly over the past decade, thanks to a combination of advanced characterization techniques and multiscale modeling approaches. These insights have laid the groundwork for rational materials design and interface engineering strategies that address longstanding issues such as poor wettability, grain boundary resistance, and chemical instability. Notably, progress in tailoring the electrolyte–electrode interface has shown great promise in extending battery life and preventing failure modes such as short-circuiting and capacity fade.

Despite these encouraging advancements, the widespread commercialization of solid-state batteries (SSBs) is still constrained by several factors. Manufacturing complexity, high material costs, and insufficient standardization across research and industry remain substantial barriers. Furthermore, integrating solid-state electrolytes into full-cell architectures, particularly with lithium metal anodes and high-voltage cathodes, continues to demand significant innovation in materials design, processing techniques, and device engineering.

Looking ahead, the future of next-generation SSEs hinges on several strategic directions. One promising avenue lies in composite electrolytes, which synergistically combine the mechanical robustness of ceramics with the flexibility and processability of polymers. These materials can provide a balanced solution to the brittleness-conductivity trade-off and are well-suited for scalable fabrication techniques such as tape casting and roll-to-roll processing.

Another critical frontier is interface optimization, where in situ and operando characterization methods will play a pivotal role in elucidating dynamic processes at the electrolyte–electrode interface. Surface coatings, artificial interphases, and advanced current collector designs are likely to evolve into essential components of future battery systems. In parallel, machine learning and high-throughput screening are expected to accelerate the discovery of novel SSE materials with tailored properties and improved performance metrics.

Additionally, system-level innovations, such as monolithic cell architectures and solid-state battery modules with integrated thermal management, will be key to translating material-level advances into commercial viability. The incorporation of recyclable and earth-abundant materials will also be vital for ensuring sustainability and supply chain resilience, particularly as the demand for electric vehicles and grid-scale storage continues to rise.

In conclusion, the pathway toward practical and commercially viable solid-state batteries is a multidisciplinary endeavor that encompasses materials science, electrochemistry, engineering, and data science. While significant challenges remain, the ongoing convergence of fundamental research and technological innovation suggests that solid-state electrolytes will play a central role in shaping the future of energy storage. Continued collaboration across academia, industry, and policy will be essential in unlocking the full potential of next-generation SSEs and ushering in a new era of safe, high-performance battery technologies.

References

1. Mauger A, Julien CM (2017) Critical review on lithium-ion batteries: are they safe? Sustainable? Ionics 23:1933–1947
2. Li J, Ma C, Chi M, Liang C, Dudney NJ (2015) Solid electrolyte: the key for high-voltage lithium batteries. Adv Energy Mater 5:1401408
3. Zhang Z et al (2018) New horizons for inorganic solid state ion conductors. Energy Environ Sci 11:1945–1976
4. Wang Y et al (2015) Design principles for solid-state lithium superionic conductors. Nat Mater 14:1026–1031
5. Kraft MA et al (2017) Influence of lattice polarizability on the ionic conductivity in the lithium superionic Argyrodites Li_6PS_5X (X = Cl, Br, I). J Am Chem Soc 139:10909–10918

6. Banik A et al (2021) On the underestimated influence of synthetic conditions in solid ionic conductors. Chem Sci 12:6238–6263
7. O'Callaghan MP, Lynham DR, Cussen EJ, Chen GZ (2006) Structure and ionic-transport properties of lithium-containing garnets $Li_3Ln_3Te_2O_{12}$ (Ln = Y, Pr, Nd, Sm−Lu). Chem Mater 18:4681–4689
8. Wang C et al (2020) Garnet-type solid-state electrolytes: materials, interfaces, and batteries. Chem Rev 120:4257–4300
9. Chen S et al (2022) Ultrafast sintering for ceramic-based all-solid-state lithium-metal batteries. Adv Mater 34:2200430
10. Wang R et al (2021) High-temperature ultrafast sintering: exploiting a new kinetic region to fabricate porous solid-state electrolyte scaffolds. Adv Mater 33:2100726
11. Wang Q et al (2024) Designing lithium halide solid electrolytes. Nat Commun 15:1050
12. Ghidiu M, Ruhl J, Culver SP, Zeier WG (2019) Solution-based synthesis of lithium thiophosphate superionic conductors for solid-state batteries: a chemistry perspective. J Mater Chem A 7:17735–17753
13. Liu Z et al (2013) Anomalous high ionic conductivity of Nanoporous β-Li3PS4. J Am Chem Soc 135:975–978
14. Ito S, Nakakita M, Aihara Y, Uehara T, Machida N (2014) A synthesis of crystalline Li7P3S11 solid electrolyte from 1, 2-dimethoxyethane solvent. J Power Sources 271:342–345
15. Yubuchi S et al (2019) An argyrodite sulfide-based superionic conductor synthesized by a liquid-phase technique with tetrahydrofuran and ethanol. J Mater Chem A 7:558–566
16. Sedlmaier SJ et al (2017) Li4PS4I: a Li^+ superionic conductor synthesized by a solvent-based soft chemistry approach. Chem Mater 29:1830–1835
17. Indrawan RF, Gamo H, Nagai A, Matsuda A (2023) Chemically understanding the liquid-phase synthesis of Argyrodite solid electrolyte Li_6PS_5Cl with the highest ionic conductivity for all-solid-state batteries. Chem Mater 35:2549–2558
18. Calpa M et al (2020) Formation mechanism of thiophosphate anions in the liquid-phase synthesis of Sulfide solid electrolytes using polar aprotic solvents. Chem Mater 32:9627–9632
19. Delnick FM, Yang G, Self EC, Meyer HMI, Nanda J (2020) Investigation of complex intermediates in solvent-mediated synthesis of thiophosphate solid-state electrolytes. J Phys Chem C 124:27396–27402
20. Manthiram A, Yu X, Wang S (2017) Lithium battery chemistries enabled by solid-state electrolytes. Nat Rev Mater 2:16103
21. Abakumov AM, Fedotov SS, Antipov EV, Tarascon J-M (2020) Solid state chemistry for developing better metal-ion batteries. Nat Commun 11:4976
22. Murugan R, Thangadurai V, Weppner W (2007) Fast lithium ion conduction in garnet-type Li7La3Zr2O12. Angew Chem Int Ed 46:7778–7781
23. Rangasamy E, Wolfenstine J, Sakamoto J (2012) The role of Al and Li concentration on the formation of cubic garnet solid electrolyte of nominal composition Li7La3Zr2O12. Solid State Ionics 206:28–32
24. Zhao Y, Daemen LL (2012) Superionic conductivity in lithium-rich anti-perovskites. J Am Chem Soc 134:15042–15047
25. Braga MH, Ferreira JA, Stockhausen V, Oliveira JE, El-Azab A (2014) Novel Li3ClO based glasses with superionic properties for lithium batteries. J Mater Chem A 2:5470–5480
26. Emly A, Kioupakis E, Van der Ven A (2013) Phase stability and transport mechanisms in Antiperovskite Li3OCl and Li3OBr superionic conductors. Chem Mater 25:4663–4670
27. Zhang Y, Zhao Y, Chen C (2013) Ab initio study of the stabilities of and mechanism of superionic transport in lithium-rich antiperovskites. Phys Rev B 87:134303
28. Li Y et al (2016) Fluorine-doped Antiperovskite electrolyte for all-solid-state lithium-ion batteries. Angew Chem Int Ed 55:9965–9968
29. Patel SV et al (2021) Tunable lithium-ion transport in mixed-halide Argyrodites Li_{6-x}PS5–xClBrx: an unusual compositional space. Chem Mater 33:1435–1443
30. Eulenberger G (1977) Die Kristallstruktur der Tieftemperaturmodifikation von Ag_8GeS_6. MonShefte Für Chem/Chem Mon 108:901–913

31. Deiseroth H-J et al (2008) Li_6PS_5X: a class of crystalline Li-rich solids with an unusually high Li^+ mobility. Angew Chem Int Ed 47:755–758
32. Lee Y et al (2021) Superionic Si-substituted lithium Argyrodite Sulfide electrolyte $Li_{6+x}Sb_{1-x}Si_xS_5I$ for all-solid-state batteries. ACS Sustain Chem Eng 9:120–128
33. Goodenough JB, Hong HY-P, Kafalas JA (1976) Fast Na^+-ion transport in skeleton structures. Mater Res Bull 11:203–220
34. Anantharamulu N et al (2011) A wide-ranging review on Nasicon type materials. J Mater Sci 46:2821–2837
35. Hayashi A, Noi K, Sakuda A, Tatsumisago M (2012) Superionic glass-ceramic electrolytes for room-temperature rechargeable sodium batteries. Nat Commun 3:856
36. Krauskopf T, Culver SP, Zeier WG (2018) Local tetragonal structure of the cubic superionic conductor Na_3PS_4. Inorg Chem 57:4739–4744
37. Kandagal VS, Bharadwaj MD, Waghmare UV (2015) Theoretical prediction of a highly conducting solid electrolyte for sodium batteries: $Na_{10}GeP_2S_{12}$. J Mater Chem A 3:12992–12999
38. Richards WD et al (2016) Design and synthesis of the superionic conductor $Na_{10}SnP_2S_{12}$. Nat Commun 7:11009
39. Duchardt M, Ruschewitz U, Adams S, Dehnen S, Roling B (2018) Vacancy-controlled Na+ Superion conduction in $Na_{11}Sn_2PS_{12}$. Angew Chem Int Ed 57:1351–1355
40. Brunklaus G, Lennartz P, Winter M (2024) Metal electrodes for next-generation rechargeable batteries. Nat Rev Electr Eng 1:79–92
41. Aurbach D et al (2000) Prototype systems for rechargeable magnesium batteries. Nature 407:724–727
42. Jaschin PW, Gao Y, Li Y, Bo S-H (2020) A materials perspective on magnesium-ion-based solid-state electrolytes. J Mater Chem A 8:2875–2897
43. Zhan Y, Zhang W, Lei B, Liu H, Li W (2020) Recent development of Mg ion solid electrolyte. Front Chem 8
44. Canepa P et al (2017) High magnesium mobility in ternary spinel chalcogenides. Nat Commun 8:1759
45. Adamu M, Kale GM (2016) Novel sol–gel synthesis of $MgZr_4P_6O_{24}$ composite solid electrolyte and newer insight into the Mg^{2+}-ion conducting properties using impedance spectroscopy. J Phys Chem C 120:17909–17915
46. Song H, Su J, Wang C (2021) Hybrid solid electrolyte interphases enabled Ultralong life Ca-metal batteries working at room temperature. Adv Mater 33:2006141
47. Skov LN, Grinderslev JB, Kjær TSS, Kristensen LR, Jensen TR (2025) Towards solid-state batteries using a calcium Hydridoborate electrolyte. Angew Chem Int Ed 64:e202500613
48. Sun X-G et al (2016) Polymer gel electrolytes for application in aluminum deposition and rechargeable aluminum ion batteries. Chem Commun 52:292–295
49. Mo Y, Ong SP, Ceder G (2012) First principles study of the $Li_{10}GeP_2S_{12}$ lithium super ionic conductor material. Chem Mater 24:15–17
50. Schwietert TK, Vasileiadis A, Wagemaker M (2021) First-principles prediction of the electrochemical stability and reaction mechanisms of solid-state electrolytes. JACS Au 1:1488–1496
51. Zhu Z, Zhu Y (2022) Developing classical interatomic potentials for solid electrolytes. Acc Mater Res 3:1101–1105
52. Xu M, Ding J, Ma E (2012) One-dimensional stringlike cooperative migration of lithium ions in an ultrafast ionic conductor. Appl Phys Lett 101:031901
53. Liu Y, Wang S, Nolan AM, Ling C, Mo Y (2020) Tailoring the cation lattice for chloride lithium-ion conductors. Adv Energy Mater 10:2002356
54. Galvez-Aranda DE, Seminario JM (2020) Solid electrolyte interphase formation between the $Li_{0.29}La_{0.57}TiO_3$ solid-state electrolyte and a Li-metal anode: an ab initio molecular dynamics study. RSC Adv 10:9000–9015
55. Jain A et al (2013) Commentary: the materials project: a materials genome approach to accelerating materials innovation. APL Materials 1:011002

56. Kirklin S et al (2015) The open quantum materials database (OQMD): assessing the accuracy of DFT formation energies. npj Comput Mater 1:15010
57. Kamaya N et al (2011) A lithium superionic conductor. Nat Mater 10:682–686
58. Xiao Y et al (2021) Lithium oxide superionic conductors inspired by garnet and NASICON structures. Adv Energy Mater 11:2101437
59. Butler KT, Davies DW, Cartwright H, Isayev O, Walsh A (2018) Machine learning for molecular and materials science. Nature 559:547–555
60. Vasylenko A et al (2021) Element selection for crystalline inorganic solid discovery guided by unsupervised machine learning of experimentally explored chemistry. Nat Commun 12:5561
61. Han G et al (2024) Superionic lithium transport via multiple coordination environments defined by two-anion packing. Science 383:739–745
62. Xu Z et al (2023) Machine learning molecular dynamics simulation identifying weakly negative effect of polyanion rotation on Li-ion migration. npj Comput Mater 9:105
63. Guo H, Wang Q, Urban A, Artrith N (2022) Artificial intelligence-aided mapping of the structure–composition–conductivity relationships of glass–ceramic lithium thiophosphate electrolytes. Chem Mater 34:6702–6712

Interfacial Reaction and Characteristics of Solid-State Batteries

Divya Rani and Pushpendra Kumar

Abstract Interfaces including solid–solid, solid–liquid, liquid–gas, and solid–gas play a crucial role in electrochemical cells, particularly in all-solid-state batteries (ASSBs). While extensive research has focused on addressing the lower ionic conductivity of solid electrolytes (SEs) in ASSBs, the high interfacial impedance caused by multiple solid–solid interfaces remains a major obstacle to their widespread commercialization. Unlike liquid electrolytes, SEs cannot effectively fill interfacial voids, often leading to inadequate physical contact between the anode, cathode, and electrolyte, even when pressure is applied during cell fabrication. Moreover, the limited electrochemical window of SEs, coupled with the chemical potential difference between SEs and electrodes, can lead to SE decomposition, resulting in the formation of the cathode–electrolyte interface (CEI) and the solid–electrolyte interface (SEI). While these interfacial layers can sometimes extend the electrochemical stability of SEs, their continued growth particularly if they conduct both ions and electrons can further increase interfacial impedance. Therefore, a detailed analysis of CEI/SEI composition and evolution is essential for improving ASSB performance. Additionally, space-charge layer formation at the SE–electrode interface reduces the local cation concentration compared to the bulk SE, further exacerbating impedance. Inhomogeneous contact between the anode and SE can also lead to non-uniform current distribution, promoting dendrite formation, even in the presence of lithophilic thin films. Therefore, a comprehensive understanding of the physicochemical and electrochemical processes at ASSB interfaces is critical for mitigating these challenges. This chapter provides an in-depth analysis of SE–cathode, SE–anode, SE–additive, and SE–collector interfaces. We will examine the key challenges associated with each interface and discuss potential strategies to enhance interfacial stability and performance.

D. Rani · P. Kumar (✉)
School of Physical Science, Jawaharlal Nehru University, New Delhi, India
e-mail: pushpendra@jnu.ac.in

A. K. Rai et al. (eds.), *Electrochemistry of Next-Generation Solid-State Batteries*, Green Energy and Technology, https://doi.org/10.1007/978-981-95-7603-6_5

1 Introduction

Solid-state batteries (SSBs) are considered a viable alternative to liquid electrolyte-based batteries, as they can potentially overcome safety concerns and energy density limitations. However, the low ionic conductivity of solid electrolytes (SEs) hinders the widespread adoption of SSBs. Significant progress has been made in developing SEs with ionic conductivities comparable to those of liquid electrolytes. In SSBs, all components are solid, necessitating the application of high pressure to ensure adequate contact between the various solid components [1]. In liquid-based batteries, ions can easily navigate between electrodes through the porous structure of the separator and electrode. However, in SSBs, ion transport is more challenging due to the solid–solid interfaces present [2]. The high interfacial impedance in SSBs limits their efficiency as energy storage systems. This elevated impedance arises from multiple factors, which will be discussed in detail in this chapter.

The application of high pressure can lead to the formation of voids and crevices at solid–solid interfaces, further increasing interfacial impedance. Another contributing factor is the mismatch in the Fermi levels of different solids, which leads to the formation of a space-charge layer. If the chemical potential of the SE does not align with that of the electrode, a thermodynamically unstable interface is formed, inevitably resulting in the development of a solid–electrolyte interphase (SEI) and a cathode–electrolyte interphase (CEI). SEI/CEI layers composed of mixed electronic and ionic conducting (MEIC) materials further passivate the electrode|SE interface, leading to an increase in impedance [3]. A comprehensive understanding of charge transfer kinetics and mechanical properties is crucial for designing and optimizing SEI/CEI structures. Using various probing and characterization techniques, this chapter will analyze the reactions occurring at the anode–electrolyte and cathode–electrolyte interfaces, as well as the decomposition products that form the SEI/CEI. Once the interfacial processes are well understood, it becomes possible to engineer SEI/CEI layers with tailored properties to enhance battery performance [4].

Numerous research groups have successfully developed solid-state electrolytes. In all-solid-state batteries (ASSBs), various types of solid electrolytes are employed, including sulfide-based, oxide-based, chloride-based, and polymer-based electrolytes. Among these, sulfide-based electrolytes exhibit the highest ionic conductivities compared to oxide-, chloride-, and polymer-based solid electrolytes (Table 1).

2 Electrochemical Stability Window

The highest occupied molecular orbital (HOMO) and lowest unoccupied molecular orbital (LUMO) of SEs must be analyzed to understand their stability against redox reactions. If the HOMO level of the SE is higher than the cathode's Fermi energy level (μ_c), oxidation occurs. Similarly, if the LUMO level is higher than the anode's Fermi level (μ_a), reduction takes place. These redox reactions lead to the formation

Table 1 Various solid electrolytes and their ionic conductivity

Solid electrolyte	Conductivity (at room temperature)	Ref. No.
LBO ($Li_{2+x}C_{1-x}B_xO_3$)	5×10^{-6} S cm^{-1}	[5]
LATP ($Li_{1+x}Al_xTi_{2-x}(PO_4)_3$)	0.48×10^{-3} S cm^{-1}	[6]
$Li_{9.54}Si_{1.74}P_{1.44}S_{11.7}Cl_{0.3}$	25×10^{-3} S cm^{-1}	[7]
LGPS ($Li_{10}GeP_2S_{12}$)	12×10^{-3} S cm^{-1}	[8]
LPS ($Li_7P_3S_{11}$)	17×10^{-3} S cm^{-1}	[9]
LPSCl (Li_6PS_5Cl)	3.15×10^{-3} S cm^{-1}	[10]
LYCl (Li_3YCl_6)	0.032×10^{-3} S cm^{-1}	[11]
LiPON ($Li_{3.6}PO_{3.4}N_{0.6}$)	2×10^{-3} S cm^{-1}	[12]
$Li_{9.54}Si_{1.74}P_{1.44}S_{11.7}Cl_{0.3}$	25×10^{-3} S cm^{-1}	[7]
$Li_{9.6}P_3S_{12}$	$\approx 55 \times 10^{-3}$ S cm^{-1}	[7]

of the SEI and CEI. Therefore, to obtain electrochemical stability, the HOMO should be lower than μ_c, and the LUMO should be higher than μ_a, oxide- and chloride-based electrolytes exhibit a wider electrochemical stability window [8].

To understand the significance of interfacial phenomena, the experiment by Kamaya et al. provides valuable insights. They constructed a Li|LGPS|Au cell and performed cyclic voltammetry (CV), obtaining an unusually high electrochemical stability window of 5–10 V. This outcome is attributed to the use of a flat metal plunger, which results in poor electrical contact with LGPS [8]. Reduced contact leads to less electron transfer and, consequently, greater stability against decomposition. However, when conductive carbon is introduced between the SE and the metal plunger, additional CV peaks appear [13]. The presence of an electronically conductive material at the electrode enhances interfacial contact, increasing decomposition. This experimental setup allows CV measurements to more accurately quantify the electrochemical stability window. It is evident that solid electrolytes with high ionic conductivity tend to undergo more decomposition, leading to lower coulombic efficiency (CE) during the first cycle. In contrast, electrolytes with lower ionic conductivity experience less degradation, yielding higher initial CE [14].

The oxidation and reduction onset potentials of solid electrolytes are determined by their anionic and cationic structures. The oxidation potential is correlated with the electron density of anions in the order $N^{3-} > S^{2-} > O^{2-} > Cl^-$, leading to a decreasing oxidation stability trend: LiYCl > LLZO > LPS > LiPON [15]. Electrolytes with higher electron density exhibit lower oxidation potentials. Similarly, transition metal-based cations, such as Y^{3+}, Zr^{4+}, and Ti^{4+}, can undergo reduction to lower oxidation states or even their metallic forms (e.g., Y^{3+} to Y^0). In Li_6PS_5Cl, phosphorus in the P^{5+} state is reduced to P/P^{3-} within the Li–In|Li_6PS_5Cl|Li_6PS_5Cl–C cell [16]. The decomposition of solid electrolytes is influenced not only by their intrinsic electrochemical stability but also by the cathode material. For instance, Li_3YCl_6 decomposes at 2.3 V when paired with $LiCoO_2$ (LCO), but when combined with nickel-cobalt-aluminum oxide (NCA), decomposition occurs within 2.5–3.6 V.

This is evidenced by a minor voltage slope during the first charge cycle, which disappears in subsequent cycles [17, 18]. The CEI formed in the first cycle extends the electrochemical stability window beyond the intrinsic thermodynamic limits of the solid electrolyte [17].

LiPON, a commonly used SE, exhibits decomposition in the first cycle, but the formation of a stable SEI and CEI enables the cell to function over an extended potential range of 0–5 V, as demonstrated by Li et al. in a Li|LiPON|$LiNi_{0.5}Mn_{1.5}O_4$ cell [17]. From computational calculation, LiPON's stability window is found to be 0.68 V to 2.63 V against the Li metal. Additionally, the decomposition products of LiPON in SEI are predicted to be Li_3P, Li_2O and Li_3N [4]. Cheng et al. have performed the cryo-high-resolution TEM at the Li|LiPON interface and fast Fourier transform pattern instantiates the copresence of Li metal, Li_2O, Li_3N, and Li_3PO_4 in the SEI, contributing to SEI stability [19]. Put et al. have designed the cell TiN|LiPON|Au, to study the decomposition of LiPON. As applied voltage reached 4.3–4.5 V LiPON starts to decompose into phosphates ($Li_4P_2O_7$ and $LiPO_3$), and ultimately it decomposes to P_4O_{10}. As applied potential increases (around 4.5 V), the formed layer of P_4O_{10} has the voltage drop present over the layer, which further shield the rest of the SE. Because of this very reason, its experimental decomposition potential is bit higher than the calculated one [20].

3 Effect of Carbon Additives

Inadequate physical contact and low ionic conductivity at the SE|cathode interface necessitate the incorporation of SE into the cathode to form a cathode@SE composite. Therefore, positive electrode in ASSB comprises of cathode-active material, SE, binder, and carbon additive. The use of carbon additive is quite necessary in cathode@SE, as cathode@SE contains electronically insulating SE, so to prevent the electronic insulation of cathode-active material, carbon additives are added [15]. Carbon additives possess an inherently large surface area and high electronic conductivity, facilitating fast kinetics; however, they also promote SE decomposition. The rate of decomposition of SE, influenced by the surface area and electronic conductivity of the carbon additive, for example Zhang et al. investigated the $LiNb_{0.5}Ta_{0.5}O_3$-coated LCO@LGPS cathode composite's decomposition mechanism, containing different carbon additives. They employed a 70:30 weight ratio of coated LCO to LGPS. Once the composite was formed, Super P, PVDF, and various carbon additives were incorporated in different ratios. A decrease in the slope of the charging curve indicated SE decomposition as shown in Fig. 1b, d. To determine the specific surface area of different carbon additives, Zhang et al. performed a BET analysis. The cathode containing Ketjen black as the carbon additive exhibited the lowest charging curve slope due to its largest surface area (1454 m^2g^{-1}) among all tested carbon additives. Additionally, the coulombic efficiency (CE) of the cell without carbon additives was higher than that of the cell with carbon additives, suggesting that SE decomposition was enhanced in the presence of carbon additives [21].

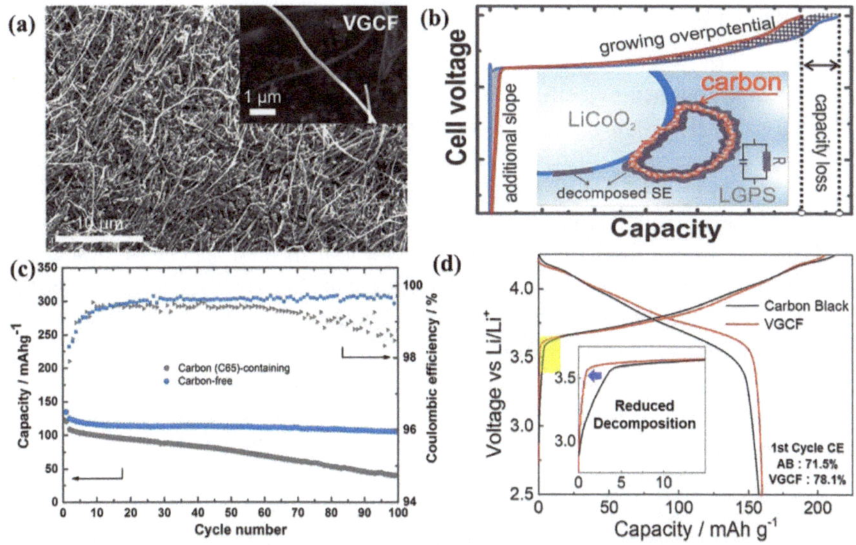

Fig. 1 **a** SEM image of VGCF [22]. **b** Comparison of the first charge curve with (blue) and without (red) carbon additives and graphical illustration of cathode–SE interface of SSB during the charging process [21]. **c** GCD performance of SSB with 5.7 wt% of C65 with and without C65 additive [21]. **d** First charge and discharge profile of Li-In|Li_6PS_5Cl|NMC811 [22]. Copyright © 2019 American Chemical Society, Copyright © 2017 American Chemical Society

The difference in the Fermi energy between the SE and cathode also contributes to the slope reduction during charging. This issue can be mitigated by applying an electronically insulating coating on either the SE or the cathode. However, excessive carbon additive content negates the benefits of such coatings.

Therefore, optimizing the carbon additive quantity is crucial when an electronically insulating coating is applied to the cathode [21]. SSBs with cathodes containing high-surface-area carbon additives exhibit electrochemical impedance spectroscopy (EIS) spectra with more well-defined semicircles compared to cathodes without carbon additives. An imperfect semicircle is attributed to surface roughness and the non-uniform distribution of current. However, the introduction of carbon improves both current and potential distribution uniformity [21]. This indicates that electrolyte decomposition kinetics are accelerated on the cathode's surface. Thus, optimizing cathode composites with lower carbon content is essential to minimize SE decomposition while maintaining high specific capacity. Research findings suggest that replacing acetylene black (AB) with vapor-grown carbon fibers (VGCF) in NMC cathodes eliminates decomposition-related features in the charge curve. The CE for the first cycle was reported as 71% for AB and 78% for VGCF (Fig. 1c) [22]. C65 exhibits a larger surface area (BET surface area of 82 m^2g^{-1}) compared to VGCF (BET surface area of 11 m^2g^{-1}). Consequently, SE decomposition in the cathode composite yields 80 mAh/g when C65 is used, whereas only 7 $mAhg^{-1}$ is obtained with VGCF [23].

4 Cathode–SE Interface

Chemical reactions at the electrode–SE interface are governed by the chemical potential difference between the electrode and the SE, which varies with the state of charge. Compared to pure $LiCoO_2$ (LCO), delithiated LCO exhibits more negative reaction energy, indicating a higher tendency for reactivity due to the elevated oxidation state of cobalt in the charged state. The reaction energies of delithiated LCO with $Li_{10}GeP_2S_{12}$ (LGPS) and Li_3PS_4 (LPS) are − 499 meV/atom and − 564 meV/atom, respectively. This is attributed to the higher oxidation states of Co, in delithiated LCO, which facilitates transition metal reduction. Theoretical calculations estimate the breakdown energies of LCO with LGPS and Li_3PS_4 as − 340 meV/atom and − 405 meV/atom, respectively, suggesting that LPS is more reactive with LCO than LGPS [4]. This behavior is analogous to that of other transition metal layered oxides, such as the NCA cathode, when paired with sulfide-based SEs. Banerjee et al. observed that delithiated NCA exhibits greater reactivity with LPSCl than lithiated NCA [17].

However, theoretical calculations using density functional theory (DFT) do not always align with experimental observations. The low elasticity and high surface energy of oxide-based SEs make it challenging to achieve intimate contact between the cathode and SE. Mechanical milling and room-temperature pressing are often insufficient for establishing proper contact. To address this, annealing at temperatures above 700 °C is employed. However, this process leads to the formation of additional interfacial compounds with poor ionic conductivity, thereby increasing the impedance of all-solid-state batteries (ASSBs) [24]. Jung et al. constructed two Li|LPSCl|NMC cells, with one cycled immediately and the other left to rest for 48 h before cycling. The rested cell exhibited higher initial capacity and reduced polarization but suffered from lower capacity retention, indicating the formation of a CEI at the NMC|Li_6PS_5Cl interface even in the absence of an applied bias. Impedance measurements as a function of time confirmed that within 48 h, the impedance increased fourfold. The increase in bulk impedance was attributed to electrode and SE degradation, while the rise in interfacial impedance was ascribed to CEI formation [25].

Several research groups have employed transmission electron microscopy (TEM) to investigate the cathode–SE interface. The interface between LCO and $Li_2S–P_2S_5$ SE was found to be amorphous [3]. Cross-sectional energy-dispersive X-ray spectroscopy (EDX) line profiles reveal mutual diffusion of Co, P, and S (Fig. 2a). Electrochemical impedance spectroscopy (EIS) and first discharge curves of cells containing Li_2SiO_3-coated LCO and uncoated LCO indicate reduced capacity and increased interfacial impedance (Fig. 2b, c). A similar observation was made at the LCO|LGPS interface [26]. DFT calculations provide insights into reaction energy diagrams between various compounds. The reaction energy between NCA and LPSCl is highly negative, indicating an unstable interface between the two materials. LiCl, Li_3PO_4, Ni_3S_4, and NiS_2 are identified as potential reaction products. Conversely, the reaction energy between $LiNbO_3$ and LPSCl is less negative, indicating a more stable interface. This makes $LiNbO_3$ an ideal coating material for LPSCl. Raman spectra of NCA/LPSCl reveal a blue shift in the LPSCl peak for both pristine and charged NCA, indicating S-to-O substitution and the formation of oxidized LPSCl [17]. The reaction energy between two compounds varies with the state of charge (degree of lithiation). Lithiated NMC exhibits highly reactive oxygen species that interact with P and S, leading to the formation of $Mn_3(PO_4)_2$, $Ni_3(PO_4)_2$, and Li_3PO_4 at the NMC|LPSCl interface [15]. De-lithiated NMC reacts with LPSCl to form $Ni_3(PO_4)_2$, Li_3PO_4, and $Mn_3(PO_4)_2$, which are electronically insulating but exhibit some ionic conductivity. Banerjee et al. charged NCA in liquid electrolyte (LE) to 4.3 V, collected the charged cathode powder, and mixed it with LPSCl. X-ray diffraction (XRD) and Raman analysis of the mixed powder revealed additional peaks corresponding to Ni_3S_4, LiCl, and Li_3PO_4, confirming that oxygen in charged NCA participates in the reaction with LPSCl. This results in a degraded LPSCl phase with ionic conductivity reduced to one-third of its original value [17].

It is important to note that charged cathodes and certain SEs, such as sulfide-based SEs, are highly sensitive to atmospheric exposure. Therefore, measurements should be conducted in an isolated environment to prevent any unwanted reactions. Another crucial aspect that demands attention is the formation of a space-charge layer (SCL).

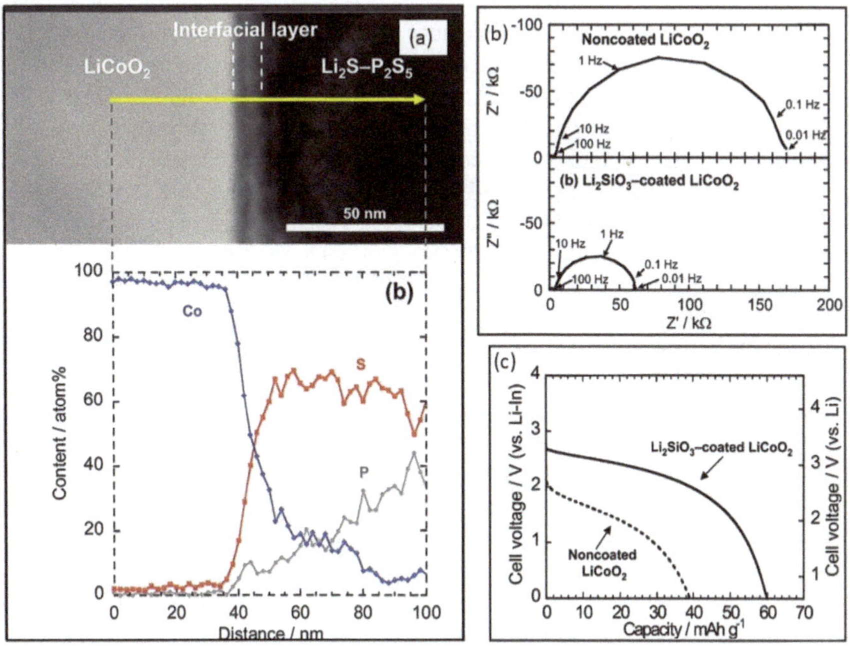

Fig. 2 **a** Cross-sectional HAADF-STEM image of Li_2SiO_3 coated-$LiCoO_2$|$Li_2S–P_2S_5$ interface, cross-sectional EDX line profile for Co, P, and S elements. **b, c** EIS and discharge curve of cells containing Li_2SiO_3 coated-$LiCoO_2$ and uncoated $LiCoO_2$ [17]. Copyright © 2009 American Chemical Society

To equilibrate the Fermi energy at the interface between the positive electrode and the solid electrolyte, band bending occurs, leading to the development of the SCL. This layer can be understood as the energy gap between the valence band maximum and the conduction band minimum, distinguishing it from the electrostatic potential observed in the electric double layer (EDL). Due to the Fermi energy mismatch, electrons migrate from the sulfide SE to the positive electrode. To maintain charge neutrality, Li ions also move from the SE to the positive electrode, resulting in the formation of a Li-ion depletion layer. An additional oxidation peak observed during charging has been attributed to the presence of the SCL. Since the SCL forms at the interface, it induces a localized electrostatic potential, leading to Li-ion deficiency on the SE side [27]. The existence of the depletion layer has been confirmed through numerical simulations. However, only in situ Kelvin probe and electron holography techniques have successfully detected the localized electrostatic potential [28].

4.1 Coating Layer on Cathode Material

By coating an electronically insulating material on the cathode, we can prevent the chemical reaction between the SE and the cathode, avoid the formation of an SCL, and ultimately reduce interfacial resistance. The density functional theory (DFT)-calculated E_{hull} value measures the phase stability of the coating material; the lower the E_{hull}, the coating on cathode is more thermodynamically stable [17]. Chemical stability is another essential characteristic that cathode coatings must possess. Pseudo-binary phase diagrams and decomposition energy calculations for various interfaces have shown that the LNO-LPSCl interface is more stable than the NCA-LPSCl interface. Before annealing the LCO@LLZO composite cathode, Kato et al. applied an Nb layer to the LCO@LLZO cathode. During annealing, Li^+ diffusion from LLZO to Nb results in the oxidation of the Nb layer, forming amorphous or poorly crystalline $LiNbO_3/Li_3Nb_4O_6$. This amorphous phase has an ionic conductivity of 10^{-6} S cm^{-1} at ambient temperature, whereas the crystalline LNO layer exhibits a significantly lower ionic conductivity of 10^{-11} S cm^{-1}. The amorphous $LiNbO_3/Li_3Nb_4O_6$ phase is also resistant to strain induced by volume changes. An all-solid-state battery (ASSB) with an Nb-coated cathode, compared to a cell using the same electrode materials but employing a liquid electrolyte (LE), undergoes CV measurements. The ASSB with the Nb-coated cathode exhibits a much narrower gap between oxidation and reduction peaks, indicating lower polarization [29]. Ohta and colleagues found that ASSBs with an LCO|$Li_4Ti_5O_{12}$ layer|LPS interface show no slope during charging. When a sufficiently thick $Li_4Ti_5O_{12}$ layer is present, SCL formation can be suppressed. This layer at the cathode–SE interface prevents space-charge layer formation [30]. Commonly used coating materials include ternary metal oxides (TMOs) such as Li_2SiO_3, $Li_4Ti_5O_{12}$, $LiTaO_3$, $LiAlO_2$, $Li_2O–ZrO_2$, and $LiNbO_3$. However, these materials are unstable above 4 V. The low electronic conductivity of TMO contributes to their sluggish kinetics, which in turn enhances long-term cycling stability. Coating materials that contain mobile charge carriers can result in higher discharge capacities; for instance, Li_2SiO_3 exhibits a higher discharge capacity than SiO_2 because it contains Li^+ charge carriers [29]. Although Li_3InCl_6 is highly stable with LCO, it does not require a coating layer on the cathode material [31].

$LiNbO_3$ is commonly used coating material in sulfide-based all-solid-state lithium-ion battery, because of its high Li^+ conductivity. But its coating is prepared by wet method using flammable alcohols; it can hinder their large-scale coating process. Jung et al. have used Li_3BO_3 (LBO), LBO coated on LCO via wet method using water. Although LBO has low Li^+ conductivity, but after wet coating LBO converts to $Li_{3-x}B_{1-x}C_xO_3$ (LBCO) which is highly Li^+ conducting (6×10^{-6} S cm^{-1} at 30 °C) material [32, 33]. After cycling grain size of LCO decreases, Zhang et al. observed this via Williamson-Hall plots derived from the X-ray pattern. Because of volumetric changes, degradation continues even after the thin coating of oxide layer on LCO. Electrochemical reaction between LCO and LGPS still persists even after coating, which results in Co diffusion at the LCO|LGPS interface, leading to LCO degradation. While charging, –S–S– bonds formed between PS_4 molecules, further leads to the formation of polymerized PS_4 ions. During this whole process, Li migrates from LGPS to LCO|LGPS interface and results in the Li depletion layer, and this hinders the mobility of Li ions. The instability of the c-LCO|thiophosphate SE interface is attributed to the Co-P exchange at the interface, which is energetically favorable. Electron energy loss spectroscopy (EELS) confirms this transfer. As shown in Fig. 3a, b, the Co signal in the coating layer is initially negligible but becomes significant after galvanostatic cycling, indicating Co diffusion to the surface. EIS data reveal SE decomposition, inferred from the enlargement of the semicircle in the high-frequency region. The increase in semicircle size in the mid- and low-frequency regions is attributed to instability at the cathode|SE interface [32]. During lithiation and delithiation, the mechanical behavior of the coating becomes crucial. Positive volume fluctuation induces stress, whereas negative volume fluctuation leads to void formation. These stresses and voids on the electrode surface cause cracking, exposing the cathode-active material to the SE, which results in continuous CEI growth [34]. Electrolytes in ASSBs do not undergo volume changes unless they participate in chemical reactions. Sakuda et al. applied a pressure of 360 MPa at room temperature to sulfide-based SE (Li_2S–P_2S_5). Due to the minimal bond energy between Li and S in Li_2S–P_2S_5, Li^+ and $PS_4{}^{3-}$ anions diffuse into the voids of the cathode material during pressing. In contrast, oxide-based SEs have a higher binding energy between Li and O atoms, making diffusion into voids more difficult under pressure. Li^+ and $PS_4{}^{3-}$ ions in Li_2S–P_2S_5 exhibit covalent bonding characteristics.

Unlike ionic compounds with strong electrostatic repulsion, the low electrostatic repulsion and bond energy in these materials support their covalent nature. Understanding Young's modulus of SEs is essential since electrode material volume changes require sustained contact between the SE and the active material. A lower Young's modulus is preferable for ASSBs. Sulfide-based SEs have a lower Young's modulus than oxide and ceramic SEs due to their lower bond dissociation energy per unit volume [35]. Because of the larger Na^+ ionic size, the bond dissociation energy between Na^+ and $PS_4{}^{3-}$ is lower than that of Li^+ and $PS_4{}^{3-}$, making

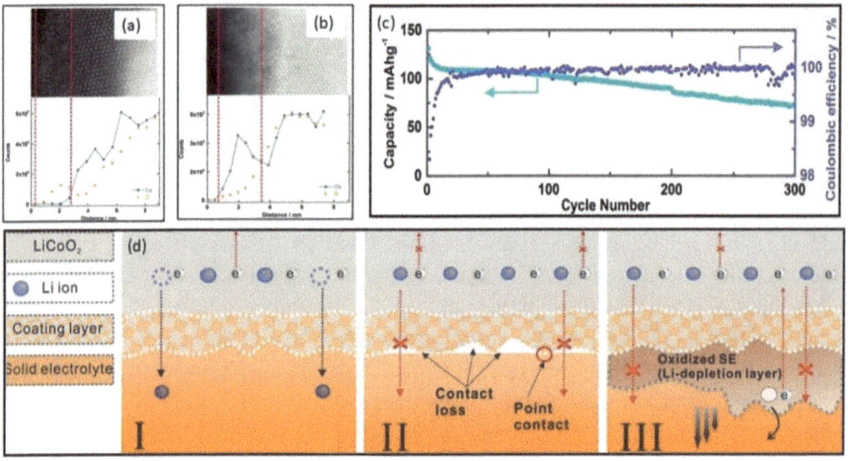

Fig. 3 **a**, **b** High-resolution ADF-STEM image of a pristine c-LCO and c-LCO after 300 GCD cycling. **c** Cycling performance of c-LCO|LGPS|LTO cell. **d** Schematic diagram of three possible situations that can occur in ASSB with coated LCO as active cathode material [32]. Copyright © 2018 American Chemical Society

$75Na_2S{:}25P_2S_5$ glass more moldable than $75Li_2S{:}25P_2S_5$. Consequently, Na^+ and PS_4^{3-} ions diffuse more easily into cathode voids in $75Na_2S{:}25P_2S_5$ glass SE [36]. Kato et al. added various lithium halides to Li_2S-P_2S_5 glasses, observing a reduction in Young's modulus with increasing lithium halide content [37]. Although ionic crystals like LiI and LiBr have a higher Young's modulus, the elastic modulus is directly related to the free volume in the glasses; a lower Young's modulus corresponds to a higher free volume. In (100 − y) $(0.75Li_2S{:}0.25P_2S_5)$.yLiI composites, the bond dissociation energies are unknown. However, the reduction in transition and crystallization temperatures with LiI addition suggests a decrease in bond energy [38]. Excessive cycling leads to voids and cracks at the SE|electrode interface, causing a loss of contact between the SE and the electrode. Due to the buried nature of the SE|electrode interface, analyzing Li-ion distribution at this interface is challenging. Simulations indicate a non-uniform Li-ion distribution, making restacking ineffective in restoring ASSB capacity [35].

It has been found that crystallographic axes exhibit varying degrees of distortion and that most cathode-active materials do not shrink uniformly over the entire composition range. $LiCoO_2$ volume expands by 2% up to a 50% state of discharge ($Li_{0.5}CoO_2$). Beyond this point, the oxidation state of oxygen ions decreases, leading to slight repulsion between oxygen layers, which in turn causes volume shrinkage in the cathode material [39]. Wu et al. utilized synchrotron radiation X-ray tomographic microscopy to observe the reaction of micro-sized Sn anodes in SSBs. Their analysis revealed that expansion and compression during cycling are reversible due to the self-healing properties of LPS SE. Image analysis also showed that cracks in the Sn electrode led to high tortuosity, which impedes Li-ion transport [40].

Several strategies can be employed to mitigate electrode material expansion and compression: (1) using a flexible polymer binder to reduce stress and compression, (2) selecting cathodes with positive volume change and anodes with negative volume change (or vice versa) to partially counteract expansion effects, (3) utilizing materials with minimal volume change, such as LTO, and (4) combining materials with opposing volume changes to compensate for overall expansion and contraction [41]. SEs like Na_3PS_4 (NPS) exhibit positive volume changes when used with NMC and LCO cathodes but show negative volume changes with Fe-based cathodes such as $NaFePO_4F$ [42].

SEs are considered ideal for addressing polysulfide shuttling and electrolyte consumption issues at the Li|liquid electrolyte interface in Li-S batteries. These challenges can be mitigated by using sulfide-based SEs, as the reaction energy between sulfur and sulfide SEs is effectively zero [43]. This non-reactive S|sulfide-based SE interface necessitates cathode composite fabrication via high-energy ball milling or high-temperature treatment. To enhance the contact area at the S|carbon and S|SE interfaces, reducing the particle size of S, Li_2S, and SE is essential. Smaller particles enable higher sulfur loading in the electrode. Nagao et al. increased the sulfur content to 50 wt% and mixed it with nanocarbon via mechanical milling at 155 °C, a temperature above sulfur's melting point but low enough to maintain its molten state with reduced viscosity. This process results in a larger molten sulfur–nanocarbon interface, increasing the contact area. The reduced particle size achieved through mechanical milling retains a capacity of 1050 mAh g^{-1} at 0.064 mA cm^{-2} for 50 cycles in a Li|Li_2S–P_2S_5|S-carbon composite cell (Fig. 4d) [44].

Suzuki et al. prepared a sulfur-carbon replica-solid electrolyte (S/CR/SE) composite using three different methods: 1. mechanical mixing, 2. liquid-phase mixing, and 3. combined mixing. Via mechanical mixing particle size are even and each part of the S/CR/SE composite is homogeneously dispersed, but potentiality of SE to penetrate in the mesopores is also quite low. While in liquid-phase mixing, there is disparity in the size of the particle but filling ratio of SE in the mesopores is relatively high. Combined mixing has the advantageous properties of both mechanical and liquid-phase mixing, and it has homogeneous dispersion of uniform particles, while maintaining the high filling ratio of SE in mesopores. The cathode prepared via liquid-phase mixing exhibited the poorest capacity retention due to the large particle size of the S/CR/SE surface. In contrast, the electrode prepared using the combined mixing method demonstrated stable capacity retention, attributed to the high ionic conductivity in the mesopores introduced by liquid-phase mixing and the significant downsizing and uniform dispersion of particles achieved through mechanical milling [45].

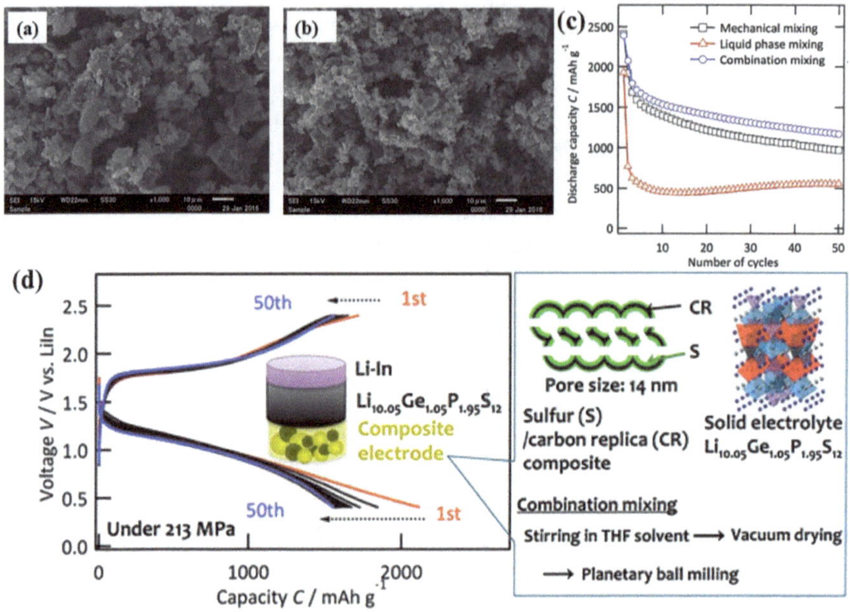

Fig. 4 **a**, **b** SEM image of S/carbon replica/SE prepared by liquid-phase mixing and combination mixing. **c** Comparison of cycling performance of various cathode composite made of mechanical mixing, liquid-phase mixing and combination mixing. **d** GCD curve of Li-In|LGPS|S-CR-SE and schematic representation of cathode preparation by combination mixing [45]. Copyright © 2018 American Chemical Society

5 Anode–SE Interface

Considering the safety issues associated with Li metal batteries, ASSBs are regarded as an effective solution to the problem. They can also suppress the issue of dendritic growth. However, the highly electropositive and reactive nature of Li poses a significant challenge in the development of ASSBs. Due to its low electronegativity, Li metal exhibits extreme reactivity. Although certain compounds demonstrate zero reaction energy with Li, for example, the reaction energy at the LiI|Li and Li_3N|Li interfaces is nearly zero, other SEs exhibit high reactivity with Li metal. These types of compounds form mixed electronic and ionic compounds (MEICs) at the Li|SE interface. Another type of interface involves the formation of an ionically conducting and electronically insulating layer following the reaction between Li metal and SE [46]. The stability of three- or four-element ionic compounds depends on the formation energy of the products generated at the Li anode.

For example, $Li_{10}GeP_2S_{12}$ decomposes into Li_2S, Li_3P, and a Li-Ge alloy, as observed by in situ XPS. The Li/LGPS interface comprises Li_2S, an insulator; a Li-Ge electronically conducting alloy; and ionically conducting Li_3P. This combination incrementally increases the Li|SE interfacial impedance, ultimately leading to cell failure [46]. Wenzel et al. studied Li_6PS_5X (X=Cl, Br, I) type argyrodites, which exhibit ionic conductivity in the range of 10^{-2} S cm^{-1} to 10^{-3} S cm^{-1} at room temperature. They derived a general expression for the interface reaction based on in situ XPS analysis.

$$Li_6PS_5X + 8Li \rightarrow 5Li_2S + Li_3P + LiX. \quad (1)$$

The primary product of this process is Li_2S, which has low ionic and electronic conductivity. In contrast, Li_3P and LiX are ionically conductive materials. Since all the compounds formed are electronically insulating, they contribute to the formation of a robust interface, supporting an extended cycle life [47]. Tatsumigo et al. synthesized $80Li_2S{:}20P_2S_5$ via mechanical milling of Li_2S and P_2S_5 precursors for 20 h. To obtain a glass-ceramic material, the ball-milled powder was subsequently heated to 230 °C and then cooled to room temperature. Electrochemical impedance spectroscopy (EIS), galvanostatic charge–discharge (GCD), and cyclic voltammetry (CV) measurements were conducted to evaluate the stability of LPS glass SEs in a Li|$80Li_2S{:}20P_2S_5$|Li cell. The impedance of the cell increased from 1000 Ω to 1200 Ω after 24 h. However, over time, the interface between Li and the LPS SE stabilized, leading to only a slight increase in impedance beyond 1200 Ω after 1500 h. CV measurements revealed two distinct peaks corresponding to lithium dissolution and deposition. No additional peaks were observed up to 10 V, indicating a wide electrochemical stability window [48]. Only a few solid electrolytes exhibit high ionic conductivity, which leads to the formation of mixed-conducting interfaces. Tian et al. initially employed Na_3PS_4 in a Na|Na_3PS_4|Na symmetric cell. The formation of Na_3P, an electronically conductive phase, at the interface caused the cell to become unstable. However, due to the kinetically stable Na-Sn and NPS contact, the Na-Sn alloy effectively replaced Na metal, enabling stable plating and stripping for up to 40 h [49].

6 Thermodynamically Stable Layer on Anode and SE

Applying a thermodynamically stable layer on Li metal and employing a high ionic conductivity SE instead of an alloy as the anode material are effective strategies to enhance battery performance. Zhang et al. spin-coated LiH_2PO_4 on Li metal using H_3PO_4 dissolved in tetrahydrofuran (THF), varying the H_3PO_4 concentration. The EIS measurements were conducted on the coated-Li|LGPS|LCO cell on the first and 14th days after preparation. Figure 5b shows that cells coated with LiH_2PO_4 at an 80% H_3PO_4 concentration exhibit reduced interfacial impedance. X-ray photoelectron spectroscopy (XPS) characterization reveals that the amount of Ge with an oxidation state of + 4 on the LGPS surface remains largely unchanged after 14 days, indicating that the LiH_2PO_4 layer effectively prevents Ge reduction. To evaluate the stability of LGPS against Li and coated-Li, Li|LGPS|Li and 80%-Li|LGPS|80%-Li cells underwent galvanostatic cycling at 0.1 mA cm^{-2} at 25 °C. A sudden voltage increase

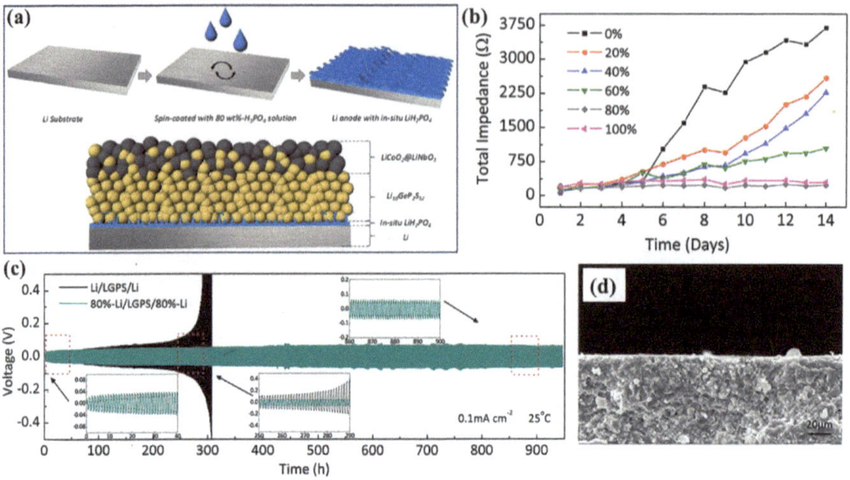

Fig. 5 **a** Schematic presentation of preparation of in situ LiH_2PO_4 on Li metal anode. **b** Impedance variation for different concentrations of LiH_2PO_4. **c** Galvanostatic cycling of symmetric cell anode|LGPS|anode. **d** SEM of LGPS-80% Li interface [50]. Copyright © 2017 American Chemical Society

occurred after 40 h in the Li|LGPS|Li cell, whereas the 80%-Li|LGPS|80%-Li cell remained stable for 950 h at − 0.05/+ 0.05 V [50].

Another study introduced a protective layer on the highly ionically conductive $Li_{3.833}Sn_{0.833}As_{0.166}S$ SE. Lithium borohydride ($LiBH_4$) and lithium iodide (LiI) were dissolved in THF in a 3:1 molar ratio to create a protective layer ($3LiBH_4$:LiI) on $Li_{3.833}Sn_{0.833}As_{0.166}S$. The pellet was soaked in the solution and then vacuum-dried at 170 °C for 1 h. To assess stability in humid air, the coated SE pellets were exposed to an 80% humid environment for 48 h, followed by X-ray diffraction (XRD) measurements. Moisture-induced intensity broadening at 16° disappeared after drying at 140 °C for 1 h. A symmetric cell was fabricated by attaching Li foil to both sides of the coated $Li_{3.833}Sn_{0.833}As_{0.166}S$ pellet and cycled at room temperature at 0.1 mA cm^{-2}. The cell containing the coated SE exhibited no significant increase in potential until 1200 minutes, whereas the uncoated $Li_{3.833}Sn_{0.833}As_{0.166}S$ SE displayed considerable fluctuations [51].

LATP ($Li_{1.3}Al_{0.3}Ti_{1.7}(PO_4)_3$) is another high ionic conductivity SE; however, it is unstable against Li metal. Liu et al. deposited an Al_2O_3 layer on LATP using atomic layer deposition (ALD). The cyclability test of Li|LATP|Li and Li|ALD@LATP|Li cells was performed at 0.01 mA cm^{-2}. The uncoated LATP cell initially exhibited a low overpotential (~ 0.1 V), but it increased from 0.1 V to 3.5 V over 300 cycles due to the decomposition of LATP at the Li-LATP interface. Al_2O_3-coated LATP (Al_2O_3 coated on LATP via atomic layer deposition) have higher overpotential (10 V), as observed in initial cycles owing to the insulating nature of Al_2O_3. But as the cycling goes on, lithiation of Al_2O_3 progresses, which decreases the overpotential to 0.9 V after 100 cycles and it remains nearly constant for next 500 cycles [52].

Hao et al. coated a 200 nm ZnO layer on both sides of LATP pellets via magnetron sputtering, formed ZnO@LATP@ZnO is sandwiched between two lithium metal pieces to form Li|ZnO@LATP@ZnO|Li symmetric cell, and this symmetric cell was tested for cyclability at 0.2 mA cm^{-2} and 0.05 mA cm^{-2}, showing no potential variation up to 200 and 2000 cycles, respectively. EIS measurement revealed that initial impedance of Li|LATP|Li symmetric cell is 80,554 Ω, which further increases to 190,000 Ω after the first cycle at 0.05 mA cm^{-2}, while in Li|ZnO@LATP@ZnO|Li symmetric cell, impedance decreases from 353 Ω to 300 Ω after first cycle at same current density.

The production of Zn and Li_2O at the LATP-Li contact, which limits side reactions, prevents an increase in impedance [54]. Mechanically, ZnO and Al_2O_3 are stiffer, possess a higher Young's modulus, and are not suitable for accommodating severe stress. In contrast, polymers or hybrid coating layers have a lower Young's modulus, making them effective in mitigating induced stress due to their flexibility. Hu et al. coated Na_3SbS_4 (NAS) SE with cellulose-poly(ethylene oxide) (CPEO), a polymer coating that not only accommodates stress caused by anode volume changes but also suppresses the formation of Na_3Sb, which results from the interaction between NAS and the Na interface. A Na|CPEO|NAS|CPEO|Na symmetric cell cycled continuously for 800 h without any potential change (Fig. 6b) [53].

However, polymer-based coatings suffer from low transference numbers and inadequate ionic conductivity, leading to anion depletion at the Li-metal interface and

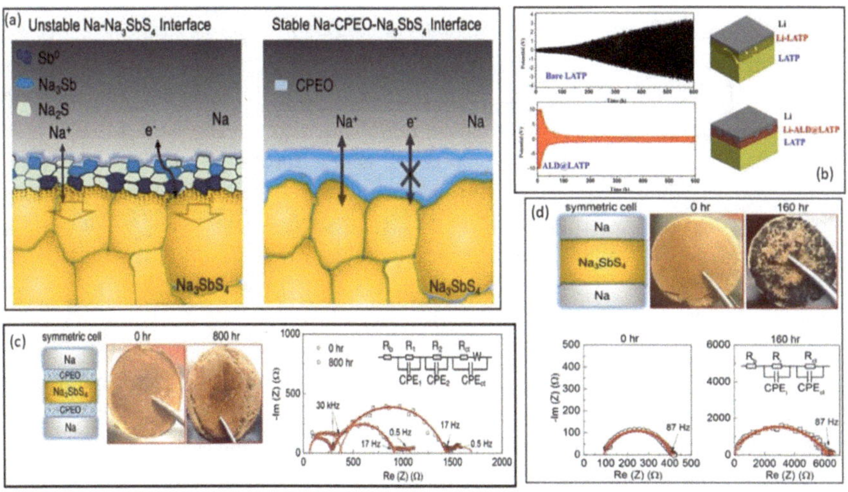

Fig. 6 **a** Schematic diagram of NAS|Na and NAS|CPEO|Na interface [53]. **b** Galvanostatic cycling of Al_2O_3 coated LATP by ALD as SE in Li|ALD@LATP|Li symmetrical cell [52]. **c** Picture CPEO coated-NAS before and after cycling at 0.1 mA cm^{-2} current density and EIS of the cell containing CPEO coated-NAS as electrolyte before and after cycling [53]. **d** CPEO coated-NAS before and after cycling at 0.1 mA cm^{-2} current density and EIS of the cell containing CPEO coated-NAS as electrolyte before and after cycling [53]. Copyright © 2018 American Chemical Society. Copyright © 2019 American Chemical Society

an irregular electric field distribution that promotes Li dendrite formation [55]. The breakdown energy of SE decreases because certain metal oxides, such as LLZO and $LiAlO_2$, exhibit strong ionic conductivity and a high activation barrier for breaking metal-oxygen bonds. Despite their high ionic conductivity (10^{-4} to 10^{-3} S cm^{-1}), the resistance at the lithium metal-SE interface remains excessively high due to reduced intimate contact, resulting in small gap formation. Inhomogeneous contact leads to an uneven electric field, making these gaps hotspots for lithium dendrite nucleation [56].

To address this issue, improving the wettability of lithium metal through the application of a lithophilic coating at the SE interface is essential. Techniques such as atomic layer deposition (ALD), molecular layer deposition (MLD), and pulsed laser deposition (PLD) are employed to deposit lithophilic materials onto the SE [57]. For instance, an amorphous silicon layer applied to niobium-doped LLZO via plasma-enhanced chemical vapor deposition (PECVD) transforms into a Li–Si alloy, reducing the resistance from 925 Ω cm^{-2} to 125 Ω cm^{-2} [58]. Numerous review articles have been published on the characterization of the anode–SE interface, as well as the properties and mechanisms governing its formation. For further reading, these references may be consulted [59, 60].

7 Summary

All-solid-state batteries (ASSBs) have the potential to replace liquid electrolyte-based batteries, as lithium metal can be employed as the anode due to its mechanical robustness. However, the high interfacial impedance remains a significant challenge to commercialization. Two primary factors contribute to the elevated impedance in ASSBs: (1) chemical reactions occurring at various interfaces and (2) inadequate physical contact between different components of the cell. The electrochemical stability window of the solid electrolyte (SE) plays a crucial role in determining its compatibility with both the anode and cathode. However, accurately assessing this stability window presents inherent challenges. The incorporation of carbon compounds between the SE and the electrode during cyclic voltammetry (CV) measurements results in an increased number of peaks, underscoring the complexity of interfacial behavior. The impact of carbon additives is particularly pronounced in the positive electrode during the charging process, where the slope on the charging curve may diminish upon their addition. Furthermore, different types of carbon additives have been observed to induce variations in the slope, emphasizing their role in interfacial dynamics.

A comprehensive investigation of the cathode–SE interface has revealed that the cathode undergoes decomposition at different energy levels depending on its charge state. The formation of a lithium-depleted layer arises due to the space-charge layer, which develops as a result of differences in the Fermi energies of the electrode and the SE. To mitigate undesired chemical reactions and space-charge layer formation at the cathode–SE interface, an electrically insulating coating is applied to either the cathode material or the SE. The mechanical properties of this coating enable it to accommodate the stress induced by the volumetric changes of the electrode materials. A detailed analysis of previous studies by various researchers on the mechanical characteristics of the cathode–electrolyte interphase (CEI) and coating layers is presented. The inherently unstable SE-Li interface is another critical challenge. Due to lithium's high reactivity, the solid–electrolyte interphase (SEI) that forms at the SE-Li interface exhibits both electronic and ionic conductivity, leading to the gradual decomposition of the SE.

The formation of mixed ionic-electronic conductors (MIECs) can be suppressed by introducing a thermodynamically stable coating on either lithium or the solid electrolyte. Finally, a detailed discussion of various research efforts concerning coating layers that enhance stable lithium metal stripping and plating on the anode is provided. These coatings effectively reduce interfacial impedance, thereby improving the overall performance and longevity of ASSBs.

References

1. Bautkinova T, Prokop M, Bystron T, Bouzek K (2025) Interface between anode porous transport layer and catalyst layer: a key to efficient, stable and competitive proton exchange membrane water electrolysis. Curr Opin Electrochem 49:101624
2. Takada K, Ohta N, Zhang L, Fukuda K, Sakaguchi I, Ma R, Osada M, Sasaki T (2008) Interfacial modification for high-power solid-state lithium batteries. Solid State Ionics 179:1333–1337
3. Sakuda A, Hayashi A, Tatsumisago M (2010) Interfacial observation between $LiCoO_2$ electrode and $Li_2S-P_2S_5$ solid electrolytes of all-solid-state lithium secondary batteries using transmission electron microscopy. Chem Mater 22:949–956
4. Zhu Y, He X, Mo Y (2016) First principles study on electrochemical and chemical stability of solid electrolyte–electrode interfaces in all-solid-state Li-ion batteries. J Mater Chem A 4:3253–3266
5. Ohta S, Komagata S, Seki J, Saeki T, Morishita S, Asaoka T (2013) All-solid-state lithium ion battery using garnet-type oxide and Li_3BO_3 solid electrolytes fabricated by screen-printing. J Power Sources 238:53–56
6. He S, Xu Y, Zhang B, Sun X, Chen Y, Jin Y (2018) Unique rhombus-like precursor for synthesis of $Li_{1.3}Al_{0.3}Ti_{1.7}(PO_4)_3$ solid electrolyte with high ionic conductivity. Chem Eng J 345:483–491
7. Kato Y, Hori S, Saito T, Suzuki K, Hirayama M, Mitsui A, Yonemura M, Iba H, Kanno R (2016) High-power all-solid-state batteries using sulfide superionic conductors. Nat Energy 1:16030
8. Kamaya N, Homma K, Yamakawa Y, Hirayama M, Kanno R, Yonemura M, Kamiyama T, Kato Y, Hama S, Kawamoto K, Mitsui A (2011) A lithium superionic conductor. Nat Mater 10:682–686
9. Mizuno F, Hayashi A, Tadanaga K, Tatsumisago M (2005) New, highly ion-conductive crystals precipitated from $Li_2S–P_2S_5$ glasses. Adv Mater 17:918–921
10. Wang S, Zhang Y, Zhang X, Liu T, Lin Y-H, Shen Y, Li L, Nan C-W (2018) High-conductivity argyrodite Li_6PS_5Cl solid electrolytes prepared via optimized sintering processes for all-solid-state lithium–sulfur batteries. ACS Appl Mater Interfaces 10:42279–42285
11. Hu L, Zhu J, Duan C, Zhu J, Wang J, Wang K, Gu Z, Xi Z, Hao J, Chen Y, Ma J, Liu J-X, Ma C (2023) Revealing the PNMA crystal structure and ion-transport mechanism of the Li_3YCl_6 solid electrolyte. Cell Rep Phys Sci 4:101428

12. López-Aranguren P, Reynaud M, Głuchowski P, Bustinza A, Galceran M, López Del Amo JM, Armand M, Casas-Cabanas M (2021) Crystalline LiPON as a bulk-type solid electrolyte. ACS Energy Lett 6:445–450
13. Swamy T, Chen X, Chiang Y-M (2019) Electrochemical redox behavior of Li ion conducting sulfide solid electrolytes. Chem Mater 31:707–713
14. Kato A, Nose M, Yamamoto M, Sakuda A, Hayashi A, Tatsumisago M (2018) Mechanical properties of sulfide glasses in all-solid-state batteries. J Ceram Soc Japan 126:719–727
15. Banerjee A, Wang X, Fang C, Wu EA, Meng YS (2020) Interfaces and interphases in all-solid-state batteries with inorganic solid electrolytes. Chem Rev 120:6878–6933
16. Tan DHS, Wu EA, Nguyen H, Chen Z, Marple MAT, Doux J-M, Wang X, Yang H, Banerjee A, Meng YS (2019) Elucidating reversible electrochemical redox of Li_6PS_5 cl solid electrolyte. ACS Energy Lett 4:2418–2427
17. Banerjee A, Tang H, Wang X, Cheng J-H, Nguyen H, Zhang M, Tan DHS, Wynn TA, Wu EA, Doux J-M, Wu T, Ma L, Sterbinsky GE, D'Souza MS, Ong SP, Meng YS (2019) Revealing nanoscale solid–solid interfacial phenomena for long-life and high-energy all-solid-state batteries. ACS Appl Mater Interfaces 11:43138–43145
18. Wang S, Bai Q, Nolan AM, Liu Y, Gong S, Sun Q, Mo Y (2019) Lithium chlorides and bromides as promising solid-state chemistries for fast ion conductors with good electrochemical stability. Angew Chem Int Ed 58:8039–8043
19. Cheng D, Wynn TA, Wang X, Wang S, Zhang M, Shimizu R, Bai S, Nguyen H, Fang C, Kim M, Li W, Lu B, Kim SJ, Meng YS (2020) Unveiling the stable nature of the solid electrolyte interphase between lithium metal and LiPON via cryogenic electron microscopy. Joule 4:2484–2500
20. Put B, Vereecken PM, Stesmans A (2018) On the chemistry and electrochemistry of LiPON breakdown. J Mater Chem A 6:4848–4859
21. Zhang W, Leichtweiß T, Culver SP, Koerver R, Das D, Weber DA, Zeier WG, Janek J (2017) The detrimental effects of carbon additives in Li_{10} GeP_2S_{12}-based solid-state batteries. ACS Appl Mater Interfaces 9:35888–35896
22. Tan DHS, Wu EA, Nguyen H, Chen Z, Marple MAT, Doux J-M, Wang X, Yang H, Banerjee A, Meng YS (2019) Elucidating reversible electrochemical redox of Li_6PS_5Cl solid electrolyte. ACS Energy Lett 4:2418–2427
23. Ates T, Keller M, Kulisch J, Adermann T, Passerini S (2019) Development of an all-solid-state lithium battery by slurry-coating procedures using a sulfidic electrolyte. Energy Storage Mater 17:204–210
24. Ohta S, Kobayashi T, Seki J, Asaoka T (2012) Electrochemical performance of an all-solid-state lithium ion battery with garnet-type oxide electrolyte. J Power Sources 202:332–335
25. Jung S-K, Gwon H, Lee S-S, Kim H, Lee JC, Chung JG, Park SY, Aihara Y, Im D (2019) Understanding the effects of chemical reactions at the cathode–electrolyte interface in sulfide based all-solid-state batteries. J Mater Chem A 7:22967–22976
26. Woo JH, Trevey JE, Cavanagh AS, Choi YS, Kim SC, George SM, Oh KH, Lee S-H (2012) Nanoscale Interface modification of $LiCoO_2$ by Al_2O_3 atomic layer deposition for solid-state Li batteries. J Electrochem Soc 159:A1120–A1124
27. Tateyama Y, Gao B, Jalem R, Haruyama J (2019) Theoretical picture of positive electrode–solid electrolyte interface in all-solid-state battery from electrochemistry and semiconductor physics viewpoints. Curr Opin Electrochem 17:149–157
28. Masuda H, Ishida N, Ogata Y, Ito D, Fujita D (2017) Internal potential mapping of charged solid-state-lithium ion batteries using in situ kelvin probe force microscopy. Nanoscale 9:893–898
29. Kato T, Hamanaka T, Yamamoto K, Hirayama T, Sagane F, Motoyama M, Iriyama Y (2014) In-situ $Li_7La_3Zr_2O_{12}/LiCoO_2$ interface modification for advanced all-solid-state battery. J Power Sources 260:292–298
30. Ohta N, Takada K, Zhang L, Ma R, Osada M, Sasaki T (2006) Enhancement of the high-rate capability of solid-state lithium batteries by nanoscale interfacial modification. Adv Mater 18:2226–2229

31. Li X, Liang J, Luo J, Norouzi Banis M, Wang C, Li W, Deng S, Yu C, Zhao F, Hu Y, Sham T-K, Zhang L, Zhao S, Lu S, Huang H, Li R, Adair KR, Sun X (2019) Air-stable Li_3InCl_6 electrolyte with high voltage compatibility for all-solid-state batteries. Energy Environ Sci 12:2665–2671
32. Zhang W, Richter FH, Culver SP, Leichtweiss T, Lozano JG, Dietrich C, Bruce PG, Zeier WG, Janek J (2018) Degradation mechanisms at the $Li_{10}GeP_2S_{12}/LiCoO_2$ cathode Interface in an all-solid-state lithium-ion battery. ACS Appl Mater Interfaces 10:22226–22236
33. Jung SH, Oh K, Nam YJ, Oh DY, Brüner P, Kang K, Jung YS (2018) Li_3BO_3–Li_2CO_3: rationally designed buffering phase for sulfide all-solid-state Li-ion batteries. Chem Mater 30:8190–8200
34. Koerver R, Zhang W, De Biasi L, Schweidler S, Kondrakov AO, Kolling S, Brezesinski T, Hartmann P, Zeier WG, Janek J (2018) Chemo-mechanical expansion of lithium electrode materials—on the route to mechanically optimized all-solid-state batteries. Energy Environ Sci 11:2142–2158
35. Sakuda A, Hayashi A, Tatsumisago M (2013) Sulfide solid electrolyte with favorable mechanical property for all-solid-state lithium battery. Sci Rep 3:2261
36. Nose M, Kato A, Sakuda A, Hayashi A, Tatsumisago M (2015) Evaluation of mechanical properties of Na_2S–P_2S_5 sulfide glass electrolytes. J Mater Chem A 3:22061–22065
37. Kato A, Yamamoto M, Sakuda A, Hayashi A, Tatsumisago M (2018) Mechanical Properties of Li_2S–P_2 S_5 Glasses with Lithium Halides and Application in All-Solid-State Batteries. ACS Appl Energy Mater 1:1002–1007
38. Koerver R, Aygün I, Leichtweiß T, Dietrich C, Zhang W, Binder JO, Hartmann P, Zeier WG, Janek J (2017) Capacity fade in solid-state batteries: interphase formation and Chemo-mechanical processes in nickel-rich layered oxide cathodes and lithium thiophosphate solid electrolytes. Chem Mater 29:5574–5582
39. Reimers JN, Dahn JR (1992) Electrochemical and in situ X-ray diffraction studies of lithium intercalation in $LixCoO_2$. J Electrochem Soc 139:2091–2097
40. Wu X, Billaud J, Jerjen I, Marone F, Ishihara Y, Adachi M, Adachi Y, Villevieille C, Kato Y (2019) Operando visualization of morphological dynamics in all-solid-state batteries. Adv Energy Mater 9:1901547
41. McDowell MT, Lee SW, Harris JT, Korgel BA, Wang C, Nix WD, Cui Y (2013) In situ TEM of two-phase lithiation of amorphous silicon nanospheres. Nano Lett 13:758–764
42. Ma J, Chen B, Wang L, Cui G (2018) Progress and prospect on failure mechanisms of solid-state lithium batteries. J Power Sources 392:94–115
43. Banerjee S, Zhang X, Wang L-W (2019) Motif-based design of an oxysulfide class of lithium superionic conductors: toward improved stability and record-high Li-ion conductivity. Chem Mater 31:7265–7276
44. Nagao M, Hayashi A, Tatsumisago M (2011) Sulfur–carbon composite electrode for all-solid-state Li/S battery with Li_2S–P_2S_5 solid electrolyte. Electrochim Acta 56:6055–6059
45. Suzuki K, Mashimo N, Ikeda Y, Yokoi T, Hirayama M, Kanno R (2018) High cycle capability of all-solid-state lithium–sulfur batteries using composite electrodes by liquid-phase and mechanical mixing. ACS Appl Energy Mater 1:2373–2377
46. Wenzel S, Randau S, Leichtweiß T, Weber DA, Sann J, Zeier WG, Janek J (2016) Direct observation of the interfacial instability of the fast ionic conductor $Li_{10}GeP_2S_{12}$ at the lithium metal anode. Chem Mater 28:2400–2407
47. Wenzel S, Sedlmaier SJ, Dietrich C, Zeier WG, Janek J (2018) Interfacial reactivity and interphase growth of argyrodite solid electrolytes at lithium metal electrodes. Solid State Ionics 318:102–112
48. Hayashi A, Hama S, Mizuno F, Tadanaga K, Minami T, Tatsumisago M (2004) Characterization of Li_2S–P_2S_5 glass-ceramics as a solid electrolyte for lithium secondary batteries. Solid State Ionics 175:683–686
49. Tian Y, Sun Y, Hannah DC, Xiao Y, Liu H, Chapman KW, Bo S-H, Ceder G (2019) Reactivity-guided interface design in Na metal solid-state batteries. Joule 3:1037–1050
50. Zhang Z, Chen S, Yang J, Wang J, Yao L, Yao X, Cui P, Xu X (2018) Interface re-engineering of $Li_{10}GeP_2S_{12}$ electrolyte and lithium anode for all-solid-state lithium batteries with ultralong cycle life. ACS Appl Mater Interfaces 10:2556–2565

51. Sahu G, Lin Z, Li J, Liu Z, Dudney N, Liang C (2014) Air-stable, high-conduction solid electrolytes of arsenic-substituted Li_4SnS_4. Energy Environ Sci 7:1053–1058
52. Liu Y, Sun Q, Zhao Y, Wang B, Kaghazchi P, Adair KR, Li R, Zhang C, Liu J, Kuo L-Y, Hu Y, Sham T-K, Zhang L, Yang R, Lu S, Song X, Sun X (2018) Stabilizing the Interface of NASICON solid electrolyte against Li metal with atomic layer deposition. ACS Appl Mater Interfaces 10:31240–31248
53. Hu P, Zhang Y, Chi X, Kumar Rao K, Hao F, Dong H, Guo F, Ren Y, Grabow LC, Yao Y (2019) Stabilizing the interface between sodium metal anode and sulfide-based solid-state electrolyte with an electron-blocking interlayer. ACS Appl Mater Interfaces 11:9672–9678
54. Hao X, Zhao Q, Su S, Zhang S, Ma J, Shen L, Yu Q, Zhao L, Liu Y, Kang F, He Y (2019) Constructing multifunctional interphase between $Li_{1.4}Al_{0.4}Ti_{1.6}(PO_4)_3$ and Li metal by magnetron sputtering for highly stable solid-state lithium metal batteries. Adv Energy Mater 9:1901604
55. Zhou W, Wang S, Li Y, Xin S, Manthiram A, Goodenough JB (2016) Plating a dendrite-free lithium anode with a polymer/ceramic/polymer sandwich electrolyte. J Am Chem Soc 138:9385–9388
56. Zhang S, Ding MS, Xu K, Allen J, Jow TR (2001) Understanding solid electrolyte Interface film formation on graphite electrodes. Electrochem Solid-State Lett 4:A206
57. Han X, Gong Y, Fu K, He X, Hitz GT, Dai J, Pearse A, Liu B, Wang H, Rubloff G, Mo Y, Thangadurai V, Wachsman ED, Hu L (2017) Negating interfacial impedance in garnet-based solid-state Li metal batteries. Nat Mater 16:572–579
58. Luo W, Gong Y, Zhu Y, Fu KK, Dai J, Lacey SD, Wang C, Liu B, Han X, Mo Y, Wachsman ED, Hu L (2016) Transition from superlithiophobicity to superlithiophilicity of garnet solid-state electrolyte. J Am Chem Soc 138:12258–12262
59. Wang R, Sun S, Xu C, Cai J, Gou H, Zhang X, Wang G (2024) The interface engineering and structure design of an alloying-type metal foil anode for lithium ion batteries: a review. Mater Horiz 11:903–922
60. Paul PP, Chen B-R, Langevin SA, Dufek EJ, Nelson Weker J, Ko JS (2022) Interfaces in all solid state Li-metal batteries: a review on instabilities, stabilization strategies, and scalability. Energy Storage Mater 45:969–1001

Reaction Dynamics and Kinetic Analysis in Solid-State Batteries

Sagar M. Mane, Komal S. Wagh, Rajneesh Kumar Mishra, Sanjay G. Chavan, Shankar S. Dhasade, and Jaewoong Lee

Abstract This chapter delves into the reaction dynamics and kinetic analysis of solid-state batteries, exploring the intricate processes that govern their performance and efficiency. Solid-state batteries (SSBs) offer promising advantages over traditional liquid electrolyte batteries, such as higher safety, enhanced energy density, and longer cycle life. However, optimizing their performance requires a thorough understanding of the reaction mechanisms and kinetic behaviors, including ion transport, interfacial reactions, and phase transformations within solid electrolytes and electrode materials. The chapter first outlines key theoretical frameworks for analyzing reaction kinetics in SSBs, including models of ionic conductivity, interfacial resistance, and activation energies that influence electrochemical processes. Emphasis is placed on various kinetic models that predict battery behavior under different operating conditions, alongside experimental techniques like electrochemical impedance spectroscopy (EIS), nuclear magnetic resonance (NMR), and neutron diffraction that allow for precise kinetic measurements. Finally, the chapter discusses challenges and opportunities in tailoring reaction kinetics through material design and interface engineering, as well as future trends in SSB development. By providing insights into the dynamic and kinetic properties of SSBs, this chapter aims to support the advancement of high-performance solid-state battery technology for next-generation energy storage.

S. M. Mane (✉) · K. S. Wagh · J. Lee
Department of Fiber System Engineering, Yeungnam University, Gyeongsan, Republic of Korea

R. K. Mishra
Department of Physics, Yeungnam University, Gyeongsan, Republic of Korea

S. G. Chavan
Department of Physics, Karmaveer Bhaurao Patil Mahavidyalaya, Pandharpur, Maharashtra, India

S. S. Dhasade
Department of Physics, Dr. Ganpatrao Deshmukh Mahavidyalaya, Sangola, Solapur, Maharashtra, India

A. K. Rai et al. (eds.), *Electrochemistry of Next-Generation Solid-State Batteries*, Green Energy and Technology, https://doi.org/10.1007/978-981-95-7603-6_6

1 Introduction

Solid-state batteries (SSBs) are increasingly recognized as a cornerstone of future energy storage technologies, driven by their potential to overcome the intrinsic limitations of conventional lithium-ion batteries (LIBs). The shift from liquid electrolytes to solid-state ionic conductors is not merely a material substitution but represents a fundamental reimagining of battery architecture, electrochemical behavior, and safety paradigms. The appeal of SSBs lies in several compelling attributes: the ability to achieve higher energy densities, enhanced intrinsic safety, improved electrochemical stability, and broader thermal operating ranges. One of the most transformative advantages of SSBs is the potential to integrate metallic lithium anodes, which are incompatible with liquid electrolyte systems due to their high reactivity and dendrite formation risks [1–4]. The use of lithium metal anodes can theoretically increase the specific energy of a battery by up to 70%, offering significant benefits for electric vehicles (EVs), aerospace, and grid storage [5, 6]. Additionally, the non-flammable nature of solid electrolytes, including ceramics and certain polymers, eliminates the fire and explosion hazards associated with organic solvents, making SSBs inherently safer. From a thermal perspective, solid electrolytes offer superior high-temperature stability, enabling battery operation at elevated temperatures without the risk of electrolyte decomposition or combustion. This characteristic also opens the door for applications in harsh or extreme environments, such as in aerospace, deep-well drilling, or military systems, where conventional batteries may fail [7–10].

However, despite these promising features and the rapid pace of materials innovation, the commercial realization of solid-state battery technology remains fraught with complex challenges. These challenges stem primarily from the unique physical and chemical behavior of solid–solid interfaces and the altered dynamics of ion transport in rigid matrices. These challenges are rooted in several interconnected problems, as illustrated in Fig. 1. (a) Interfacial Instability—unlike liquid systems, where the electrolyte can easily wet and conform to electrode surfaces, solid electrolytes must form an intimate contact with electrodes to facilitate efficient ion transport. However, mismatches in thermal expansion coefficients, chemical potential, and mechanical properties can lead to interfacial delamination, void formation, or the growth of resistive interphase layers. These interfacial instabilities often lead to increased impedance, reduced cyclability, and, in extreme cases, cell failure [11–15]. (b) Ionic conductivity challenge—Although significant strides have been made in the development of high-conductivity Li-ion and Na-ion solid electrolytes, such as sulfide-based materials (e.g., $Li_{10}GeP_2S_{12}$, Na_3Ps_4, $Na_{10}SnP_2S_{12}$), garnet-type oxides (e.g., LLZO), and NASICON-type ceramic electrolyte (e.g., $Na_3Zr_2Si_2PO_{12}$), their ionic conductivities still generally lag behind those of liquid electrolytes at room temperature [16–19]. Moreover, many promising solid electrolytes are sensitive to moisture or air, which complicates manufacturing and handling. Enhancing ionic mobility while maintaining electrochemical and environmental stability remains a central materials design challenge. (c) Mechanical integrity and stress evolution—solid electrolytes must act as both ion conductors and mechanical supports, often under dynamic operating conditions. During charge and discharge, volume changes in electrodes, especially with lithium metal or alloying-type anodes, can exert significant stresses on the solid electrolyte. This can result in cracking, pulverization, or dendrite penetration, all of which degrade performance and pose safety risks [20–24]. The coupled chemo-mechanical behavior of SSB components is a particularly active area of research, with implications for both short-term reliability and long-term durability. (d) Reaction dynamics and kinetic complexity—perhaps the most nuanced and least understood set of challenges lies in the reaction dynamics and kinetics that govern the performance of solid-state batteries. In liquid systems, ion transport, and charge transfer are often well described by classical models assuming isotropic, homogeneous environments. In contrast, solid-state systems are characterized by: (i) heterogeneous ionic transport through polycrystalline materials with grain boundaries and anisotropic diffusion paths. (ii) Localized and nonlinear charge transfer at solid–solid interfaces with variable contact areas. (iii) Formation of interphases with distinct ionic/electronic transport properties and evolving morphology. (iv) Stress-driven phenomena such as interfacial delamination or dendritic growth, which alter local reaction rates [25–31].

The confluence of these factors underscores the reality that the development of solid-state batteries is not simply a materials problem, but a deeply multidisciplinary challenge. It involves navigating the complex interactions between materials selection, microstructural engineering, interface design, mechanical reinforcement, and electrochemical modeling. Each of these dimensions contributes to the overall behavior of the battery, and each must be optimized in concert to achieve commercially viable performance. In conventional batteries, reaction kinetics have been extensively studied and optimized through decades of research. In contrast, solid-state systems remain less well understood due to the heterogeneous nature of solid–solid contacts and the presence of multiple kinetic bottlenecks. Therefore, a deep understanding of the reaction dynamics and kinetic processes within solid-state batteries is

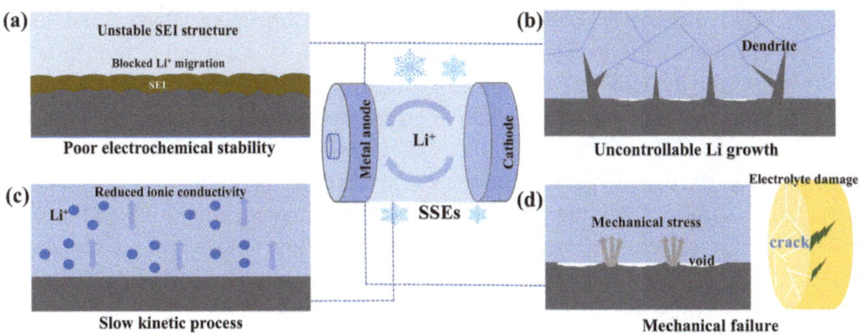

Fig. 1 Schematic illustrating issues in the solid-state batteries, **a** poor electrochemical stability, **b** dendrite formation due to uncontrolled Li growth, **c** kinetic process due to less ionic conductivity, **d** mechanical failures due to stress and cracks in electrolytes. Reproduced with permission [27] Copyright 2025, RSC

essential for overcoming these barriers. This chapter aims to provide a comprehensive overview of the fundamental reaction dynamics and kinetic analysis methodologies relevant to solid-state battery systems. It integrates insights from electrochemistry, materials science, solid-state physics, and computational modeling, with an emphasis on understanding how the kinetics of ion transport, charge transfer, and interfacial reactions evolve under different operating conditions and within different material frameworks.

2 Fundamentals of Reaction Dynamics in SSBs

2.1 Solid-State Electrochemical Reactions

To understand the reaction dynamics and kinetics in solid-state batteries (SSBs), it's essential to begin with the fundamental principles of solid-state electrochemistry. These principles underpin how ions and electrons behave within solid materials, how electric fields drive transport, and how chemical and physical interactions at interfaces influence overall battery performance. Electrochemical reactions within SSBs occur at the interfaces between electrodes and solid electrolytes. The absence of a liquid medium fundamentally alters the nature of ion and electron transfer. Instead of solvated ion transport through a fluid electrolyte, ions must migrate through a rigid, often polycrystalline, solid matrix. This introduces a series of challenges that are less significant in liquid systems. In solid media, ions traverse the structure through site-to-site hopping, which inherently restricts their mobility compared to the more fluid ion motion in liquids. A critical impediment to electrochemical activity in SSBs is the limited contact at solid–solid interfaces, where insufficient wettability raises interfacial resistance. At the atomic scale, ionic transport is governed by thermally activated hopping events, where ions move between energetically favorable sites such as interstitials or vacancies in the host lattice. In crystalline materials, ordered lattice structures provide periodic potential energy landscapes for ion migration. In amorphous materials, the lack of long-range order leads to more complex transport pathways but can sometimes enable higher conductivities due to a broader distribution of low-barrier sites. The nature of the electrochemical reaction also differs from liquid systems. In SSBs, the charge transfer process involves the simultaneous movement of ions and electrons at solid interfaces. This coupled transport can be hindered by electronic or ionic isolation, necessitating mixed-conducting interphases or composite electrode designs. Furthermore, solid-state reactions are often governed by diffusion-limited kinetics [32–36]. As ions move through solids much more slowly than through liquids, any compositional gradient or defect structure can significantly influence the reaction front. This effect is particularly pronounced during charge/discharge cycles, where kinetic asymmetry can lead to irreversible structural transformations or phase separation [3]. Multiple interdependent factors influence the electrochemical analysis of solid-state batteries. In this context, Joshi et al. have compiled a comprehensive set

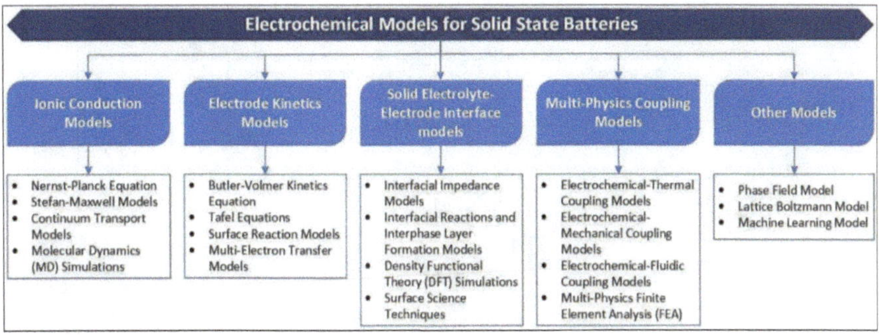

Fig. 2 Various electrochemical models for analyzing the overall kinetic process in the solid-state batteries. Reproduced with permission [1] Copyright 2025, ELSEVIER

of models that facilitate the evaluation of electrochemical reactions occurring within solid-state battery systems, as illustrated in Fig. 2 [1].

2.2 *Reaction Dynamics in Solid-State Batteries*

Reaction dynamics in solid-state batteries (SSBs) refer to the time-dependent processes involved in ion and electron transport, interfacial charge transfer, and chemical transformation within the battery during operation. These dynamic phenomena are fundamentally different from those in conventional liquid-based systems, due to the rigid nature of solid components, the complexity of interfaces, and the limited mobility of species in solids.

2.2.1 Ion and Electron Transport in Solids

In solid-state batteries (SSBs), lithium-ion mobility is fundamentally governed by the nature of the solid electrolyte, which may be crystalline, amorphous, or a composite of both. Solid-state ionics provide a foundational framework for understanding the migration of ions and mixed ion-electron processes in solid materials [37–40]. This theoretical approach serves as a bridge between the fundamental science of ionic conductors and the practical development of solid-state batteries. As described by Funke et al. [41, 42], the evolution of material properties spanning atomic to macroscopic levels helps elucidate how ion transport behaviors arise from structural and chemical characteristics. Considering the Li^+ ion transportation through solid matrices, Yu et al. [25] derived the ion motion at various levels, schematically represented in Fig. 3.

At scales below the microscale, lithium-ion transport is governed by atomistic vibrations. In solid electrolytes, lattice and molecular dynamics influence the ion's

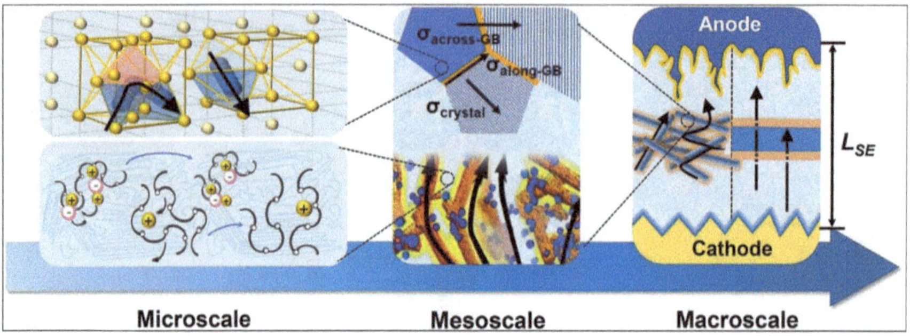

Fig. 3 Schematic illustrating the way of ion transport mechanism in solid-state batteries from a microscale to a macroscale view. Reproduced with permission [25] Copyright 2025, KeAi

ability to overcome energy barriers during hopping events. These events involve interactions such as ion-dipole complexation and decomplexation, where the pathway selection is determined by local atomic configurations. In inorganic solids, cations preferentially follow routes of minimal energy, but the precise motion down to each "hop" is dictated by diverse structural influences, making ion jumps highly dependent on the material's local architecture. On the microscopic scale, where materials exhibit quantum-level effects, lithium transport is highly dependent on whether the medium is crystalline or molecular in nature. In the case of crystalline electrolytes, it relies on framework arrangements, coordination environments, lattice volume, defects, and local disorder. These factors determine jump distances, activation energies, and hopping frequencies. In the case of molecular crystals, including supramolecular assemblies and liquid crystal frameworks, allow lithium movement via externally introduced lithium ions, with 3D lattice arrangements enabling channel formation. On the other hand, polymer-based electrolytes support lithium migration through segmental motion influenced by the coexistence of crystalline and amorphous phases. These systems allow tuning of transport properties by manipulating intermolecular interactions and modifying structural order [43–46]. At the mesoscale, transport is governed by heterogeneity, such as grain boundaries, voids, and composite interfaces. These structural discontinuities, common in polycrystalline and hybrid materials, introduce unique transport phenomena. At the macroscopic scale relevant to the entire device transport phenomena become averaged and can be described by continuum models. Here, the cumulative effects of atomic and mesoscopic processes manifest as observable properties such as conductivity, capacity, and efficiency [47–50].

2.2.2 Interfaces in Solid-State Batteries

The interface between the solid electrolyte and electrode materials is one of the most pivotal and complex regions in solid-state battery (SSB) systems. Unlike liquid electrolyte-based batteries, where the electrolyte conforms naturally to the electrode

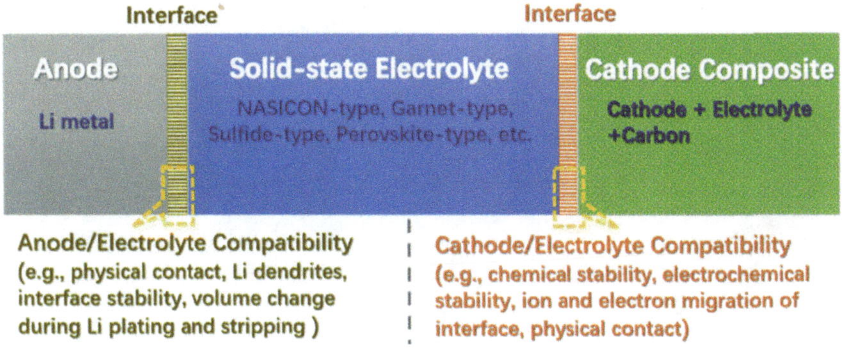

Fig. 4 Interface issues related to the solid-state batteries. Reproduced with permission [11] Copyright 2019, RSC

surface, the solid–solid interfaces in SSBs lack fluidity, often leading to poor physical contact, increased resistance, and dynamic evolution of interfacial chemistry and mechanics. These interfaces critically influence the overall reaction dynamics, rate capability, and long-term cycling stability of SSBs. Charge transfer at solid–solid interfaces involves the movement of lithium ions across the boundary between the electrode and solid electrolyte, accompanied by electron transfer through the external circuit. The efficiency of this process hinges on several interdependent factors as illustrated in Fig. 4: (a) interfacial structure and the atomic arrangement and crystallographic orientation at the interface determine the availability of continuous ion-conducting pathways. Poor structural alignment or lattice mismatch between the electrode and electrolyte can lead to disordered regions or barriers that impede Li^+ (Na^+) transfer. (b) Chemical compatibility—chemical reactions at the interface can produce unwanted interphase compounds (e.g., Li_2CO_3, LiOH, Na_2S, Na_3P, or transition metal oxides) that reduce ionic conductivity. Incompatible interfaces may undergo spontaneous redox reactions that degrade both the electrolyte and electrode materials, especially under voltage extremes. (c) Contact quality—the degree of physical intimacy between surfaces affects the contact area and, by extension, the charge transfer kinetics. Interfacial voids or rough surfaces reduce the effective area for ion exchange, increasing interfacial impedance and leading to current constriction effects [11, 14, 51–54].

2.3 *Kinetics of Ion Transport*

Ion transport kinetics are at the heart of solid-state battery (SSB) performance. The ability of ions such as Li^+, Na^+, or other mobile species to migrate through the solid electrolyte and into or out of electrode materials directly determines the rate capability, efficiency, and long-term stability of the battery. Unlike in liquid electrolytes, where ion motion is relatively unrestricted, solid-state materials impose structural

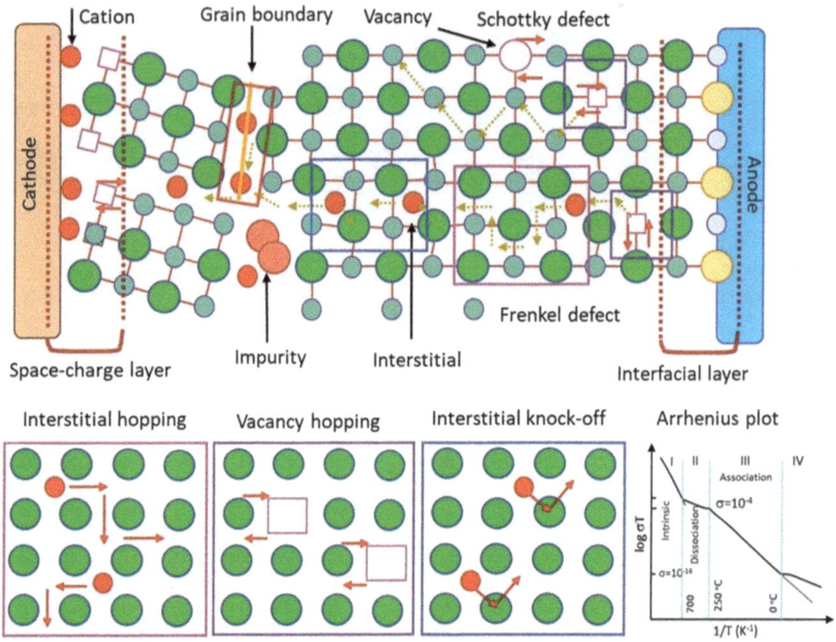

Fig. 5 Kinetics of ion migration in the solid-state electrolytes derived based on the ion Frenkel and Schottky defects and grain boundaries (vacancy, interstitial, and interstitial knock-off hopping). Reproduced with permission [55] Copyright 2023, Wiley-VCH GmbH

and energetic barriers that influence how easily and quickly ions can move. In crystalline solid-state electrolytes, ionic conductivity is largely governed by the presence and distribution of structural defects and available ion-hosting sites. One common strategy to enhance ion mobility is doping with cations of different valencies. Higher-valence dopants typically lead to the creation of cation vacancies or interstitial anions, while lower-valence substitutions can introduce cation interstitials or anion vacancies. At interfaces and grain boundaries, lattice distortions induced by potential gradients can further influence carrier distribution and migration behavior. Ion transport mechanisms such as those driven by Schottky and Frenkel defects (Fig. 5) rely on combinations of vacancy and interstitial-based movement, including more complex processes like di-vacancy hopping or collective transport events [55]. This section explores the key factors governing ionic transport kinetics.

2.3.1 Ionic Conductivity and Transport Pathways

The ionic conductivity σ of a material quantifies how easily ions can move through it. In solid electrolytes, this is determined by both the number of mobile ions and the ease with which they hop between available sites in the crystal lattice. The general relationship is [32].

$$\sigma = nq\mu, \tag{1}$$

where n is the number of charge carriers (ion concentration), q is the charge of the carrier, μ is the mobility of the ion. Transport pathways in solids are governed by the crystal structure, defect chemistry, and microstructure. Materials with open frameworks (e.g., perovskites, garnets, NASICON) provide channels or interstitial sites for fast ion diffusion. In solid-state batteries (SSBs), ion transport mechanisms can be broadly categorized into two fundamental domains: (a) Bulk transport, which involves ionic migration within the homogeneous interior of individual components such as the solid electrolyte or electrode materials; and (b) Interfacial transport, which pertains to ion transfer across junctions such as at the boundary between the solid electrolyte and the electrode, or across grain interfaces within polycrystalline structures. While bulk ion conduction can be enhanced through strategic material engineering approaches, including compositional tuning, structural optimization, and elemental doping, the transport across interfaces remains significantly more complex [4].

2.3.2 Grain Boundary Effects

Grain boundaries (GBs) play a critical yet complex role in ion transport within polycrystalline solid-state electrolytes (SSEs). Their influence varies depending on structural and chemical characteristics. Some amorphous GBs enhance lithium mobility due to high defect density, while many crystalline systems exhibit significant resistance at these interfaces. Notably, studies on materials like LISICON and Na_3PS_4 show that ion diffusion across GBs is often slower than through the bulk, contributing to lower overall conductivity in polycrystalline forms. Additionally, GBs serve as preferential sites for the accumulation of point defects, such as vacancies and dopant ions, which are essential for enabling fast lithium-ion movement in symmetric crystal structures. The redistribution of these defects due to GB segregation can significantly alter transport properties. Ultimately, GBs influence ion (Li^+/Na^+) conduction through both reduced diffusivity and defect dynamics, with transport occurring through multiple pathways, including across, along, and around GBs and their junctions, making their impact on ionic kinetics difficult to isolate and quantify [56, 57].

2.3.3 Activation Energy and Temperature Dependence

Ion migration in solids follows an activated hopping mechanism that requires ions to overcome an energy barrier between adjacent sites. The rate of ion transport is temperature-dependent and can be modeled by the Arrhenius equation [3, 55, 58].

$$\sigma(T) = \sigma_0 \exp\left(-\frac{E_a}{kT}\right), \tag{2}$$

where E_a is the activation energy, k is the Boltzmann constant, and T is the absolute temperature. Lower activation energies result in higher conductivities at a given temperature. Therefore, much research in SSB development focuses on: Reducing Ea through doping or structural modifications, increasing site connectivity for faster percolation, and enhancing the entropy of the system (e.g., through disorder engineering). The pre-exponential factor σ_0 is also important, representing the intrinsic mobility and concentration of carriers, and is influenced by vibrational dynamics and defect density. Ion-host interactions, lattice polarizability, and the phase of the material are key barriers to the activation energy. Ion-host interaction, such as electrostatic attraction between lithium ions and surrounding anions (e.g., oxygen in oxide electrolytes or sulfur in sulfides) and lattice polarizability with more polarizable anions (e.g., S^{2-} in sulfide electrolytes), facilitates lower activation energies compared to less polarizable ones (e.g., O^{2-} in oxide-based materials). If the material phase is amorphous or glassy, electrolytes generally exhibit lower activation barriers due to the absence of long-range order, providing a more flexible and isotropic environment for ion migration.

2.3.4 Blocking Effects and Space-Charge Layers

Space-charge layers (SCLs) arise at the interface between two materials with differing chemical potentials, leading to an imbalanced charge distribution. In solid-state batteries, these layers form when mobile ions move toward regions of lower chemical potential, but electrons, due to the insulating nature of solid electrolytes, remain largely immobile. This creates localized charge accumulation near the interface. The behavior and impact of these space-charge regions depend on several material properties, including electric permittivity, ion concentration, and interfacial voltage. The buildup continues until electrochemical equilibrium is reached, where chemical driving forces are balanced by the internal electric field. Depending on the ionic transport mechanism, space-charge layers can either improve or impair battery performance. For instance, when increased lithium vacancies enhance ion conduction, SCLs support better ionic transport. Conversely, in systems where these vacancies hinder conduction, the effect is detrimental. These interactions are especially relevant under open-circuit conditions, where the interface begins to evolve as soon as the electrode and electrolyte are brought into contact. In sum, space-charge layers represent a complex interfacial phenomenon that can influence ionic mobility and battery efficiency in both beneficial and adverse ways, depending on the interplay between material characteristics and electrochemical conditions [59, 60].

2.3.5 Dendrite Formation and Kinetic Competition

Dendrite growth remains a critical obstacle for solid-state lithium batteries, posing safety and reliability concerns. These metallic filaments form over repeated charging cycles and, if unchecked, can breach the electrolyte layer and cause internal short circuits. Compared to liquid-based systems, solid electrolytes offer enhanced resistance to dendrite intrusion and eliminate flammability and leakage risks associated with liquid components. Despite this resilience, dendrites are not absent in solid-state systems. Poor contact at the electrode–electrolyte interface is a major contributor, often exacerbated by mechanical defects like cracks or voids. Solutions such as implementing interfacial buffer layers show potential, but the underlying processes driving dendrite growth through solid materials, especially given their relative mechanical strength, are still being actively studied [61, 62].

3 Reaction Dynamics and Kinetics Analysis

Analyzing the kinetics of reactions and transport phenomena in solid-state batteries (SSBs) is essential for understanding their performance, failure mechanisms, and for guiding material and interface design. Due to the complexity of solid–solid interfaces and the often buried or nanoscale nature of key processes, specialized experimental techniques are required. These techniques must be capable of probing ion transport, interfacial phenomena, and reaction dynamics across different length and time scales. This section reviews the most important experimental tools and methodologies used to study the kinetics in SSBs, categorizing them into electrochemical, spectroscopic, microscopic, and modeling-assisted methods.

3.1 Electrochemical Techniques for Kinetics Analysis

Electrochemical techniques play a foundational role in analyzing the kinetics of ion and charge transport in solid-state batteries (SSBs) [50]. These methods are indispensable for quantifying diffusion coefficients, interfacial resistances, and charge transfer rates, all of which are essential for understanding reaction mechanisms and performance bottlenecks.

3.1.1 Electrochemical Impedance Spectroscopy (EIS)

Electrochemical impedance spectroscopy is one of the most powerful tools for investigating ionic and electronic transport processes in SSBs. EIS provides frequency-dependent information about bulk resistance, interfacial impedance, and grain boundary effects. Electrochemical impedance spectroscopy (EIS) involves analyzing the response of a system to a small alternating voltage or current over a range of frequencies. When a stable electrochemical cell is subjected to a low-amplitude sinusoidal input at a specific angular frequency (ω), it generates an output signal of the same frequency. The ratio of input to output defines the system's impedance at that frequency. By varying the frequency and collecting corresponding impedance values, a full impedance spectrum is obtained [63, 64]. Liu et al. [65] used impedance modeling to understand the reaction kinetics in the solid-state thin-film batteries. This study reveals that the high-frequency semicircle in impedance spectra corresponds to bulk resistance linked to ion migration, while the mid- and low-frequency features relate to charge transfer resistance and space-charge layer capacitance. A larger free enthalpy difference reduces charge transfer resistance but increases the electrostatic potential drop. Lithium-ion diffusion primarily affects bulk resistance without significantly influencing interfacial impedance. These findings are supported by experimental impedance measurements.

3.1.2 Galvanostatic/Potentiostatic Intermittent Titration Technique (GITT)/(PITT)

GITT is used to evaluate lithium diffusion coefficients by applying intermittent current pulses and monitoring voltage response during relaxation periods. GITT applies a constant current pulse followed by an open-circuit relaxation. The voltage response during relaxation is used to estimate the chemical diffusion coefficient of lithium [66].

$$D = \frac{4}{\pi}\left(\frac{\tau}{\Delta E/\Delta E_s}\right)^2\left(\frac{L^2}{T}\right), \tag{3}$$

where ΔE is the voltage drop during the current pulse, ΔE_s is the steady-state voltage change, L is the diffusion length, and τ is the pulse duration. Similar to GITT, PITT applies small voltage steps instead of current and measures the corresponding current response over time. This current-time decay behavior provides critical insights into diffusion mechanisms and interfacial charge transfer kinetics. In a typical PITT experiment, a stepwise voltage perturbation is applied to the battery cell, and the transient current response is recorded at each step. The system is allowed to relax after each step until the current approaches zero or reaches a steady state. This procedure is repeated over a voltage range of interest, generating a series of current-time profiles. These profiles can be mathematically modeled to extract kinetic parameters such as chemical diffusion coefficients, ionic conductivity in solid phases, activation energy for ion migration and interfacial resistance, and charge transfer dynamics. Since the technique directly examines the response to potential changes, it can effectively isolate processes that are rate-limited by ionic movement rather than electronic or capacitive effects.

3.1.3 Cyclic Voltammetry

Cyclic voltammetry (CV) is an essential electrochemical technique where the current response is measured as the electrode potential is swept at a constant rate over a defined voltage range. This method reveals redox processes in lithium-ion batteries by identifying characteristic peaks corresponding to oxidation and reduction reactions. It is widely used to assess the electrochemical stability of various cell components, including electrolytes, electrodes, and binders. CV also enables the investigation of key atomic-scale events such as solid–electrolyte interphase (SEI) formation, lithium-ion intercalation and deintercalation, as well as structural and compositional changes in electrode materials [50].

3.2 Spectroscopic and Surface Analysis Techniques

Spectroscopic tools provide chemical and electronic structure information, which is essential for understanding reaction dynamics and interphase formation.

3.2.1 X-Ray Photoelectron Spectroscopy (XPS)

Surface reactions at the electrode–electrolyte interface are a key factor in the degradation of solid electrolytes. X-ray photoelectron spectroscopy (XPS) is a highly effective surface analysis method used to detect chemical changes, surface irregularities, and compositional variations near the interface. Ex situ XPS is commonly employed to study coating layers on solid electrolytes and to investigate the mechanisms of interfacial reactions. While ex situ studies are informative, in situ or operando XPS measurements remain technically difficult due to the challenge of accessing buried interfaces in solid-state battery architectures.

Quérel et al. [67] and Williams et al. [68] This work presents an operando XPS approach where a thin sodium (Na^0) layer is deposited onto a NaSICON-type solid electrolyte ($Na_{3.4}Zr_2Si_{2.4}P_{0.6}O_{12}$ or NZSP) while simultaneously monitoring the interphase composition in real time as illustrated in Fig. 6a–c. The study confirms the formation of a solid–electrolyte interphase (SEI) at the Na^0|NZSP interface and, notably, showcases this method's capability to investigate the kinetics of interphase development. Vardar et al. [69] used XPS depth profiling for understanding the temperature-dependent decomposition process of $LiCoO_2$ (LCO) in the matrix of the $Li_7La_3Zr_2O_{12}$ (LLZO). This kind of decomposition process yields different products such as Li_2CO_3, $LaCoO_3$, and $La_2Zr_2O_7$ and results in hindering of ion transportation process, resulting in the capacity reduction of the solid-state batteries. Davis et al. [70] used operando XPS to verify the dynamics of the electronically insulating and conducting solid–electrolyte interphase (SEI) for solid electrolytes Li_6PS_5Cl and $Li_{10}GeP_2S_{12}$ before and after Li plating. It is noted that, in Li_6PS_5Cl-based systems,

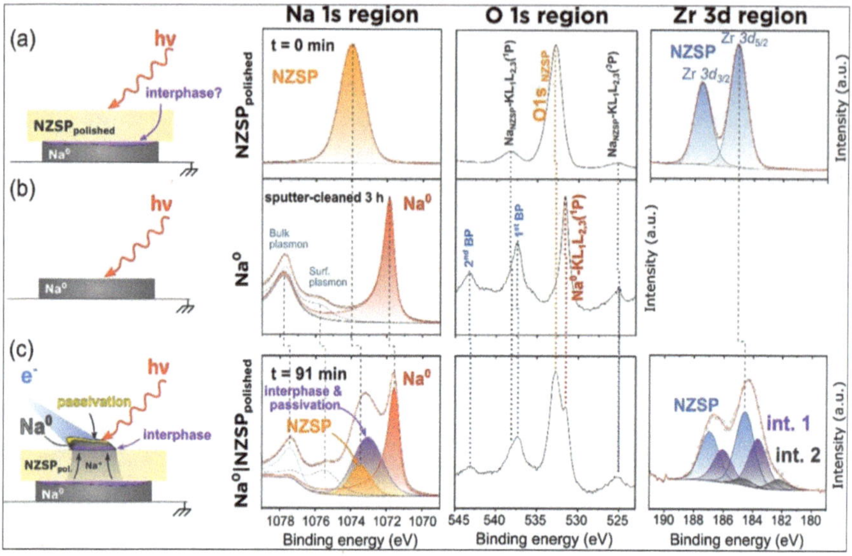

Fig. 6 Real-time monitoring of the formation of the interphase layer upon addition of Na^0 to NaSICON-type solid electrolyte ($Na_{3.4}Zr2Si_{2.4}P_{0.6}O_{12}$ or NZSP), **a** fitting model and XPS spectra of polished NZSP, **b** fitting model and XPS spectra of Na^0, and **c** fitting model and XPS spectra of Na^0|NZSP polished interface at 91 min of plating. Reproduced with permission [67] Copyright 2023, ACS

the SEI enables lithium metal to nucleate and plate effectively after SEI formation. However, when the SEI is electronically conductive, lithium plating is hindered due to ongoing decomposition of the solid electrolyte, leading to reduced Faradaic efficiency. Wenzel et al. [71] assessed the chemical reactivity of $Li_{10}GeP_2S_{12}$ with lithium, utilizing in situ XPS combined with time-resolved electrochemical measurements as illustrated in Fig. 7. Through this combined analysis, they have confirmed that the breakdown of $Li_{10}GeP_2S_{12}$ results in an interphase containing Li_3P, Li_2S, and a lithium–germanium alloy.

3.2.2 Raman Spectroscopy

Raman spectroscopy is a complementary vibrational technique used to probe the local bonding environment and vibrations within materials. These methods provide vital information about the chemical and structural characteristics of solid-state battery (SSB) components, particularly at interfaces. This technique is useful for detecting phase transitions, identifying interphase species, and monitoring degradation reactions. Kuwata et al. [72] performed in situ Raman assessment of the $Li_xMn_2O_4$ cathode during cycling performance of a thin-film solid-state battery assembled with $Li/Li_3PO_4/LiMn_2O_4$ to analyze the ion dynamics. Characteristic Raman signals corresponding to the α, β, and λ phases of $Li_xMn_2O_4$ were identified. A noticeable

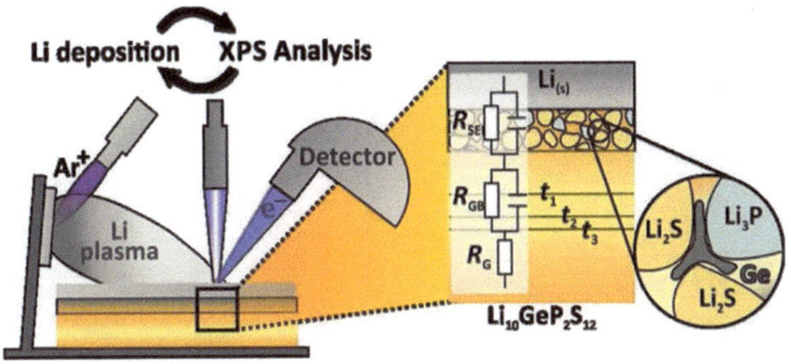

Fig. 7 Schematic of in situ X-ray photoelectron spectroscopy coupled with electrochemical impedance spectroscopy used to analyze Li kinetics in $Li_{10}GeP_2S_{12}$ solid-state electrolyte. Reproduced with permission [71] Copyright 2016, ACS

hysteresis in Raman intensity was observed between charge and discharge from the front surface, while minimal hysteresis appeared when measurements were taken from the back, highlighting directional dependence in structural response. Ji et al. [73] investigated the Na-ion reaction pathway under thermodynamic and kinetic effects. This study offers an in-depth analysis of the sulfur cathode redox behavior in all-solid-state sodium batteries (ASSSBs) by integrating electrochemical tests (CV, GCD, and GITT), in situ Raman spectroscopy, ex situ X-ray absorption spectroscopy (XAS), and theoretical modeling. It reveals a thermodynamically driven three-step solid–solid reaction pathway ($S_8 \rightarrow Na_2S_5/Na_2S_4 \rightarrow Na_2S_2 \rightarrow Na_2S$) at very low current rates or elevated temperatures (≥ 90 °C), reflected in a three-plateau voltage profile. However, under standard testing conditions (≤ 60 °C, $\geq$ C/20), kinetic constraints suppress the Na_2S_2 intermediate, leading to a two-step reaction as shown in Fig. 8a, b. First-principles calculations support this by showing only a minimal energy difference between Na_2S_2 and the two-phase Na_2S_4–Na_2S system, explaining its kinetic exclusion. This work provides clarity on both thermodynamic and kinetic factors in the multi-electron conversion processes of sulfur cathodes. Similarly, Cao et al. [74] used similar techniques to compare the Li-ion pathway during redox reactions in the all-solid-state lithium battery. It is noted that the intermediate phase formation of Li_2S_2 takes place during the charge and discharge process, which is due to the conversion between S_8 and Li_2S electrolyte. Luo et al. [75] validated the improvement in the interfacial (intercalation/deintercalation) ion dynamics of Na^+ in a sodium solid-state battery using in situ Raman and EIS analysis, which constitutes a hybrid solid polymer electrolyte.

3.2.3 X-Ray Absorption Spectroscopy

X-ray absorption spectroscopy (XAS) is a powerful analytical technique used to investigate the local chemical and electronic structure of elements within materials. It

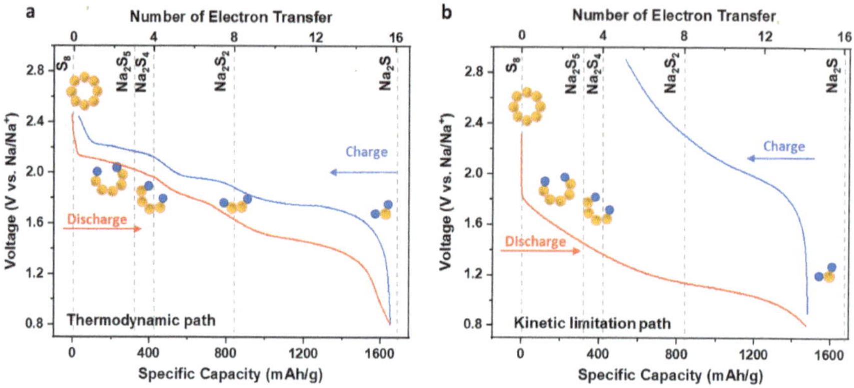

Fig. 8 Reaction pathways followed by Na ion in solid-state sodium battery **a** higher temperature ($\geq$ 90 °C) and **b** at lower temperatures ($\leq$ 60 °C). Reproduced with permission [73] Copyright 2024, RSC

is especially useful in solid-state battery (SSB) research for probing buried interfaces and bulk electrode or electrolyte materials at the atomic level. XAS is divided into two primary regions: X-ray Absorption Near Edge Structure (XANES) and Extended X-ray Absorption Fine Structure (EXAFS), both of which provide complementary information. Solid-state batteries often face interfacial challenges due to complex microstructures and chemical inconsistencies. While ion transport has traditionally been attributed to direct contact at the interface, Lou et al. [76] used in-operando transmission X-ray microscopy, combined with XANES, to probe the electrochemical kinetics at the cathode of the solid-state battery. This study reveals that internal conditions within polycrystalline ($LiNi_{0.6}Mn_{0.2}Co_{0.2}O_2$) particles also significantly influence ion movement. Even with limited initial contact, lithium ions can diffuse uniformly, driven by chemical potential gradients. However, repeated cycling can disturb the internal environment, leading to uneven charge distribution and rapid capacity loss. These findings emphasize the critical role of internal particle environments in governing interfacial transport and battery performance. The interface kinetics between the cathode and electrolyte and the involved chemical changes were illustrated in Fig. 9a–c Operando XAS studies in sodium-ion batteries have largely focused on cathode materials, offering insights into redox behavior and structural changes [77].

For example, operando XANES and EXAFS at the Co K-edge on $Na_{0.74}CoO_2$ revealed that Na insertion is coupled with the Co^{3+}/Co^{4+} redox transition and a shortening of the c-axis in the hexagonal structure [78]. Similarly, investigations on $Na_{0.85}Li_{0.17}Ni_{0.21}Mn_{0.64}O_2$ showed, via Ni and Mn K-edge XANES, that Mn stays as Mn^{4+} while Ni undergoes reversible valence changes between Ni^{3+} and Ni^{4+}, with EXAFS detecting local structural distortions around these cations during cycling [79]. Further work by Sottmann et al. [80] on $Na_{1.32}Mn[Fe(CN)_6]_{0.83}\cdot zH_2O$ linked the redox activity of Mn^{2+}/Mn^{3+} and Fe^{2+}/Fe^{3+} to distinct voltage plateaus using simultaneous XAS and XRD. Lin et al. [81] explored TiS_2 cathodes in Na-ion batteries

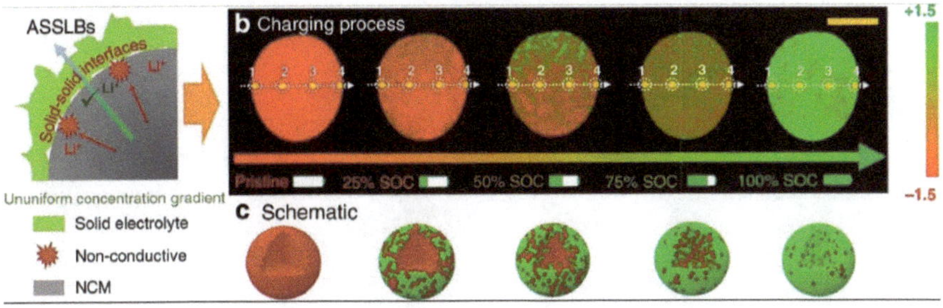

Fig. 9 Analysis of Li kinetics during the charging process using 2D TXM-XANES, **a** solid–solid interface kinetic model representation, **b** mapping of a single particle cathode during the charging process using 2D TXM-XANES, and **c** solid-state electrochemistry process during charging. Reproduced with permission [76] Copyright 2020, Nature Communications

and showed that both Ti and S undergo redox during sodiation. Using MCR-ALS to analyze XANES data, they identified an intermediate phase ($Na_{0.55}TiS_2$) alongside the initial TiS_2 and the final $NaTiS_2$, confirmed through EXAFS and ex situ XRD. These studies demonstrate how operando XAS unveils redox mechanisms and intermediate states during battery operation. Xu et al. [82] used XANES supported by computer modeling to understand the electrochemical reactions of the $Na_2Ti_3O_7$ anode of the Na-ion solid-state battery. This work presents an in-depth analysis of $Na_2Ti_3O_7$ as a low-voltage anode material for sodium-ion batteries. Applying a carbon coating improves both cycle life and coulombic efficiency by enhancing conductivity and limiting SEI development. By utilizing the anode's low voltage, a full cell with a high overall voltage is achieved. A key discovery is the self-relaxation of the fully sodiated $Na_4Ti_3O_7$ phase, attributed to its structural instability, as supported by computational modeling. XANES confirms that Ti^{4+}/Ti^{3+} acts as the redox pair during operation.

3.2.4 Nuclear Magnetic Resonance Spectroscopy

Nuclear magnetic resonance (NMR) spectroscopy is a powerful analytical technique that provides unique insights into the reaction dynamics and kinetic processes in solid-state batteries (SSBs). Unlike surface-sensitive methods such as XPS or spectroscopy techniques that target electronic structures (like XAS), NMR offers local and element-specific information in the bulk and interfacial regions, particularly useful for investigating ion mobility, structural evolution, and phase transformations during battery operation [25]. Yu et al. [29] in their research work utilized two-dimensional exchange NMR spectroscopy (2D-EXSY) to quantitatively investigate lithium-ion exchange at the interface between a solid electrolyte (Li_6PS_5Br) and an electrode material (Li_2S). The technique, enabled by distinct NMR chemical shifts, reveals how ion transport across the interface evolves throughout electrode preparation and cycling, as illustrated in Fig. 10a–l. Findings show that nanoscaling Li_2S and

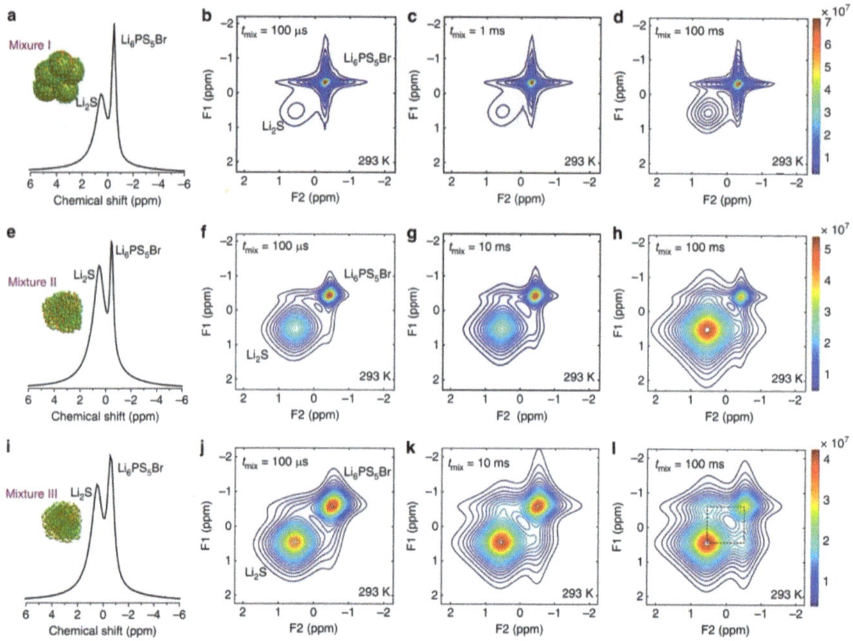

Fig. 10 Exchange analysis of Li ion between Li_2S cathode and Li_6PS_5Br solid-state electrolyte using nuclear magnetic resonance spectroscopy, **a** 1D MAS spectra of Li_6PS_5Br–Li_2S cathode mixtures when Li_2S is micron-sized, **b–d** Li-Li exchange spectra at different mixing times for the initial mixture, **e** 1D MAS spectra of Li_6PS_5Br–Li_2S cathode mixtures when Li_2S is nano-sized, **f–h** Li–Li exchange spectra at different mixing times for this mixture, **i** 1D MAS spectra of Li_6PS_5Br–Li_2S cathode mixtures when Li_2S is nano-sized is completely mixed with Li_6PS_5Br, and **j–l** Li–Li exchange spectra at different mixing times for this mixture. Reproduced with permission [29] Copyright 2017, Nature Communications

ensuring strong contact with the electrolyte are critical for effective charge transfer. Initially, interfacial conductivity is significantly lower than bulk conductivity due to the limited contact area. After cycling, lithium-ion mobility at the interface declines sharply, likely due to structural changes and diffusion barriers caused by side product formation, as illustrated in Fig. 11a–d. These results highlight the importance of maintaining interface stability during cycling and showcase exchange NMR as a powerful tool for directly probing interfacial transport phenomena.

3.3 Microscopy and Imaging Techniques

Microscopy (SEM, TEM, AFM) provides crucial insight into the nanoscale structure, morphology, and composition of materials in solid-state batteries. These techniques

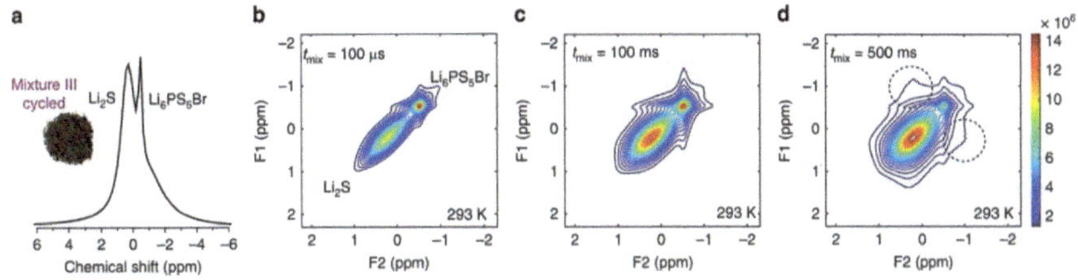

Fig. 11 After cycling exchange analysis of Li ion between Li_2S cathode and Li_6PS_5Br solid-state electrolyte using nuclear magnetic resonance spectroscopy, **a** 1D MAS spectra of Li_6PS_5Br–Li_2S cathode mixtures after cycling, and **b–d** Li–Li exchange spectra at different mixing times for the mixture after cycling. Reproduced with permission [29] Copyright 2017, Nature Communications

are pivotal for understanding degradation mechanisms, ion transport pathways, interfacial evolution, and mechanical failure, all of which influence the kinetics and reaction dynamics of SSBs. SEM is widely used to examine surface morphology, grain structure, crack formation, and porosity in solid-state battery components, especially the electrolyte and electrode layers. When integrated with SEM, EDS enables elemental mapping to analyze composition and interdiffusion at the electrode|electrolyte interface. It is especially useful for visualizing chemical gradients and interphase formation. TEM allows ultra-high-resolution imaging of internal structures, grain boundaries, crystallographic defects, and solid–solid interfaces at the atomic scale. AFM provides three-dimensional surface topography and mechanical properties such as stiffness and adhesion forces at the nanoscale. Together, these microscopy techniques offer a comprehensive toolkit to analyze physical and chemical phenomena underpinning reaction dynamics and kinetics in solid-state batteries [83].

3.4 Modeling and Simulation of Reaction Kinetics

As solid-state battery (SSB) technologies advance, understanding their complex reaction mechanisms requires robust theoretical tools that complement experimental efforts. Modeling and simulation have emerged as indispensable approaches to dissect ion transport, charge transfer, interfacial evolution, and degradation phenomena across multiple scales—from atomic-level interactions to full-cell performance. These computational techniques enable predictive design of novel materials, extraction of kinetic parameters, and identification of failure pathways. Atomistic modeling methods such as Density Functional Theory (DFT), Ab Initio Molecular Dynamics (AIMD), and Classical Molecular Dynamics (MD) offer insight into ion migration barriers, defect energetics, and dynamic ion behavior. At the mesoscale, phase-field modeling and kinetic Monte Carlo (KMC) simulations provide understanding of phenomena like dendrite growth, phase transformation, and interfacial reactions. On a continuum level, multiphysics models simulate macroscopic ion transport, electrochemical kinetics, thermal behavior, and mechanical responses under real-world operating conditions. The challenging electrode–electrolyte interfaces are increasingly studied using a multiscale approach that combines thermodynamic, electrochemical, and mechanical modeling to understand contact degradation and interphase formation. Furthermore, machine learning (ML) and data-driven models are gaining traction for predicting material properties, analyzing spectroscopy data, and guiding optimization of cell components [84]. The integration of modeling with experimental observation through parameter fitting, inverse modeling, and computational screening enhances the accuracy and impact of both domains, accelerating innovation and development in SSB technology.

4 Conclusion and Future Aspects

Understanding the reaction mechanisms that govern the behavior of solid-state batteries is crucial for designing high-performance systems. The interplay between ionic conductivity, electrochemical stability, and mechanical integrity dictates how effectively SSBs perform under varying operating conditions. The future of SSBs relies on a deeper understanding of the kinetic processes that govern charge and discharge cycles. Real-time monitoring, advanced computational models, and in-depth studies of interfacial reactions will be crucial to advancing the field. Additionally, research into the aging processes and degradation pathways will provide valuable insights into improving the lifespan and efficiency of SSBs.

References

1. Joshi A, Mishra DK, Singh R, Zhang J, Ding Y (2025) A comprehensive review of solid-state batteries. Appl Energy 386:125546
2. Machín A, Morant C, Márquez F (2024) Advancements and challenges in solid-state battery technology: an in-depth review of solid electrolytes and anode innovations. Batteries 10:29
3. Jiang C, Li H, Wang C (2017) Recent progress in solid-state electrolytes for alkali-ion batteries. Sci Bull (Beijing) 62:1473–1490
4. Kim JG, Son B, Mukherjee S, Schuppert N, Bates A, Kwon O, Choi MJ, Chung HY, Park S (2015) A review of lithium and non-lithium based solid state batteries. J Power Sources 282:299–322
5. Liu W, Placke T, Chau KT (2022) Overview of batteries and battery management for electric vehicles. Energy Rep 8:4058–4084
6. Zhao G, Wang X, Negnevitsky M (2022) Negnevitsky, iScience connecting battery technologies for electric vehicles from battery materials to management. Iscience 25:103744
7. Liang J, Luo J, Sun Q, Yang X, Li R, Sun X (2019) Recent progress on solid-state hybrid electrolytes for solid-state lithium batteries. Energy Storage Mater 21:308–334
8. Li C, Yu Wang Z, Jiang He Z, Jiao Li Y, Mao J, Hua Dai K, Yan C, Chao Zheng J (2021) An advance review of solid-state battery: challenges, progress and prospects, sustainable. Mater Technol 29:e00297

9. Mageto T, Bhoyate SD, de Souza FM, Mensah-Darkwa K, Kumar A, Gupta RK (2022) Developing practical solid-state rechargeable Li-ion batteries: concepts, challenges, and improvement strategies. J Energy Storage 55:105688
10. Xia S, Wu X, Zhang Z, Cui Y, Liu W (2019) Practical challenges and future perspectives of all-solid-state lithium-metal batteries. Chem 5:753–785
11. Du M, Liao K, Lu Q, Shao Z (2019) Recent advances in the interface engineering of solid-state Li-ion batteries with artificial buffer layers: challenges, materials, construction, and characterization, energy. Environ Sci 12:1780–1804
12. Zhao Y, Wang C, Dai Y, Jin H (2021) Homogeneous Na^+ transfer dynamic at Na/$Na_3Zr_2Si_2PO_{12}$ interface for all solid-state sodium metal batteries. Nano Energy 88:106293
13. Xiao Y, Wang Y, Bo SH, Kim JC, Miara LJ, Ceder G (2020) Understanding interface stability in solid-state batteries. Nat Rev Mater 5:105–126
14. Wang S, Xu H, Li W, Dolocan A, Manthiram A (2018) Interfacial chemistry in solid-state batteries: formation of interphase and its consequences. J Am Chem Soc 140:250–257
15. Takada K, Ohno T, Ohta N, Ohnishi T, Tanaka Y (2018) Positive and negative aspects of interfaces in solid-state batteries. ACS Energy Lett 3:98–103
16. Ai S, Wu X, Wang J, Li X, Hao X, Meng Y (2024) Research Progress on solid-state electrolytes in solid-state lithium batteries: classification, ionic conductive mechanism, interfacial challenges. Nanomaterials 14:1773
17. Hu S, Li YF, Yang R, Yang Z, Wang L (2018) Structure and ionic conductivity of $Li_7La_3Zr_2{-}xGexO_{12}$ garnet-like solid electrolyte for all solid state lithium ion batteries. Ceram Int 44:6614–6618
18. Xu J, Tao H, Deng Z, Yang X, Fan LZ (2024) Simultaneously enhancing ionic conductivity and interfacial stability by Fe_2O_3 for solid-state sodium metal batteries. J Mater 10:1243–1251
19. He S, Xu Y, Chen Y, Ma X (2020) Enhanced ionic conductivity of an F--assisted $Na_3Zr_2Si_2PO_{12}$ solid electrolyte for solid-state sodium batteries. J Mater Chem A 8:12594–12602
20. Yu HC, Taha D, Thompson T, Taylor NJ, Drews A, Sakamoto J, Thornton K (2019) Deformation and stresses in solid-state composite battery cathodes. J Power Sources 440:227116
21. Lewis JA, Tippens J, Cortes FJQ, McDowell MT (2019) Chemo-mechanical challenges in solid-state batteries. Trends Chem 1:845–857
22. Kalnaus S, Dudney NJ, Westover AS, Herbert E, Hackney S (2023) Solid state batteries: the critical role of mechanics. Science 381:1300
23. Liang F, Wang S, Liang Q, Zhong A, Yang C, Qian J, Song H, Chen R (2024) Insight into all-solid-state Li–S batteries: challenges, advances, and engineering design. Adv Energy Mater 14:2401959
24. Kasemchainan J, Zekoll S, Spencer Jolly D, Ning Z, Hartley GO, Marrow J, Bruce PG (2019) Critical stripping current leads to dendrite formation on plating in lithium anode solid electrolyte cells. Nat Mater 18:1105–1111
25. Yu W, Deng N, Feng Y, Feng X, Xiang H, Gao L, Cheng B, Kang W, Zhang K (2024) Understanding multi-scale ion-transport in solid-state lithium batteries. eScience 5:100278
26. Burov AS, Boev AO, Abakumov AM, Aksyonov DA (2024) Mechanism of Li^+ charge transfer at Li/$Li_7La_3Zr_2O_{12}$ interfaces: a density functional theory study. Phys Rev B 109:045305
27. Xu H, Song X, Gu Y, Fan J, Liu J, Wang S (2025) Failure mechanisms and design strategies for low-temperature solid-state metal batteries. J Mater Chem A 13:10388–10414
28. Tappan BA, Geng K, Vivona D, Wang D, Mankus D, Lytton-Jean A, Bresser D, Shao-Horn Y (2025) Investigation of charge transfer kinetics in multilayer PEO/LLZO solid-state batteries. ACS Appl Mater Interfaces 17:18255–18267
29. Yu C, Ganapathy S, Eck ERHV, Wang H, Basak S, Li Z, Wagemaker M (2017) Accessing the bottleneck in all-solid state batteries, lithium-ion transport over the solid-electrolyte-electrode interface. Nat Commun 8:1086
30. Weiss M, Simon FJ, Busche MR, Nakamura T, Schröder D, Richter FH, Janek J (2020) From liquid- to solid-state batteries: ion transfer kinetics of Heteroionic interfaces. Electrochem Energy Rev 3:221–238

31. Lou S, Yu Z, Liu Q, Wang H, Chen M, Wang J (2020) Multi-scale imaging of solid-state battery interfaces: from atomic scale to macroscopic scale. Chem 6:2199–2218
32. Sotoudeh M, Baumgart S, Dillenz M, Döhn J, Forster-Tonigold K, Helmbrecht K, Stottmeister D, Groß A (2024) Ion mobility in crystalline battery materials. Adv Energy Mater 14:2302550
33. Mutter D, Pantano DA, Elsässer C, Urban DF (2025) Diffusion behavior of Li ions in crystalline and amorphous Li–Zr–O and Li–Nb–O phases. Solid State Ionics 421:116787
34. Zsoldos E, Cormier MME, Phattharasupakun N, Liu A, Dahn JR (2023) How to measure solid state Lithium-ion diffusion using the Atlung method for intercalant diffusion. J Electrochem Soc 170:040511
35. Baker DR, Verbrugge MW, Bower AF (2016) Thermodynamics, stress, and Stefan-Maxwell diffusion in solids: application to small-strain materials used in commercial lithium-ion batteries. J Solid State Electrochem 20:163–181
36. Park JH, Choi YS, Byeon YW, Ahn JP, Lee JC (2019) Diffusion kinetics governing the diffusivity and diffusion anisotropy of alloying anodes in Na-ion batteries. Nano Energy 65:104041
37. Naik KG, Vishnugopi BS, Mukherjee PP (2022) Kinetics or transport: whither Goes the solid-state battery cathode? ACS Appl Mater Interfaces 14:29754–29765
38. Wang MJ, Kazyak E, Dasgupta NP, Sakamoto J (2021) Transitioning solid-state batteries from lab to market: linking electro-chemo-mechanics with practical considerations. Joule 5:1371–1390
39. Knauth P, Tuller HL (2002) Solid-state ionics: roots, status, and future prospects. J Am Ceram Soc 85:1654–1680
40. Linford RG (1988) Applications of solid state ionics for batteries. Solid State Ionics 28–30:831–840
41. Funke K, Cramer C, Wilmer D (2005) Concept of mismatch and relaxation for self-diffusion and conduction in ionic materials with disordered structures, diffusion in condensed matter. Springer, pp 857–893
42. Funke K (2013) Solid state ionics: from Michael faraday to green energy-the European dimension. Sci Technol Adv Mater 14:043502
43. Yu Z, Zhang X, Fu C, Wang H, Chen M, Yin G, Huo H, Wang J (2021) Dendrites in solid-state batteries: ion transport behavior, advanced characterization, and interface regulation. Adv Energy Mater 11:2003250
44. Fu ZH, Chen X, Zhang Q (2023) Review on the lithium transport mechanism in solid-state battery materials. WIREs Comput Mol Sci 13:e1621
45. Zhao C, Liu L, Qi X, Lu Y, Wu F, Zhao J, Yu Y, Hu YS, Chen L (2018) Solid-state sodium batteries. Adv Energy Mater 8:1703012
46. De Gol A, Dermenci KB, Farkas L, Berecibar M (2024) Electro-chemo-mechanical degradation in solid-state batteries: a review of microscale and multiphysics modeling. Adv Energy Mater 14:2403255
47. Miao X, Guan S, Ma C, Li L, Nan CW (2023) Role of interfaces in solid-state batteries. Adv Mater 35:2206402
48. Lou S, Zhang F, Fu C, Chen M, Ma Y, Yin G, Wang J (2021) Interface issues and challenges in all-solid-state batteries: lithium, sodium, and beyond. Adv Mater 33:2000721
49. Zhu H, Huang Y, Zhu H, Wang L, Lan S, Xia X, Liu Q (2020) In situ probing multiple-scale structures of energy materials for Li-ion batteries. Small Methods 4:1900223
50. Liu X, Zhang L, Yu H, Wang J, Li J, Yang K, Zhao Y, Wang H, Wu B, Brandon NP, Yang S (2022) Bridging multiscale characterization technologies and digital modeling to evaluate lithium battery full lifecycle. Adv Energy Mater 12:2200889
51. Tikekar MD, Choudhury S, Tu Z, Archer LA (2016) Design principles for electrolytes and interfaces for stable lithium-metal batteries. Nat Energy 1:16114
52. Jia L, Zhu J, Zhang X, Guo B, Du Y, Zhuang X (2024) Li–solid electrolyte interfaces/interphases in all-solid-state Li batteries. Electrochem Energy Rev 7:12
53. Sagane F, Abe T, Iriyama Y, Ogumi Z (2005) Li^+ and Na^+ transfer through interfaces between inorganic solid electrolytes and polymer or liquid electrolytes. J Power Sources 146:749–752

54. Wenzel S, Leichtweiss T, Weber DA, Sann J, Zeier WG, Janek J (2016) Interfacial reactivity benchmarking of the sodium ion conductors Na_3PS_4 and sodium β-alumina for protected sodium metal anodes and sodium all-solid-state batteries. ACS Appl Mater Interfaces 8:28216–28224
55. Shinde SS, Wagh NK, Kim SH, Lee JH (2023) Li, Na, K, Mg, Zn, Al, and Ca anode interface chemistries developed by solid-state electrolytes. Adv Sci 10:2304235
56. Shen K, Wang Y, Zhang J, Zong Y, Li G, Zhao C, Chen H (2020) Revealing the effect of grain boundary segregation on Li ion transport in polycrystalline anti-perovskite Li_3ClO: a phase field study. Phys Chem Chem Phys 22:3030–3036
57. Milan E, Pasta M (2023) The role of grain boundaries in solid-state Li-metal batteries. Mater Futures 2:013501
58. Zhou D, He YB, Liu R, Liu M, Du H, Li B, Cai Q, Yang QH, Kang F (2015) In situ synthesis of a hierarchical all-solid-state electrolyte based on nitrile materials for high-performance lithium-ion batteries. Adv Energy Mater 5:1500353
59. De Klerk NJJ, Wagemaker M (2018) Space-charge layers in all-solid-state batteries; important or negligible? ACS Appl Energy Mater 1:5609–5618
60. Dobhal G, Walsh TR, Tawfik SA (2022) Blocking directional lithium diffusion in solid-state electrolytes at the interface: first-principles insights into the impact of the space charge layer. ACS Appl Mater Interfaces 14:55471–55479
61. Antony Jose S, Gallant A, Gomez PL, Jaggers Z, Johansson E, LaPierre Z, Menezes PL (2025) Solid-state lithium batteries: advances, challenges, and future perspectives. Batteries 11:90
62. Sun H, Celadon A, Cloutier SG, Al-Haddad K, Sun S, Zhang G (2024) Lithium dendrites in all-solid-state batteries: from formation to suppression. Batter Energy 3:20230062
63. Vadhva P, Hu J, Johnson MJ, Stocker R, Braglia M, Brett DJL, Rettie AJE (2021) Electrochemical impedance spectroscopy for all-solid-state batteries: theory, methods and future outlook. ChemElectroChem 8:1930–1947
64. Zhang L, Dai Y, Li C, Dang Y, Zheng R, Wang Z, Wang Y, Cui Y, Arandiyan H, Shao Z, Sun H, Zhuang Q, Liu Y (2024) Recent advances in electrochemical impedance spectroscopy for solid-state batteries. Energy Storage Mater 69:103378
65. Liu Y, Yu WB, Xu BX (2022) Impedance modelling of all-solid-state thin film batteries: influence of the reaction kinetics. J Mater Chem A Mater 10:313–325
66. Horner JS, Whang G, Ashby DS, Kolesnichenko IV, Lambert TN, Dunn BS, Talin AA, Roberts SA (2021) Electrochemical modeling of GITT measurements for improved solid-state diffusion coefficient evaluation. ACS Appl Energy Mater 4:11460–11469
67. Quérel E, Williams NJ, Seymour ID, Skinner SJ, Aguadero A (2023) Operando characterization and theoretical modeling of metal|electrolyte interphase growth kinetics in solid-state batteries. Part I: experiments. Chem Mater 35:853–862
68. Williams NJ, Quérel E, Seymour ID, Skinner SJ, Aguadero A (2023) Operando characterization and theoretical modeling of metal|electrolyte interphase growth kinetics in solid-state batteries. Part II: modeling. Chem Mater 35:863–869
69. Vardar G, Bowman WJ, Lu Q, Wang J, Chater RJ, Aguadero A, Seibert R, Terry J, Hunt A, Waluyo I, Fong DD, Jarry A, Crumlin EJ, Hellstrom SL, Chiang YM, Yildiz B (2018) Structure, chemistry, and charge transfer resistance of the interface between $Li_7La_3Zr_2O_{12}$ electrolyte and $LiCoO_2$ cathode. Chem Mater 30:6259–6276
70. Davis AL, Kazyak E, Liao DW, Wood KN, Dasgupta NP (2021) Operando analysis of interphase dynamics in anode-free solid-state batteries with sulfide electrolytes. J Electrochem Soc 168:070557
71. Wenzel S, Randau S, Leichtweiß T, Weber DA, Sann J, Zeier WG, Janek J (2016) Direct observation of the interfacial instability of the fast ionic conductor $Li_{10}GeP_2S_{12}$ at the lithium metal anode. Chem Mater 28:2400–2407
72. Kuwata N, Matsuda Y, Okawa T, Hasegawa G, Kamishima O, Kawamura J (2022) Ion dynamics of the $Li_xMn_2O_4$ cathode in thin-film solid-state batteries revealed by in situ Raman spectroscopy. Solid State Ionics 380:115925

73. Ji T, Tu Q, Zhao Y, Wierzbicki D, Plisson V, Wang Y, Wang J, Burch KS, Yang Y, Zhu H (2024) Three-step thermodynamic vs. two-step kinetics-limited sulfur reactions in all-solid-state sodium batteries, energy. Environ Sci 17:9255–9267
74. Sun X, Cao D, Bak S-M, Ji T, Du Y, Geiwitz M, Burch K, Zhu H (2023) Understanding mechanisms of sulfur in all-solid-state batteries through operando and theoretical studies. Angew Chem Int Ed Engl 62:e202302363
75. Luo C, Li Q, Shen D, Zheng R, Huang D, Chen Y (2021) Enhanced interfacial kinetics and fast Na^+ conduction of hybrid solid polymer electrolytes for all-solid-state batteries. Energy Storage Mater 43:463–470
76. Lou S, Liu Q, Zhang F, Liu Q, Yu Z, Mu T, Zhao Y, Borovilas J, Chen Y, Ge M, Xiao X, Lee WK, Yin G, Yang Y, Sun X, Wang J (2020) Insights into interfacial effect and local lithium-ion transport in polycrystalline cathodes of solid-state batteries. Nat Commun 11:5700
77. Ghigna P, Quartarone E (2021) Operando x-ray absorption spectroscopy on battery materials: a review of recent developments. J Phys Energy 3:032006
78. Ding JJ, Zhou YN, Sun Q, Yu XQ, Yang XQ, Fu ZW (2013) Electrochemical properties of P2-phase $Na_{0.74}CoO_2$ compounds as cathode material for rechargeable sodium-ion batteries. Electrochim Acta 87:388–393
79. Karan NK, Slater MD, Dogan F, Kim D, Johnson CS, Balasubramanian M (2014) Operando structural characterization of the Lithium-substituted layered sodium-ion cathode material P2-$Na_{0.85}Li_{0.17}Ni_{0.21}Mn_{0.64}O_2$ by X-ray absorption spectroscopy. J Electrochem Soc 161:A1107–A1115
80. Sottmann J, Bernal FLM, Yusenko KV, Herrmann M, Emerich H, Wragg DS, Margadonna S (2016) In operando synchrotron XRD/XAS investigation of sodium insertion into the prussian blue analogue cathode material $Na_{1.32}Mn[Fe(CN)_6]_{0.83}zH_2O$. Electrochim Acta 200:305–313
81. Lin CH, Topsakal M, Sun K, Bai J, Zhao C, Dooryhee E, Northrup P, Gan H, Lu D, Stavitski E, Chen-Wiegart YCK (2020) Operando structural and chemical evolutions of TiS_2 in Na-ion batteries. J Mater Chem A Mater 8:12339–12350
82. Xu J, Ma C, Balasubramanian M, Meng YS (2014) Understanding $Na_2Ti_3O_7$ as an ultra-low voltage anode material for a Na-ion battery. Chem Commun 50:12564–12567
83. Nomura Y, Yamamoto K, Hirayama T, Ohkawa M, Igaki E, Hojo N, Saitoh K (2018) Quantitative operando visualization of electrochemical reactions and Li ions in all-solid-state batteries by STEM-EELS with hyperspectral image analyses. Nano Lett 18:5892–5898
84. Bacon S, Babar S, Dent M, Foster A, Baboo JP, Zhang T, Watts JF, Lekakou C (2024) Reaction kinetics of lithium-sulfur batteries with a polar Li-ion electrolyte: modeling of liquid phase and solid phase processes. Phys Chem Chem Phys 26:19247–19256

Solid-State Battery Design, Manufacturing, and Battery Performance Evaluation

Shubha Dubey, Kumud Dubey, Manoj Goswami, Netrapal Singh, Surender Kumar, and N. K. Gaur

Abstract Solid-state batteries (SSBs) present a next-generation energy storage solution by replacing flammable liquid electrolytes with solid counterparts, offering enhanced safety, thermal stability, and energy density. Their compact design, extended shelf life, and compatibility with lithium metal anodes make them particularly attractive for electric vehicles, portable electronics, and large-scale grid applications. Despite their promise, several challenges must be addressed to enable the widespread adoption of SSBs. These include long-term performance stability, power output limitations, and economic feasibility. Overcoming these challenges requires innovations in materials science and engineering to improve both the reliability and affordability of SSBs. Critical areas of improvement include enhancing specific energy, energy density, specific power, energy efficiency, and long-term energy retention. Achieving these objectives will require the development of novel materials with superior ionic and electronic conductivities, along with advanced cell architectures that support these materials. This chapter provides a comprehensive overview of the key materials, design strategies, and manufacturing processes essential for optimizing SSB performance. It highlights promising solid electrolyte candidates, such as sulfide-based systems, and explores scalable methods including roll-to-roll processing and 3D-printed architectures. Electrochemical performance metrics, aging mechanisms, and emerging tools like AI-guided material discovery are also discussed. By integrating materials science with engineering innovation, this work outlines a strategic roadmap for the practical realization and commercial adoption of solid-state battery technologies.

Keywords Solid-state batteries (SSBs) · Solid electrolytes · High-energy–density storage · Battery performance optimization · Energy storage applications

S. Dubey (✉) · K. Dubey · N. K. Gaur
Department of Physics, Barkatullah University, Bhopal, India
e-mail: shubha.dubey4@gmail.com

M. Goswami · N. Singh · S. Kumar
CSIR—Advanced Materials and Processes Research Institute (AMPRI), Bhopal, India

Academy of Scientific and Innovative Research (AcSIR), Ghaziabad, India

A. K. Rai et al. (eds.), *Electrochemistry of Next-Generation Solid-State Batteries*, Green Energy and Technology, https://doi.org/10.1007/978-981-95-7603-6_7

1 Introduction

Solid-state batteries (SSBs) are a revolutionary type of energy storage technology that utilize solid electrolytes instead of the conventional liquid electrolytes found in traditional batteries, such as lithium-ion batteries. The key differentiator lies in the electrolyte phase: in SSBs, the electrolyte is a solid material that facilitates the movement of ions between the battery's anode and cathode during charge and discharge cycles. This innovation eliminates the need for flammable and volatile liquid electrolytes, significantly improving safety and energy density [1]. While liquid electrolyte-based batteries rely on liquid to conduct ions between the anode and cathode, solid-state batteries use materials such as oxides, sulfides, and polymers that conduct ions in their solid form. These solid electrolytes enable the possibility of using high-energy–density anodes, such as lithium metal, without the risk of dendrite formation, a significant safety issue in traditional lithium-ion cells [2]. Recent breakthroughs in SSB development have accelerated their potential for commercialization. Companies like Toyota have demonstrated functional prototypes, while Quantum Scope has made significant progress in developing high-performance solid-state lithium-metal batteries, attracting major investment and industry interest.

Mechanisms of Ion Transport in Solid Electrolytes

Solid-state batteries offer numerous advantages over their liquid counterparts. Some of the most compelling reasons to explore SSB technology include:

- **Safety**: Traditional lithium-ion batteries are prone to leaking, overheating, and even combustion under certain conditions due to the liquid electrolytes and organic solvents they use. SSBs, on the other hand, are inherently safer since solid electrolytes are non-flammable, reducing the risk of fires and explosions [3].
- **Higher Energy Density**: SSBs can achieve higher energy densities because solid electrolytes allow for the use of lithium metal anodes, which can store more energy than conventional graphite anodes. This leads to longer battery life and more compact designs [4].
- **Longer Cycle Life**: The solid-state structure tends to reduce degradation over time, providing longer-lasting performance compared to traditional batteries, which are limited by liquid-electrolyte-induced wear [5].
- **Faster Charging**: With the right design, solid-state batteries can potentially offer faster charge times due to better ion mobility and reduced internal resistance [6].

Applications and Market Trends

The advancement of solid-state battery technology has the potential to transform a range of industries:

- **Electric Vehicles (EVs)**: Solid-state batteries are seen as a game-changer in the electric vehicle sector due to their higher energy density, longer cycle life, and improved safety features. With automakers such as Toyota, BMW, and Volkswagen investing heavily in this technology, SSBs could lead to faster charging and more affordable EVs with longer ranges [7].

- **Grid Energy Storage**: The scalability of solid-state batteries makes them highly attractive for large-scale energy storage applications, particularly for storing renewable energy generated by solar and wind power. Their longer lifespan and greater safety make them a viable option for stabilizing energy grids [8].
- **Portable Electronics**: The compact size and high energy density of SSBs open the door to portable electronic devices, offering longer-lasting power with faster charging times [9].
- **Aerospace and Medical Devices**: The lightweight and efficient energy storage provided by solid-state batteries make them ideal for aerospace applications, where weight is a critical factor. Additionally, their safety and reliability make them suitable for powering medical implants and devices that demand high-performance, long-lasting energy solutions [10].

Current Challenges and Motivation for Research

Despite the exciting possibilities, several challenges persist in the development of solid-state batteries, which motivates ongoing research and innovation:

- **Low Ionic Conductivity**: Many solid electrolytes exhibit low ionic conductivity at room temperature, limiting the rate at which ions can move between the electrodes, which in turn impacts battery performance [11].
- **Interface Issues**: The interfaces between the solid electrolyte and the electrodes can have high resistance, leading to inefficient ion transfer and diminished performance over time. In particular, dendrite formation in lithium metal anodes remains a significant challenge that can cause short-circuiting and safety concerns [12].
- **Scalability and Cost**: While solid-state batteries promise enhanced performance, the materials and processing techniques used in their production are often more expensive than traditional batteries. Scaling up production while maintaining cost-effectiveness remains an ongoing challenge [13].

This chapter explores the design principles, manufacturing processes, and performance evaluation methods for solid-state batteries, with a focus on overcoming these challenges and advancing the technology toward widespread commercial use.

1.1 Design of Solid-State Batteries

The design of solid-state batteries (SSBs) involves a delicate balance of materials selection, electrochemical properties, and manufacturing techniques to achieve high performance, safety, and long cycle life. The main components of SSBs—electrode materials, solid electrolytes, and interfacial components—must work together harmoniously to optimize the overall battery performance. Below is an overview of the key components involved in the design of SSBs.

1.1.1 Key Components of SSBs

Electrode Materials

Electrode materials are fundamental in determining the energy density, cycle life, and safety of solid-state batteries. The selection of cathode and anode materials depends on factors such as electrochemical stability, conductivity, and compatibility with solid electrolytes [14].

Cathode materials	Properties	Challenges
Lithium transition metal oxides (e.g., $LiCoO_2$, NCM, NCA)	High energy density and good cycle life [15]	Stability issues at high voltages
Polyanionic compounds (e.g., $LiFePO_4$, $LiMnPO_4$)	Excellent structural stability and safety, moderate energy density [16]	Moderate energy density
Sulfide-based cathodes (e.g., Li_2S, FeS_2)	High theoretical capacities [17]	Sulfur dissolution and compatibility issues with solid electrolytes
Anode materials	Properties	Challenges
Lithium metal	Highest theoretical capacity [18]	Prone to dendrite formation leading to short circuits
Silicon-based anodes (Si-Li Alloys)	Higher capacity than graphite, but prone to significant volume expansion [19]	Volume expansion causes mechanical stress and performance degradation
Transition metal-based anodes (e.g., TiO_2, Sn-based)	More stable and compatible with solid electrolytes, lower energy densities [20]	Lower energy density compared to lithium and silicon-based anodes

Solid Electrolytes

Solid electrolytes play a vital role in SSBs, serving as the medium for ion conduction while providing structural support to the battery. The choice of solid electrolyte depends on factors such as ionic conductivity, chemical stability, mechanical properties, and compatibility with electrode materials.

Solid electrolyte type	Examples	Properties	Challenges
Oxide-based electrolytes	LLTO ($Li_3xLa_{2/3-x}TiO_3$), LLZO ($Li_7La_3Zr_2O_{12}$), NASICON (Li_1 + $xAl_xTi_{2-x}(PO_4)_3$)	High ionic conductivity, excellent stability (e.g., against lithium metal) [21–23]	Difficult processing, moderate conductivity for NASICON, requires doping
Sulfide-based electrolytes	$Li_{10}GeP_2S_{12}$, $Li_7P_3S_{11}$	Among the highest ionic conductivity materials [24]	Sensitive to moisture, requires encapsulation for stability
Polymer-based electrolytes	PEO-LiTFSI, PVDF-based electrolytes	Mechanical flexibility and ease of processing, suitable for flexible batteries [25]	Lower ionic conductivity at room temperature

Interfacial Stability and Challenges

Interfacial stability is a critical aspect of solid-state battery performance. The interfaces between the electrode and electrolyte, as well as between the solid electrolyte and current collectors, significantly influence the ion transport and long-term stability of the battery.

- **High Interfacial Resistance**: Poor contact between the electrode materials and the electrolyte can lead to high interfacial resistance, which limits ion transport and overall efficiency [26].
- **Dendrite Suppression in Lithium Metal Anodes**: Dendrite formation in lithium metal anodes is a significant issue that can lead to internal short circuits and safety hazards. Addressing this requires strategies such as interface engineering and the use of protective coatings to suppress dendrite growth [27].
- **Electrode/Electrolyte Contact Optimization**: Improving the interface between electrodes and electrolytes is essential for enhancing ionic conductivity and battery performance. Researchers are exploring various techniques, including composite electrode design and interface coatings, to optimize these contacts [28].

These components and challenges underscore the complexities involved in designing efficient and stable solid-state batteries. Advances in material science, electrolyte development, and interfacial engineering will be key to addressing these issues and enhancing the commercial viability of SSBs.

1.2 Alternative Solid-State Battery Chemistries

1.2.1 Solid Electrolytes for Sodium-Ion Solid-State Batteries

Solid electrolytes are a cornerstone of sodium-ion solid-state battery (SISB) technology, offering critical advantages in terms of thermal stability, non-flammability, and resistance to leakage and dendritic growth. Unlike liquid electrolytes, which pose significant safety hazards and suffer from limited electrochemical stability windows, solid-state electrolytes enable the development of more robust and compact battery architectures.

Types of Solid Electrolytes

Solid electrolytes for SISBs can be broadly categorized into oxide-based, sulfide-based, and polymer-based systems, each possessing unique electrochemical, mechanical, and processing characteristics. Among them, NASICON-type oxides and sulfide-based electrolytes have emerged as the most promising candidates due to their high ionic conductivities and compatibility with sodium metal and high-voltage cathodes [29, 30].

A comparative overview of key solid electrolytes used in SISBs is presented in Table 1.

Key Attributes and Performance Metrics

- **NASICON-Type Electrolytes ($Na_3Zr_2Si_2Po_{12}$)**: These materials possess a robust three-dimensional framework that facilitates high Na^+ conductivity (~10^{-3} $S{\cdot}cm^{-1}$). They also offer good compatibility with oxide cathodes but require high-temperature sintering for densification, which can complicate integration with temperature-sensitive components [30].
- **Sulfide-Based Electrolytes (Na_3PS_4, Na_3SbS_4)**: These materials provide superior ionic conductivities, often exceeding 10^{-3} $S{\cdot}cm^{-1}$, and exhibit low activation energies for ion transport. Their soft mechanical nature also promotes intimate contact with electrodes. However, their sensitivity to moisture necessitates handling under inert atmospheres and may require encapsulation strategies.
- **Polymer Electrolytes**: Systems based on polyethylene oxide (PEO) and sodium salts offer ease of processing, flexibility, and scalability. Yet, their low room-temperature conductivity and limited electrochemical window (~2–4 V) constrain their use in high-performance applications.
- **Composite Electrolytes**: Hybrid systems combining ceramic fillers with polymers offer a synergistic balance of conductivity, flexibility, and stability. Composite designs such as NASICON/PEO or Na_3PS_4/gel polymers are being actively explored to mitigate the limitations of individual components and improve electrode–electrolyte interfacial compatibility.

Table 1 Overview of solid electrolyte materials for sodium-ion solid-state batteries

Type	Material	Ionic conductivity ($S\ cm^{-1}$)	Electrochemical stability window (V)	Key advantages	Major challenges
Oxide-based	$Na_3Zr_2Si_2PO_{12}$ (NASICON)	~ 10^{-3}	~ 2–4.5	High thermal stability, stable interface	Brittle, poor flexibility, high sintering temperature
	$Na_{11}Sn_2PS_{12}$	~ 3×10^{-3}	~ 2–4.5	High Na^+ conductivity, good processability	Air sensitivity, moisture reactivity
Sulfide-based	Na_3PS_4	~ 10^{-3} to 10^{-2}	~ 1.5–4.0	High conductivity, good electrode interface	Highly sensitive to moisture and air
	Na_3SbS_4	~ 10^{-3}	~ 1.5–4.5	Improved stability over Na_3PS_4	Requires protective coatings
Polymer-based	PEO–NaX (X = CF_3SO_3, ClO_4)	~ $10^{-6}$to 10^{-4}	~ 2–4	Flexible, low-cost, processable	Low conductivity at room temperature
Composite	NASICON/ PEO, Na_3PS_4/ Polymer	~ 10^{-4} to 10^{-3}	~ 2–4.5	Flexible, enhanced interface contact	Limited long-term stability

Recent Progress and Strategies for Enhancement Recent advancements include the development of:

- **In situ-formed interphases** to stabilize the interface between the solid electrolyte and sodium metal.
- **Glass–ceramic and doped NASICON systems** to enhance bulk conductivity.
- **Grain boundary engineering** to reduce interfacial resistance.
- **Polymer-sulfide blends** for flexible thin-film battery configurations.

Challenges and Outlook

Despite the progress, significant challenges remain. These include the **mechanical brittleness** of oxides, **poor ambient stability** of sulfides, **limited conductivity of polymers**, and **electrochemical incompatibility** at interfaces. Addressing these issues will require a combination of advanced synthesis methods, nanoscale interface engineering, and real-time in situ characterization techniques to monitor interfacial dynamics.

In the future, **tailored solid electrolyte architectures**, potentially using **multi-layered or functionally graded structures**, are expected to play a pivotal role in unlocking the full potential of SISBs for grid-scale and high-power applications.

1.2.2 Potassium-Ion Solid-State Batteries (KISBs)

Potassium-ion solid-state batteries (KISBs) have recently garnered significant interest as promising candidates for sustainable and cost-effective energy storage, particularly for large-scale grid applications. Potassium (K), an alkali metal with high natural abundance and low cost, offers distinct electrochemical advantages such as a lower standard electrode potential (−2.93 V vs. SHE) and similar redox behavior to lithium, making it suitable for high-energy–density battery systems. Unlike conventional potassium-ion batteries that rely on flammable organic liquid electrolytes, KISBs utilize solid-state electrolytes, thereby improving battery safety, reducing flammability risks, and enabling the use of potassium metal as an anode due to suppressed dendritic growth under optimized conditions.

Key Features and Components of KISBs

Solid Electrolytes

The development of high-performance potassium-ion-conducting solid electrolytes is crucial to realizing efficient KISBs. K-β-alumina and polymer-based electrolytes have shown considerable promise, with ionic conductivities ranging from 10^{-4} to 10^{-3} S·cm^{-1} at ambient temperatures [31]. Composite systems incorporating both inorganic ceramics and polymers are being actively explored to enhance mechanical properties, processability, and interfacial compatibility.

Electrode Materials

A broad range of electrode materials has been investigated for use in KISBs. Table 2 provides a comparative overview of typical cathode and anode materials used in KISBs.

Applications

KISBs are being considered for applications **in stationary energy storage systems, backup power units, and portable electronic devices**, where cost, safety, and resource abundance are key criteria. Due to potassium's wide availability and compatibility with aluminum current collectors (unlike lithium), these batteries are well-suited for scalable and cost-sensitive deployment [32].

Table 2 Electrode materials for potassium-ion solid-state batteries

Electrode type	Material	Theoretical capacity (mAh g^{-1})	Key features	Challenges
Cathode	K_xMnO_2 (layered or tunnel)	~ 120–150	Abundant, good cycling stability	Structural instability during cycling
	$K_2FeFe(CN)_6$ (Prussian blue analog)	~ 130–160	Open framework, low cost	Moisture sensitivity
	$KVPO_4F$, $KFePO_4$	~ 110–140	Stable frameworks, good safety	Limited rate performance
Anode	Graphite	~ 279	High reversibility, stable SEI	Volume expansion, slow diffusion
	Potassium metal	~ 685	Highest theoretical capacity	Dendrite formation, interfacial instability

Recent Advances in KISB Technology

- **Hybrid Solid Electrolytes**: Composite systems combining inorganic ceramic phases with polymer matrices (e.g., PEO-KFSI, K-β-Alumina–polymer blends) have led to improvements in flexibility, mechanical stability, and ionic conductivity.
- **Interface Engineering**: Coating layers and surface functionalization have been employed to stabilize the solid–electrolyte interface, especially when employing potassium metal anodes, mitigating dendrite formation and improving long-term cycling.
- **Dendrite Suppression Strategies**: Techniques such as 3D porous anode hosts and electrolyte additives have been proposed to suppress dendritic growth and ensure uniform K^+ deposition.

Challenges and Future Outlook

Despite recent progress, KISBs face several challenges that must be addressed to achieve commercial viability:

- **Lack of high-conductivity solid electrolytes** with wide electrochemical windows and long-term stability.
- **Poor compatibility with potassium metal anodes**, particularly under high current densities.
- **Mechanical degradation** of electrodes during alloying/dealloying cycles.

Future research must focus on:

- **Designing next-generation solid electrolytes** with tailored transport and mechanical properties,
- **Optimizing electrode–electrolyte interfaces** using artificial interphases and buffer layers,
- **Developing high-rate cathode materials** with stable structures under large K^+ intercalation.

The successful implementation of these strategies could position KISBs as a key technology for post-lithium energy storage, particularly in large-scale and low-cost applications where resource sustainability is a priority.

Solid Electrolytes for Potassium-Ion Solid-State Batteries

A wide range of solid electrolytes have been explored for KISBs, each exhibiting unique electrochemical characteristics, advantages, and limitations. The properties of representative solid electrolytes are summarized in Table 3.

Table 3 Representative solid electrolytes for potassium-ion solid-state batteries

Electrolyte type	Material	Ionic conductivity ($S\ cm^{-1}$ at RT)	Electrochemical stability Window (V)	Key features	Challenges
Oxide-based ceramics	K-β-Alumina	~ 10^{-3}	~ 2.5–3.0	High K^+ conductivity, good thermal stability	Moisture sensitivity, brittle nature
Phosphate-based	$KTi_2(PO_4)_3$ (NASICON-type)	~ 10^{-4}	~ 3.0–4.0	Stable framework, low electronic conductivity	Limited K^+ diffusivity, synthesis complexity
Halide-based	K_3YCl_6, K_3LaCl_6	~ 10^{-4} to 10^{-3}	Up to ~ 4.5	Low activation energy for K^+ transport	Limited research, hygroscopic behavior
Polymer-based	PEO + KFSI	~ 10^{-5} to 10^{-4}	~ 3.0–4.0	Flexible, easy processing	Low room-temp conductivity, poor mechanical strength
Composite electrolytes	K-β-Alumina/ PEO/K-salt blends	~ 10^{-4} to 10^{-3}	~ 3.5–4.5	Improved interface contact, flexibility	Interfacial degradation, thermal instability

Detailed Comparison of Solid Electrolyte Types

Among the various types of solid electrolytes explored for KISBs, oxide-based ceramics, particularly K-β-alumina, have demonstrated the highest potassium-ion conductivity, typically on the order of ~ 10^{-3} S·cm^{-1} at room temperature. Their rigid crystalline frameworks provide well-defined conduction pathways for K^+ ions and are thermally stable, making them suitable for high-temperature operations. However, their brittle nature and sensitivity to ambient moisture pose significant challenges for device fabrication and long-term durability.

Phosphate-based materials, such as NASICON-type $KTi_2(PO_4)_3$, offer moderate conductivity (~10^{-4} S·cm^{-1}) and better chemical stability than sulfide analogs. Their open 3D framework allows reasonable K^+ diffusion, but sluggish ion mobility at room temperature limits their utility without high-temperature operation or further structural modification.

Halide-based electrolytes (e.g., K_3YCl_6 and K_3LaCl_6) have recently emerged as a new class with promising conductivity values (~10^{-4} to 10^{-3} S·cm^{-1}) and broader electrochemical windows, extending up to ~ 4.5 V. These materials exhibit low migration barriers for K^+ transport, making them attractive for high-voltage cathodes. However, their hygroscopic nature and instability in ambient conditions require stringent handling protocols.

Polymer-based electrolytes, particularly PEO doped with potassium salts (e.g., KFSI), offer ease of processing, flexibility, and good interfacial contact with electrodes. Nevertheless, their ionic conductivities at room temperature remain low (~10^{-5} to 10^{-4} S·cm^{-1}), and they often suffer from limited electrochemical and mechanical stability.

To address these limitations, composite electrolytes combining ceramic and polymer phases have gained attention. Blends such as K-β-alumina/PEO composites exhibit improved conductivity (~10^{-4} to 10^{-3} S·cm^{-1}), enhanced flexibility, and better interfacial properties. These hybrid systems aim to harness the mechanical robustness of ceramics and the interfacial adaptability of polymers. However, achieving homogeneous dispersion and long-term chemical stability at the interface remains an ongoing research challenge.

In summary, while no single electrolyte system currently meets all the ideal criteria—high conductivity, wide voltage window, flexibility, chemical and thermal stability—composite approaches offer the most practical pathway for near-term development of high-performance KISBs.

1.2.3 Zinc-Ion Solid-State Batteries (ZISBs)

Zinc-ion solid-state batteries (ZISBs) have attracted significant attention in recent years due to their high theoretical capacity (820 mAh/g), safety, and environmental friendliness. Zinc, being less reactive compared to alkali metals like lithium and sodium, poses fewer risks related to dendrite formation, making it an ideal candidate for safe and efficient energy storage. ZISBs leverage solid electrolytes to enhance battery safety, thermal stability, and longevity, distinguishing them from conventional liquid electrolyte-based systems.

Key Features and Components

- **Solid Electrolytes**: Zinc-ion conduction in solid-state systems is facilitated by **Zn-based** NASICON and gel-polymer electrolytes, which provide ionic conductivities of approximately 10^{-3} S·cm^{-1}. These electrolytes offer not only high ionic conductivity but also the mechanical flexibility and electrochemical stability necessary for safe, long-term battery performance [33].
- **Electrode Materials**:
  - **Cathodes**: Common cathode materials include MnO_2, V_2O_5, and Prussian blue analogs, which exhibit stable cycling performance and good voltage stability (~1.8 V vs. Zn). These materials are chosen for their high voltage stability and capacity retention.
  - **Anodes**: Zinc metal and Zn-alloy composites are often used as anodes. These materials enable high energy density but face challenges related to dendrite formation, which can cause short circuits and battery failure over time.
- **Applications**: ZISBs are ideal for grid storage, wearable electronics, and emergency power systems, where their high energy density, long cycle life, and safety features make them suitable for stationary and portable power applications [34].
- **Recent Advances**:
  - Flexible solid-state ZISBs are under development, aiming to incorporate gel-polymer electrolytes for enhanced mechanical flexibility and higher performance.
  - The development of protective coatings for zinc anodes has also been a major focus to prevent zinc corrosion and to improve long-term cycling stability.
- **Challenges**:
  - Zinc dendrite formation is one of the primary issues limiting the performance of ZISBs. Dendrites can pierce the solid electrolyte, causing short circuits and leading to premature battery failure.
  - Corrosion of the zinc anode and side reactions with the electrolyte can significantly reduce the lifetime of the battery and impact its capacity retention.

Solid Electrolytes for Zinc-Ion Solid-State Batteries (ZISBs):

The development of effective solid electrolytes is critical to the overall performance and safety of ZISBs. Solid electrolytes ensure efficient zinc-ion transport while providing thermal stability and mechanical integrity. Several types of electrolytes are currently being researched and optimized for ZISBs, each offering distinct advantages and challenges.

1. **Zn-Based NASICON Electrolytes**:
    - **NASICON ($Na_3Zr_2Si_2PO_{12}$)**-type materials have been successfully modified to support zinc-ion conduction. $ZnTi_2(PO_4)_3$ is one such promising material, demonstrating high ionic conductivity (~10^{-3} $S{\cdot}cm^{-1}$) and electrochemical stability within the voltage range required for ZISBs.
    - **Advantages**: These materials offer excellent ionic conductivity and structural stability, making them ideal for high-performance ZISBs. Their high-temperature stability and long-term cycling performance further enhance their appeal.
    - **Challenges**: Brittleness and moisture sensitivity are key issues with NASICON-based electrolytes, and careful processing is required to prevent degradation over time.
2. **Gel-Polymer Electrolytes (GPEs)**:
    - Gel-polymer electrolytes, commonly based on polyethylene oxide (PEO) doped with zinc salts such as $ZnCl_2$ or $ZnSO_4$, offer a flexible and easily processable alternative to ceramic electrolytes. These electrolytes can be engineered for improved ionic conductivity and mechanical properties.
    - **Advantages**: GPEs provide excellent mechanical flexibility, which is ideal for applications in wearable electronics or flexible energy storage devices. They are lightweight, easy to manufacture, and offer good ionic conductivity at elevated temperatures.
    - **Challenges**: The primary challenge of GPEs is their low conductivity at room temperature, which can be improved by optimizing the electrolyte formulation. They are also prone to moisture degradation, limiting their long-term stability.
3. **Zinc-Ion Conductive Polymers**:
    - Researchers are exploring polymer-based electrolytes specifically designed for zinc-ion conduction, using polymers such as PVDF, PAN, and PEO blended with zinc salts. These materials provide excellent mechanical properties and design flexibility.
    - **Advantages**: These polymers allow for customized battery designs and are easy to process, making them suitable for tailored applications. They also offer mechanical compliance and lightweight designs.
    - **Challenges**: These electrolytes often have low ionic conductivity at room temperature and can face interface stability issues when paired with zinc metal anodes. Additional research is needed to optimize their performance.

4. **Hybrid Electrolytes (Composite Electrolytes)**:
   - Hybrid electrolytes combine ceramic electrolytes (such as Zn-based NASICON) with polymer matrices like PEO or PVDF. These composite materials aim to combine the high ionic conductivity of ceramic electrolytes with the flexibility and processability of polymers.
   - **Advantages**: These electrolytes offer the best of both worlds, with flexibility from the polymer component and high ionic conductivity from the ceramic phase. They are also easier to process and manufacture.
   - **Challenges**: Ensuring uniform dispersion of the ceramic phase within the polymer matrix is crucial, as poor dispersion can hinder ionic conductivity. Additionally, achieving interface compatibility between the two phases remains a significant challenge.
5. **Halide-Based Electrolytes**:
   - **New halide-based electrolytes**, such as $ZnCl_2$-doped materials or zinc-ion conducting sulfides, are being explored for their enhanced electrochemical stability and potential for higher ionic conductivity.
   - **Advantages**: These materials show promise for higher ionic conductivity and better electrochemical stability compared to traditional polymer-based electrolytes.
   - **Challenges**: Moisture sensitivity and manufacturing complexity are significant challenges for halide-based electrolytes, and further development is needed to improve their long-term performance.

Summary of Electrolyte Characteristics

Electrolyte type	Ionic conductivity	Advantages	Challenges
Zn-based NASICON	~ 10^{-3} S·cm^{-1}	High conductivity, electrochemical stability	Brittleness, moisture sensitivity, handling difficulty
Gel-polymer electrolytes (GPEs)	~ 10^{-3} S·cm^{-1} (elevated)	Flexibility, ease of fabrication, lightweight	Low conductivity at room temperature, degradation
Zinc-ion conductive polymers	~ 10^{-4} to 10^{-3} S·cm^{-1}	Design flexibility, easy processing	Low conductivity, interface stability with Zn anode
Hybrid electrolytes	~ 10^{-3} to 10^{-4} S·cm^{-1}	Combine benefits of ceramics and polymers	Interface compatibility, uniform dispersion of phases
Halide-based electrolytes	~ 10^{-4} to 10^{-3} S·cm^{-1}	High stability, potential for higher conductivity	Moisture sensitivity, under development

Thus, the development of solid electrolytes for zinc-ion solid-state batteries (ZISBs) is critical for enhancing their performance, safety, and longevity. Significant advancements have been made in Zn-based NASICON, gel-polymer electrolytes, and hybrid systems. Each electrolyte type has distinct advantages, such as high ionic conductivity and flexibility, but also faces challenges related to moisture sensitivity, interface stability, and room-temperature conductivity. Overcoming these challenges will be key to realizing the full potential of ZISBs in grid storage, wearable electronics, and other energy storage applications.

1.2.4 Other Multivalent Systems

In addition to sodium, potassium, and zinc-ion batteries, multivalent ion-based solid-state batteries, such as magnesium-ion solid-state batteries (Mg-SSBs) and calcium-ion solid-state batteries (Ca-SSBs), are emerging as promising candidates for future energy storage technologies. These systems offer the potential to address some of the key challenges associated with conventional lithium-ion batteries, such as cost, safety, and sustainability.

Magnesium-Ion Solid-State Batteries (Mg-SSBs)

Magnesium-ion solid-state batteries (Mg-SSBs) are gaining attention as a potential alternative to lithium-based systems. Magnesium is an earth-abundant element, and its high volumetric capacity and dendrite-free deposition offer significant advantages over lithium-ion technology. Mg-SSBs are also considered safer because magnesium does not form dendrites that could lead to short-circuiting.

Key Features and Components

- **Solid Electrolytes**: Mg-β-alumina and MgS-based electrolytes are the primary materials under investigation for Mg-SSBs, with ionic conductivities of up to 10^{-3} S/cm [35]. These electrolytes offer both high ionic conductivity and stability, which are crucial for maintaining the long-term efficiency and safety of the battery.
- **Electrode Materials**:
  - **Cathodes**: Magnesium-ion cathode materials include $MgMn_2O_4$, $MgCoO_2$, and Chevrel phases (Mo_6S_8). These materials are being explored for their high capacity and stability during cycling.
  - **Anodes**: Magnesium metal is typically used as the anode, with the advantage of high volumetric energy density.

- **Recent Advances**:
  - Electrolyte doping strategies are being developed to enhance the ionic conductivity of solid electrolytes, improving the overall performance of the battery.
  - The development of high-capacity Mg-S cathodes is underway, with the aim of increasing the energy density of Mg-SSBs while maintaining long cycle life.
- **Challenges**:
  - Slow diffusion kinetics of Mg^{2+} ions pose a challenge for the charging/discharging speed of Mg-SSBs. This can limit the performance and practicality of these batteries, especially in applications requiring fast charge times.
  - The limited availability of suitable cathode materials capable of supporting the high energy density and stable cycling of magnesium-ion systems remains a significant obstacle.

Calcium-Ion Solid-State Batteries (Ca-SSBs)

Calcium-ion solid-state batteries (Ca-SSBs) are in the early stages of research but show promise due to their high energy density and the abundant availability of calcium, making them a potentially cost-effective alternative to lithium-ion systems.

Key Features and Components

- **Solid Electrolytes**: Ca-NASICON and CaF_2-based materials are being explored as solid electrolytes for Ca-SSBs, with conductivities ranging from 10^{-5} to 10^{-3} S/cm [36]. These materials show potential for high ionic conductivity and electrochemical stability under the operating conditions of Ca-SSBs.
- **Electrode Materials**:
  - **Cathodes**: $CaTiO_3$, $CaCo_2O_4$, and $CaMn_2O_4$ are potential cathode materials that exhibit high energy density and stable cycling performance. The choice of cathode material is crucial for ensuring high capacity and cycle life in Ca-SSBs.
  - **Anodes**: As with other multivalent systems, calcium metal can be used as an anode, providing high volumetric energy. However, dendrite formation during cycling remains a challenge.
- **Recent Advances**:
  - The development of layered calcium oxides has been a key area of research, with some materials exhibiting stable cycling and improved capacity retention during prolonged charge–discharge cycles.
  - Novel electrolyte compositions are being explored to improve the mobility of Ca^{2+} ions, enabling more efficient ion transport and better performance in high-energy applications.

- **Challenges**:
  - High operating temperatures are often required for effective Ca^{2+} ion transport. This can make Ca-SSBs less suitable for some portable or low-temperature applications unless new electrolytes or temperature management techniques are developed.
  - Stability concerns in solid electrolytes, particularly at higher voltages, remain a barrier to the widespread use of Ca-SSBs. Research is needed to improve interface stability and prevent degradation during cycling.

Summary of Multivalent Solid-State Battery Systems

Battery type	Solid electrolyte	Electrode materials	Recent advances	Challenges
Magnesium-ion solid-state batteries (Mg-SSBs)	Mg-β-alumina, MgS-based materials, conductivity ~ 10^{-3} S/cm	$MgMn_2O_4$, $MgCoO_2$, Chevrel phases (Mo_6S_8)	Electrolyte doping, high-capacity Mg-S cathodes	Slow Mg^{2+} ion diffusion, limited cathode material options
Calcium-ion solid-state batteries (Ca-SSBs)	Ca-NASICON, CaF_2-based materials, conductivity ~ 10^{-5} to 10^{-3} S/cm	$CaTiO_3$, $CaCo_2O_4$, $CaMn_2O_4$	Layered calcium oxides, novel electrolyte compositions	High operating temperatures, stability of solid electrolytes

1.2.5 Comparative Overview of Performance Characteristics in Solid-State Battery Systems

The following table presents a comparative overview of the performance characteristics across various solid-state battery (SSB) systems, including sodium-ion (SISBs), potassium-ion (KISBs), zinc-ion (ZISBs), magnesium-ion (Mg-SSBs), and calcium-ion (Ca-SSBs) batteries. The table summarizes key features such as solid electrolytes, electrode materials, recent advancements, challenges, and critical performance metrics, including energy density, capacity, and cycle life. This comparison provides a clear understanding of the relative strengths and limitations of each system, highlighting their potential applications and areas for future development in energy storage technologies.

Battery type	Solid electrolyte	Electrode materials	Recent advances	Challenges	Performance
Sodium-ion solid-state batteries (SISBs)	NASICON-type ($Na_3Zr_2Si_2PO_{12}$), Sulfide-based (e.g., $Li_4Ti_5O_{12}$)	Cathodes: $NaFePO_4$, $Na_3V_2(PO_4)_3$, Layered oxides (Na_xCoO_2); Anodes: Hard carbon, Na metal, titanium-based compounds	Composite NASICON/ polymer electrolytes for flexibility; Surface coatings on cathodes	Lower energy density (~100–150 Wh/kg), stability at high temperature	**Energy Density**: ~ 100–150 Wh/kg **Capacity**: Up to 160 mAh/g **Cycle Life**: Moderate (~1000 cycles)
Potassium-ion solid-state batteries (KISBs)	K-β-Alumina, Polymer-based electrolytes	Cathodes: K_xMnO_2, $K_2FeFe(CN)_6$, Prussian blue analogs; Anodes: Graphite, potassium metal, alloy-based materials	Hybrid solid electrolytes (inorganic-polymer); Interface engineering for K metal anodes	Limited high-performance solid electrolytes, dendrite growth in K metal anodes	**Energy Density**: ~ 100–180 Wh/kg **Capacity**: ~ 100 mAh/g **Cycle Life**: Moderate (~500–1000 cycles)
Zinc-ion solid-state batteries (ZISBs)	Zn-based NASICON, Gel-polymer electrolytes	Cathodes: MnO_2, V_2O_5, Prussian blue analogs; Anodes: Zn metal, Zn-alloy composites	Flexible solid-state Zn-ion batteries; Protective coatings to suppress corrosion	Zinc dendrite formation, corrosion and side reactions reducing lifespan	**Energy Density**: ~ 150 Wh/kg **Capacity**: ~ 200 mAh/g **Cycle Life**: Moderate (~500–1500 cycles)
Magnesium-ion solid-state batteries (Mg-SSBs)	Mg-β-alumina, MgS-based electrolytes	Cathodes: $MgMn_2O_4$, $MgCoO_2$, Chevrel phases (Mo_6S_8); Anodes: Magnesium metal	Electrolyte doping strategies to improve ionic conductivity; High-capacity Mg-S cathodes	Slow Mg^{2+} ion diffusion kinetics, limited suitable cathode materials	**Energy Density**: ~ 150–300 Wh/kg **Capacity**: ~ 300 mAh/g (theoretical) **Cycle Life**: Low to Moderate (~500 cycles)
Calcium-ion solid-state batteries (Ca-SSBs)	Ca-NASICON, CaF_2-based materials	Cathodes: $CaTiO_3$, $CaCo_2O_4$, $CaMn_2O_4$; Anodes: Calcium metal	Layered calcium oxides with stable cycling; Novel electrolyte compositions for improved Ca^{2+} mobility	High operating temperatures required, stability of solid electrolytes	**Energy Density**: ~ 100–250 Wh/kg **Capacity**: ~ 150 mAh/g **Cycle Life**: Low to Moderate (~300–800 cycles)

Performance Summary

- **Sodium-Ion Solid-State Batteries (SISBs)**: These batteries offer a moderate energy density (~ 100–150 Wh/kg), with capacities of up to 160 mAh/g. They exhibit a moderate cycle life of about 1000 cycles.
- **Potassium-Ion Solid-State Batteries (KISBs)**: KISBs have an energy density range of 100–180 Wh/kg, with a typical capacity of ~ 100 mAh/g. Their cycle life is also moderate, ranging from 500–1000 cycles.
- **Zinc-Ion Solid-State Batteries (ZISBs)**: ZISBs show a higher energy density (~150 Wh/kg) and a capacity of about 200 mAh/g, with a cycle life that can reach 500–1500 cycles.
- **Magnesium-Ion Solid-State Batteries (Mg-SSBs)**: These batteries have a high theoretical energy density (~ 150–300 Wh/kg) and capacity (up to 300 mAh/g), though their cycle life is still under development and is generally low to moderate (~ 500 cycles).
- **Calcium-Ion Solid-State Batteries (Ca-SSBs)**: Ca-SSBs offer moderate energy density (~ 100–250 Wh/kg) and capacity (~ 150 mAh/g). However, their cycle life is relatively low to moderate (~300–800 cycles), and their performance is heavily influenced by the need for high operating temperatures.

This comparison provides a comprehensive view of the performance metrics for each solid-state battery system, allowing for an understanding of their potential suitability for various applications, from portable electronics to large-scale energy storage.

2 Manufacturing of Solid-State Batteries

2.1 Material Synthesis and Processing

The fabrication of solid-state batteries (SSBs) involves precise material synthesis and processing techniques to ensure high ionic conductivity, mechanical stability, and interfacial compatibility.

2.1.1 Solid Electrolyte Fabrication

The solid electrolytes used in SSBs can be synthesized using various methods, including:

1. **Solid-State Reaction Method**: High-temperature calcination (typically > 800 °C) of precursor oxides to form crystalline solid electrolytes, such as garnet-type $Li_7La_3Zr_2O_{12}$ (LLZO) and NASICON-type $Li_{1+x}Al_xTi_{2-x}(PO_4)_3$ [37].

Table 4 Electrode processing techniques in solid-state batteries

Technique	Description	Advantages	Applications
Spray coating	Deposition of thin electrode layers via atomized solution spray	Uniform coating, scalable	Thin-film electrodes, flexible devices
Dip coating	Immersion of substrates in electrode slurry followed by drying	Precise thickness control	Laboratory-scale fabrication
Tape casting	Casting slurry on a flat surface using a doctor blade	High throughput, good thickness uniformity	Commercial bulk-type SSBs
Screen printing	Pushing ink through patterned mesh	Patternable, multilayer-compatible	Printed energy storage devices
Electrophoretic deposition	Electric field-driven deposition of charged particles	High packing density, good adhesion	Thin-film SSBs

2. **Sol–Gel Synthesis**: A low-temperature wet chemistry method that enhances homogeneity and allows fine control over particle size and morphology [38].
3. **Mechanochemical Synthesis**: Ball milling at room temperature to induce solid-state reactions, leading to nanostructured electrolytes with high ionic conductivity [39].

2.1.2 Electrode Coating Techniques

Coating techniques ensure uniformity, adhesion, and electrical contact in solid-state battery electrodes. Common methods include:

- **Spray Coating**: Thin and uniform layers of active material are deposited onto current collectors.
- **Dip Coating**: Substrates are immersed in an electrode material solution to achieve precise thickness control.
- **Tape Casting**: A widely used technique for large-scale production that produces uniform films of controlled thickness [40] (Table 4).

2.1.3 Composite Electrolyte Fabrication

Composite solid electrolytes, integrating inorganic fillers (e.g., LLZO, LATP) with polymer matrices (e.g., PEO, PVDF), improve ionic conductivity while maintaining mechanical flexibility. Techniques include:

1. **Solution Casting**: Dissolving polymer and electrolyte filler in solvent and casting into films.

2. **Hot-Pressing**: Applying heat and pressure to consolidate the composite structure and enhance interfacial contact [41].

2.2 Cell Assembly Techniques

2.2.1 Thin-Film Solid-State Battery Fabrication

Thin-film solid-state batteries are fabricated using advanced deposition techniques:

- **Atomic Layer Deposition (ALD)**: Provides precise control of ultrathin electrolyte and electrode layers, ensuring uniform thickness and defect-free films [42].
- **Pulsed Laser Deposition (PLD)**: Uses high-energy laser pulses to ablate target materials, forming thin films with high purity and density.
- **Chemical Vapor Deposition (CVD)**: Enables large-area deposition of solid electrolytes and electrodes with excellent adhesion.

2.2.2 Bulk-Type Solid-State Battery Fabrication

Bulk SSBs are manufactured using:

- **Cold Pressing and Sintering**: Powder-based electrolyte materials are compacted and sintered at high temperatures to improve grain boundary conduction.
- **Hot Isostatic Pressing (HIP)**: Applies heat and isotropic pressure to enhance material densification and reduce porosity.
- **Spark Plasma Sintering (SPS)**: A rapid sintering process utilizing pulsed electric currents to produce dense and conductive electrolyte structures [43].

2.3 Scaling-Up Manufacturing for Commercialization

Scaling up SSB manufacturing requires adapting existing lithium-ion production infrastructure while introducing new techniques for solid-state components.

- **Roll-to-Roll Manufacturing**: Enables continuous processing of thin-film solid electrolytes and electrodes, enhancing production efficiency.
- **Integration into Existing Li-Ion Production Lines**: Adapting current gigafactory setups to accommodate solid-state processing reduces costs and accelerates commercialization.
- **Cost Analysis and Economic Feasibility**: Optimizing materials and processing to balance performance and affordability remains a challenge in large-scale SSB production [44].

Table 5 Encapsulation techniques in solid-state batteries

Technique	Description	Benefits	Applications
Hermetic sealing	Metal or ceramic sealing to isolate cell from moisture/air	Protects sensitive electrolytes like sulfides	All-solid-state pouch cells
Multilayer thin-film coating	ALD or CVD of barrier films (e.g., Al_2O_3, SiN)	Excellent moisture resistance, ultrathin	Thin-film batteries, flexible electronics
Laminated polymer films	Flexible, heat-sealable polymer encapsulants	Lightweight, cost-effective	Wearables, flexible SSBs
Vacuum packaging	Enclosing battery in vacuum-sealed pouches	Prevents gas ingress, preserves electrolyte	Research cells, pilot-scale formats

2.4 Encapsulation and Packaging

Packaging is crucial to ensure the longevity and safety of solid-state batteries by preventing degradation and protecting against environmental factors.

- **Prevention of Electrolyte Degradation from Moisture**: Many solid electrolytes, especially sulfides, are highly sensitive to moisture, necessitating hermetic sealing techniques.
- **Thin-Film Encapsulation Techniques**: Use of multilayer barrier coatings (e.g., Al_2O_3, SiN) to enhance moisture resistance.
- **Flexible and Bendable Battery Designs**: Novel packaging materials and structures enable the development of flexible SSBs for wearable electronics and biomedical applications [45] (Table 5).

2.5 Process Flow for Solid-State Battery Manufacturing

To visually represent the stepwise development of solid-state batteries from raw material synthesis to final packaging, the following flow chart summarizes the key stages involved in the complete manufacturing process (Fig. 1).

3 Battery Performance Evaluation

3.1 Electrochemical Characterization

- **Capacity and Energy Density Measurements**

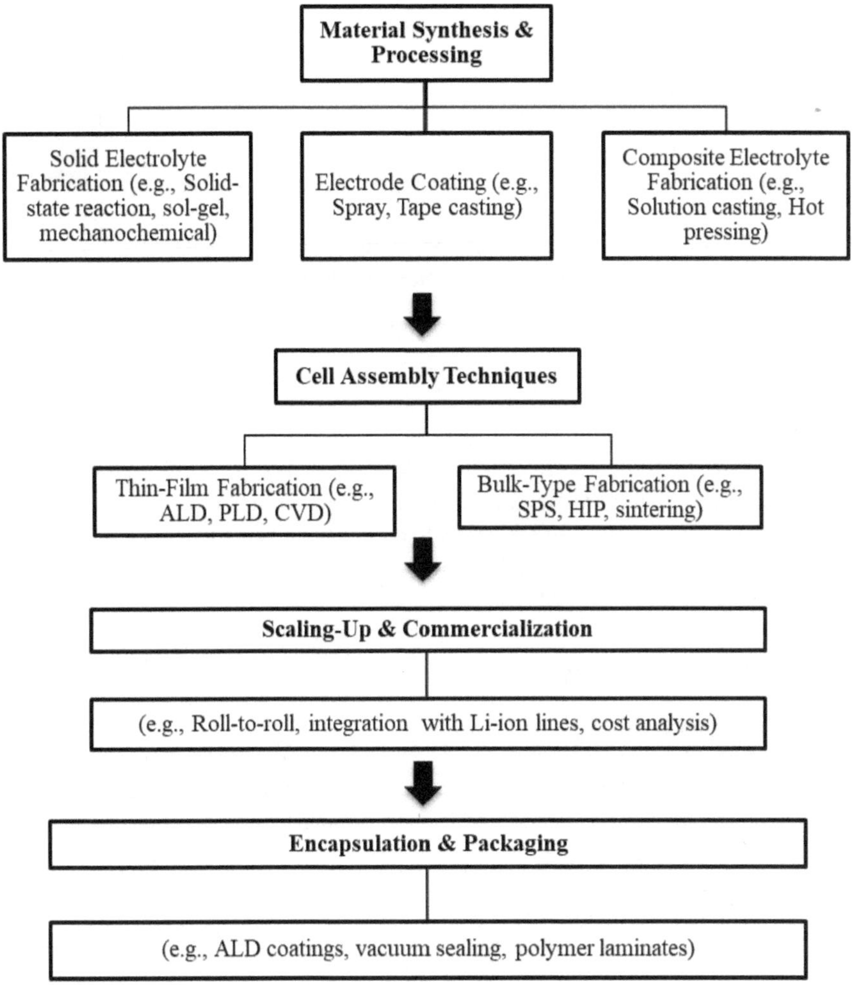

Fig. 1 Flow chart of solid-state battery manufacturing process

- **Galvanostatic Charge–Discharge Cycling**: Galvanostatic charge–discharge cycling is one of the most commonly used techniques to evaluate the electrochemical performance of solid-state batteries. This method involves applying a constant current (either charge or discharge) to the battery and recording the corresponding voltage change over time. By analyzing the charge and discharge curves, researchers can determine the battery's capacity, energy density, and the ability to retain capacity over multiple cycles. This technique is useful for evaluating the efficiency of energy storage and assessing the degradation of the battery during repeated cycles. Key parameters such as specific capacity (mAh/g) and energy density (Wh/kg) are measured to provide insight into the battery's overall performance and its potential for real-world applications.

  - **Rate Capability Tests (C-rate Analysis)**: C-rate analysis involves testing the battery at different charge and discharge rates to evaluate its performance under various conditions. The C-rate refers to the speed at which a battery is charged or discharged relative to its capacity. For example, a C-rate of 1C means the battery is charged or discharged in 1 h. Higher C-rates assess the battery's power output and its ability to function at faster charge/discharge speeds without significant loss in performance. The results provide insights into the high-power capabilities of the battery and help identify any limitations in performance at higher currents, such as a decrease in capacity or efficiency.

- **Electrochemical Impedance Spectroscopy (EIS)**
  - EIS is an essential tool for analyzing the electrochemical properties of solid-state batteries, particularly the resistance at the electrode/electrolyte interface. By applying a small alternating current and measuring the resulting voltage response, EIS can provide detailed information about the ionic and electronic resistances, charge transfer processes, and the overall efficiency of the battery. Interfacial resistance, which arises from poor contact between the electrolyte and electrodes, can significantly reduce the battery's performance and lifespan. EIS also helps detect degradation in battery components over time, providing valuable data for improving battery design and materials.

- **Cyclic Voltammetry (CV)**
  - Cyclic Voltammetry is a versatile electrochemical technique used to study redox reactions and the reversibility of ion transport in the solid electrolyte. In a CV experiment, the battery's potential is swept in a triangular waveform, and the resulting current is recorded. This process reveals important information about the kinetics of electrochemical reactions, such as the reversibility and stability of the redox reactions occurring at the electrodes. CV can also help identify unwanted side reactions and degradation processes, as well as assess the stability of the electrolyte over cycles. For solid-state batteries, understanding the redox behavior is crucial for optimizing electrolyte compositions and ensuring long-term stability.

3.2 Mechanical and Thermal Stability

- **Thermal Stability Testing**
  - **Thermal Runaway Analysis**: Thermal runaway is a critical safety concern for batteries, particularly under extreme conditions such as overcharging, overheating, or short circuits. It refers to an uncontrollable increase in temperature, which can lead to battery failure, fire, or explosion. Thermal runaway analysis is performed by subjecting solid-state batteries to elevated temperatures to simulate real-world abuse conditions. This helps assess the battery's

thermal stability and its ability to safely handle high-temperature scenarios. Researchers monitor the temperature rise, internal pressure, and any signs of degradation or chemical reactions that could trigger thermal runaway.
  - **Differential Scanning Calorimetry (DSC)**: DSC is a thermal analysis technique that measures the heat flow associated with phase transitions in the battery materials. By heating or cooling the battery or its components, DSC helps identify critical thermal events such as melting, crystallization, or phase changes in the solid electrolyte and electrodes. This information is vital for ensuring the thermal stability of solid-state batteries and for designing materials that can withstand high operating temperatures without degradation. DSC can also help determine the heat capacity of the materials, which is essential for understanding the battery's behavior under varying temperature conditions.

- **Mechanical Stress and Fracture Analysis**
  - **In Situ Mechanical Testing**: Solid-state batteries are often subjected to mechanical stresses during fabrication, cycling, and operation. In situ mechanical testing involves applying compression, tension, or bending forces to the battery components while monitoring their structural integrity. This technique helps identify failure mechanisms in solid electrolytes, such as cracking or delamination, which can lead to short circuits or loss of performance. Real-time observation of mechanical stress effects during testing provides valuable insights into how solid-state batteries can be designed to be more resilient under mechanical strain.
  - **Stress–Strain Behavior of Solid Electrolytes**: Solid electrolytes must possess sufficient mechanical strength to withstand the internal stresses caused by volume changes during charge and discharge cycles. Stress–strain analysis is used to determine how solid electrolytes behave under different mechanical loads, including their elastic modulus, tensile strength, and fracture toughness. Understanding the mechanical properties of solid electrolytes is essential for designing batteries that can maintain structural integrity and performance over long lifespans.

3.3 Longevity and Aging Mechanisms

- **Electrochemical Stability Over Cycling**
  - Long-term cycling tests are essential for evaluating the longevity of solid-state batteries. These tests involve repeated charge and discharge cycles to assess capacity fade, energy retention, and the stability of the battery over extended use. Solid-state batteries, like all batteries, experience a gradual decline in performance as they age. Cycling tests help quantify this degradation and identify any changes in battery characteristics, such as increased internal resistance

or decreased capacity. Electrochemical stability testing is crucial for determining the usable lifespan of a battery and understanding how materials or design improvements can enhance longevity.

- **Interface Degradation and Lithium Plating**
  - One of the most significant challenges for solid-state batteries, particularly those with lithium metal anodes, is the formation of lithium dendrites. Dendrites are spiky structures that grow from the anode and can pierce the electrolyte, causing short circuits or failure. Over time, dendrite formation can lead to rapid capacity loss and potential safety hazards. Interface degradation, including the breakdown of the electrolyte–electrode interface, is another common issue that can contribute to battery failure. Studying the mechanisms of dendrite formation and interface degradation is crucial for developing strategies to mitigate these issues and extend battery life.
- **Strategies to Extend Battery Life**
  - Several strategies have been proposed to enhance the cycle life of solid-state batteries. These include the use of protective coatings on electrodes to prevent dendrite formation, the development of advanced electrolyte formulations with improved stability, and interface engineering to optimize contact between the electrodes and electrolyte. By addressing the degradation mechanisms that lead to capacity fade and short circuits, researchers can significantly improve the longevity of solid-state batteries. Surface coatings, such as ceramic or polymer layers, can also help reduce electrolyte decomposition and prevent the formation of harmful side products (Table 6).

4 Challenges and Future Perspectives

4.1 Current Challenges in Solid-State Batteries

Material-Level Challenges

- **Low Ionic Conductivity in Some Solid Electrolytes:** Despite advancements in solid electrolytes, many materials still exhibit lower ionic conductivity compared to their liquid counterparts, which limits their practical applications [56, 57].
- **Stability Issues with Lithium Metal Anodes:** Lithium metal, a preferred anode material, suffers from dendrite formation, which can lead to short circuits and reduced battery lifespan [58].

Table 6 Summary of battery performance evaluation techniques

Category	Technique	Purpose	Key parameters/ insights	Refs.
Electrochemical characterization	Galvanostatic charge–discharge	Evaluate capacity, energy density, and cycle life	Specific capacity (mAh/g), energy density (Wh/kg), efficiency	[46]
	Rate capability tests (C-rate analysis)	Assess high-rate performance	Capacity retention at varying C-rates	[47]
	Electrochemical Impedance Spectroscopy (EIS)	Measure ionic/ electronic resistance and interfacial behavior	Charge transfer resistance, interfacial impedance	[48]
	Cyclic Voltammetry (CV)	Analyze redox reactions and kinetics	Redox peaks, electrochemical reversibility	[49]
Thermal stability	Thermal runaway analysis	Assess safety under high-temperature conditions	Onset temperature for degradation, self-heating behavior	[50]
	Differential Scanning Calorimetry (DSC)	Determine thermal transitions and decomposition temperatures	Phase transitions, heat capacity, enthalpy changes	[51]
Mechanical stability	In situ mechanical testing	Evaluate stress effects on battery components	Crack propagation, interfacial delamination, real-time degradation	[52]
	Stress–strain analysis	Determine mechanical strength of solid electrolyte	Young's modulus, tensile strength, fracture toughness	[53]
Longevity and aging	Long-term cycling	Measure stability over extended cycling	Capacity fade, Coulombic efficiency over time	[54]
	Interface stability testing	Monitor interfacial degradation, lithium plating	Resistance growth, dendrite formation, loss of contact	[55]

Manufacturing Challenges

- **Difficulty in Large-Scale Fabrication:** The fabrication of solid-state batteries (SSBs) involves complex processes such as sintering and thin-film deposition, which are challenging to scale up for commercial production [59].
- **High Costs of Solid-State Processing:** The need for specialized manufacturing environments and materials increases the production cost of SSBs compared to conventional lithium-ion batteries [60].

4.2 Strategies for Performance Enhancement

- **High-Conductivity Electrolyte Development**: Research is focused on developing advanced electrolytes such as doped garnet-type $Li_7La_3Zr_2O_{12}$ (LLZO) and sulfide-based electrolytes to enhance ionic conductivity [61].
- **Advanced Interfacial Engineering**: Techniques like atomic layer deposition (ALD) and artificial solid electrolyte interphase (SEI) coatings are being explored to improve electrolyte–electrode contact and reduce interfacial resistance [45].
- **3D-Structured Electrodes for Better Ion Transport**: The use of 3D architectures in electrode design can increase the active surface area and enhance ion diffusion pathways, improving overall battery performance [62].

4.3 Next-Generation Solid-State Battery Technologies

AI-Assisted Material Discovery

- **Machine Learning for Electrolyte and Electrode Screening**: AI-driven algorithms can accelerate the discovery of new materials with optimized properties for solid-state batteries, reducing experimental costs and time [63].

3D-Printed Solid-State Batteries

- **Additive Manufacturing for Complex Architectures**: 3D printing enables the fabrication of intricate battery architectures with enhanced ion transport and mechanical stability, paving the way for customizable SSB designs [27].

Hybrid Solid-State Batteries

- **Solid–Liquid Hybrid Designs for Better Performance**: A combination of solid and liquid electrolytes can mitigate some limitations of purely solid-state designs, offering improved ionic conductivity while maintaining safety advantages [64].

5 Conclusion

Solid-state batteries (SSBs) have emerged as a revolutionary alternative to conventional lithium-ion batteries, offering enhanced safety, higher energy density, and longer cycle life. This chapter has explored the fundamental aspects of SSB design, manufacturing processes, performance evaluation, and current challenges.

6 Summary of Key Advancements in Solid-State Battery Technology

Recent advancements in solid-state battery technology have led to significant improvements in ionic conductivity, interfacial stability, and mechanical robustness. Key developments include:

- The introduction of high-conductivity solid electrolytes such as garnet-type $Li_7La_3Zr_2O_{12}$ (LLZO) and sulfide-based electrolytes with superior ionic transport properties [65].
- Strategies to mitigate dendrite formation, including interface engineering and artificial solid electrolyte interphases [66].
- Manufacturing techniques such as atomic layer deposition (ALD) and 3D printing that enable precise electrode and electrolyte structuring for improved performance [58].

6.1 Comparison of Different SSB Chemistries and Their Potential Applications

Different solid-state battery chemistries present unique advantages and challenges, making them suitable for diverse applications:

- **Lithium-Based SSBs**: High energy density and efficiency, ideal for electric vehicles (EVs) and portable electronics [67].
- **Sodium-Ion Solid-State Batteries (SISBs)**: Cost-effective and abundant raw materials, suitable for grid-scale energy storage [68].
- **Potassium-Ion Solid-State Batteries (KISBs)**: Lower manufacturing costs and potential for stationary applications [69].
- **Zinc-Ion Solid-State Batteries (ZISBs)**: Safe and environmentally friendly, with applications in flexible and wearable electronics [70].

6.2 Roadmap for Future Research and Industrial Adoption

The commercialization of solid-state batteries requires continued research and development efforts in multiple areas:

- **Enhancing Electrolyte performance**: Developing novel solid electrolytes with improved ionic conductivity and stability [71].
- **Manufacturing Scalability**: Advancing cost-effective fabrication techniques for mass production [72].
- **Sustainability Considerations**: Investigating eco-friendly and recyclable materials to minimize environmental impact [73].

- **Integration with Emerging Technologies**: Exploring hybrid battery systems and AI-driven material discovery for next-generation energy storage solutions [74].

While challenges remain, solid-state batteries hold immense potential for transforming the energy storage landscape. With continued advancements, they are expected to play a pivotal role in future electric mobility, renewable energy integration, and portable electronics.

References

1. Vadhva P, Hu J, Johnson MJ, Stocker R, Braglia M, Brett DJL, Rettie AJE (2021) Electrochemical impedance spectroscopy for all-solid-state batteries: theory, methods and future outlook. ChemElectroChem 8:1930–1947
2. Tan DHS, Meng YS, Jang J (2022) Scaling up high-energy-density sulfidic solid-state batteries: a lab-to-pilot perspective. Joule 6:1755–1769
3. Kammampata SP, Basappa RH, Ito T, Yamada H, Thangadurai V (2019) Microstructural and electrochemical properties of alkaline earth metal-doped Li garnet-type solid electrolytes prepared by solid-state sintering and spark plasma sintering methods. ACS Appl Energy Mater 2:1765–1773
4. Liu M, Guan X, Liu H, Ma X, Wu Q, Ge S, Zhang H, Xu J (2022) Composite solid electrolytes containing single-ion lithium polymer grafted garnet for dendrite-free, long-life all-solid-state lithium metal batteries. Chem Eng J 445:136436
5. Lee J, Lee T, Char K, Kim KJ, Choi JW (2021) Issues and advances in scaling up sulfide-based all-solid-state batteries. Acc Chem Res 54:3390–3402
6. Kim JY, Jung S, Kang SH, Lee MJ, Jin D, Shin DO, Lee YG, Lee YM (2021) All-solid-state hybrid electrode configuration for high-performance all-solid-state batteries: comparative study with composite electrode and diffusion-dependent electrode. J Power Sour 518:230736
7. Elgrishi N, Rountree KJ, McCarthy BD, Rountree ES, Eisenhart TT, Dempsey JL (2017) A practical Beginner's guide to cyclic voltammetry. J Chem Educ 95:197–206
8. Dai Y, Panahi A (2024) Thermal runaway process in lithium-ion batteries: a review. Next Energy 6:100186
9. Zhang K, Li N, Li X, Huang J, Chen H, Jiao S, Song W (2022) Understanding enhanced ionic conductivity in composite Solid-State electrolyte in a wide frequency range of 10–2–1010 Hz. Adv Sci 9:2200213
10. Chang H, Zhang X, Li W, Liu H, Hu H, Liu Z, Liu W, Jin Y (2024) Recent developments in composite solid electrolytes for all-solid-state lithium batteries and their applications in deep-sea environment. Next Mater 6:100307
11. Sasaki Y, Song S, Hori S, Matsui N, Nomoto K, Suzuki K, Hirayama M, Song I, Jang Y, Kanno R (2024) Synthesis and structural characterization of lithium ionic conductors LI_2MS_3 (M=SI, $SI_{0.5}GE_{0.5}$, GE) and $LI_{16}GE_5S_{18}$. Chemi Mater 36:7176–7185
12. Xu L, Tang S, Cheng Y, Wang K, Liang J, Liu C, Cao Y-C, Wei F, Mai L (2018) Interfaces in solid-state lithium batteries. Joule 2(10):1991–2015
13. Mao Y, Liu J, Chen W, Zhang W, Sun C (2025) Recent advances in garnet-based electrolytes for solid-state lithium metal batteries: interfacial challenges and engineering strategies. Mater. Horizons
14. Zhu X, Jiang W, Zhao S, Huang R, Ling M, Liang C, Wang L (2022) Exploring the concordant solid-state electrolytes for all-solid-state lithium-sulfur batteries. Nano Energy 96:107093
15. Chen X, Vereecken PM (2018) Solid and solid-like composite electrolyte for lithium ion batteries: engineering the ion conductivity at interfaces. Adv Mater Interfaces 6:1800899
16. Liu L, Feng Y, Wu W (2018) Recent progress in printed flexible solid-state supercapacitors for portable and wearable energy storage. J Power Sources 410–411:69–77

17. Wang C, Liang J, Zhao Y, Zheng M, Li X, Sun X (2021) All-solid-state lithium batteries enabled by sulfide electrolytes: from fundamental research to practical engineering design. Energy Environ Sci 14:2577–2619
18. Chen L, Qiu X, Bai Z, Fan L-Z (2020) Enhancing interfacial stability in solid-state lithium batteries with polymer/garnet solid electrolyte and composite cathode framework. J Energy Chem 52:210–217
19. Parbin A, Rafiuddin N (2024) Advancing solid-state lithium phosphate electrolyte: synthesis and characterization of Li_2NaPO_4–ZrO_2 nanocomposites for solid-state batteries. J Solid State Chem 332:124580
20. Zhang T, He W, Zhang W, Wang T, Li P, Sun Z, Yu X (2020) Designing composite solid-state electrolytes for high performance lithium ion or lithium metal batteries. Chem Sci 11:8686–8707
21. Li J, Chen X, Muhammad S, Roy S, Huang H, Yu C, Ullah Z, Wang Z, Zhang Y, Wang K, Guo B (2024) Development of solid polymer electrolytes for solid-state lithium battery applications. Mater Today Energy 43:101574
22. Zheng Y, Yao Y, Ou J, Li M, Luo D, Dou H, Li Z, Amine K, Yu A, Chen Z (2020) A review of composite solid-state electrolytes for lithium batteries: fundamentals, key materials and advanced structures. Chem Soc Rev 49:8790–8839
23. Indu MS, Alexander GV, Sreejith OV, Abraham SE, Murugan R (2021) Lithium garnet-cathode interfacial chemistry: inclusive insights and outlook toward practical solid-state lithium metal batteries. Mater Today Energy 21:100804
24. Nzereogu PU, Oyesanya A, Ogba SN, Ayanwunmi SO, Sobajo, Chimsunum VC, Ayanwunmi VO, Amoo MO, Adefemi OT, Chukwudi CC (2024) Solid-state lithium-ion battery electrolytes: revolutionizing energy density and safety. Hybrid Adv 8:100339
25. Takada K (2009) Secondary batteries–lithium rechargeable systems—lithium–ion. In: Electrolytes: solid oxide. Elsevier eBooks, pp 328–336
26. Xia S, Wu X, Zhang Z, Cui Y, Liu W (2018) Practical challenges and future perspectives of All solid-state lithium-metal batteries. Chem 5:753–785
27. Yang H, Wu N (2022) Ionic conductivity and ion transport mechanisms of solid-state lithium-ion battery electrolytes: a review. Energy Sci Eng 10:1643–1671
28. Wang Q, Bai N, Wang Y, He X, Zhang D, Li Z, Sun Q, Sun H, Wang B, Wang G, Fan L-Z (2024) Optimization and progress of interface construction of ceramic oxide solid-state electrolytes in Li-metal batteries. Energy Storage Mater 71:103589
29. Sung J, Heo J, Kim D-H, Jo S, Ha Y-C, Kim D, Ahn S, Park J-W (2024) Recent advances in all-solid-state batteries for commercialization. Mater Chem Front 8:1861–1887
30. Ahmad H, Kubra KT, Butt A, Nisar U, Iftikhar FJ, Ali G (2023) Recent progress, challenges, and perspectives in the development of solid-state electrolytes for sodium batteries. J Power Sources 581:233518
31. Yang Q, Wang A, Luo J, Tang W (2021) Improving ionic conductivity of polymer-based solid electrolytes for lithium metal batteries. Chin J Chem Eng 43:202–215
32. Dolla TH, Ajayi SO, Sikeyi LL, Mathe MK, Palaniyandy N (2024) Advances in solid-state batteries fabrication strategies for their manufacture. J Energy Storage 106:114737
33. Sutherland LJ, Weerasinghe HC, Simon GP (2021) A review on emerging barrier materials and encapsulation strategies for flexible perovskite and organic photovoltaics. Adv Energy Mater 11(34):2101383
34. Xiang Y, Li X, Cheng Y, Sun X, Yang Y (2020) Advanced characterization techniques for solid state lithium battery research. Mater Today 36:139–157
35. Lee D, Lee H, Song T, Paik U (2022) Toward high rate performance solid-state batteries. Adv Energy Mater 12:2200948
36. Hori S, Kanno R, Sun X, Song S, Hirayama M, Hauck B, Dippon M, Dierickx S, Ivers-Tiffée E (2022) Understanding the impedance spectra of all-solid-state lithium battery cells with sulfide superionic conductors. J Power Sources 556:232450
37. Bhuvanendran N, Ravichandran S, Xu Q, Maiyalagan T, Su H (2022) A quick guide to the assessment of key electrochemical performance indicators for the oxygen reduction reaction: a comprehensive review. Int J Hydrogen Energy 47:7113–7138

38. McKerracher RD, Guzman-Guemez J, Wills RGA, Sharkh SM, Kramer D (2021) Advances in prevention of thermal runaway in Lithium-Ion batteries. Adv Energy Sustain Res 2:2000059
39. Yamada H, Ito T, Basappa RH, Bekarevich R, Mitsuishi K (2017) Influence of strain on local structure and lithium ionic conduction in garnet-type solid electrolyte. J Power Sources 368:97–106
40. Cheng Z, Pan H, Li F, Duan C, Liu H, Zhong H, Sheng C, Hou G, He P, Zhou H (2022) Achieving long cycle life for all-solid-state rechargeable Li-I_2 battery by a confined dissolution strategy. Nat Commun 13:125
41. Clausnitzer M, Ihrig M, Cressa L, Hein S, Finsterbusch M, Eswara S, Kuo LY, Danner T, Kaghazchi P, Fattakhova-Rohlfing D, Guillon O, Latz A (2024) Impact of degradation mechanisms at the cathode/electrolyte interface of garnet-based all-solid-state batteries. Energy Storage Mater 67:103262
42. Liu M, Song A, Zhang X, Wang J, Fan Y, Wang G, Tian H, Ma Z, Shao G (2025) Interfacial lithium-ion transportation in solid-state batteries: challenges and prospects. Nano Energy 136:110749
43. Nikfarjam N, Coman PT, Free C, Ziehl P, Sadati M, White RE (2024) Advancing ionic conductivity in solid electrolytes: Insights from polymerization-induced phase separation and microstructural optimization. J Energy Storage 93:112287
44. Qu J, Liu J, Leng G, Wang Y, Wang S, Shen S, Yin H, Hou Z, Duan B (2023) Overcoming the obstacles of lithium-metal anodes for high-energy batteries. Electrochem Commun 153:107537
45. Huang KJ, Ceder G, Olivetti EA (2021) Manufacturing scalability implications of materials choice in inorganic solid-state batteries. Joule 5:564–580
46. Wang S, Demopoulos GP (2024) High-conductive polymer-in-porous garnet solid electrolyte structure for all-solid-state lithium batteries enabled by molecular engineering. Energy Storage Mater 71:103604
47. Hatzell KB, Zheng Y (2021) Prospects on large-scale manufacturing of solid state batteries. MRS Energy Sustain 8:33–39
48. Hu X, Zhu Y, Fu Y-F, Fan Y, Chen Y, Kim D, Lu X, Yu B, Chen YI (2024) 3D-printed topological-structured electrodes with exceptional mechanical properties for high-performance flexible Li-ion batteries. Energy Storage Mater 70:103560
49. Qi B, Hong X, Jiang Y, Shi J, Zhang M, Yan W, Lai C (2024) A review on engineering design for enhancing interfacial contact in solid-state lithium-sulfur batteries. Nano-Micro Letters 16:71
50. Song J, Shang W, Zhou W, Li J, Wu X, Li W, Zhang R, Liu S (2025) From issues to solutions: 3D printing for overcoming challenges in liquid- and solid-state batteries. Energy Storage Materi 104342
51. Woolley HM, Vargas-Barbosa NM (2022) Hybrid solid electrolyte-liquid electrolyte systems for (almost) solid-state batteries: why, how, and where to? J Mater Chem A 11:1083–1097
52. Owens BB (2000) Solid state electrolytes: overview of materials and applications during the last third of the Twentieth Century. J Power Sources 90:2–8
53. Wang C, Zhang K, Cui Y, Li Q, Ma T, Li F, Qiu H, Jin D (2025) Enhancement of dendrite-free lithium metal anode performance through LiF-rich protective layer for lithium metal batteries. J Phys Chem Lett 16:3240–3248
54. Pei M, Shi H, Yao F, Liang S, Xu Z, Pei X, Wang S, Hu Y (2021) 3D printing of advanced lithium batteries: a designing strategy of electrode/electrolyte architectures. J Mater Chem A 9:25237–25257
55. Gao X, Zheng M, Yang X, Sun R, Zhang J, Sun X (2022) Emerging application of 3D-printing techniques in lithium batteries: from liquid to solid. Mater Today 59:161–181
56. Du H, Zhang X, Yu H (2024) Design of high-energy-density lithium batteries: liquid to all solid state. eTransportation 23:100382
57. Thairiyarayar CB, Huang TF, Chang JK, Li J, Kheawhom S, Liu W-R (2025) Advances in sodium all-solid-state batteries: Insights into sulfide solid electrolytes and their applications. J Taiwan Inst Chem Eng 106110

58. Sada K, Darga J, Manthiram A (2023) Challenges and Prospects of Sodium-Ion and Potassium-Ion batteries for mass production. Adv Energy Mater 13:2302321
59. Guo Y, Lim GJH, Verma V, Cai Y, Chua R, Lim JJN, Srinivasan M (2023) Solid state zinc and aluminum ion batteries: challenges and opportunities. Chemsuschem 16:e202202297
60. Chen S, Wen K, Fan J, Bando Y, Golberg D (2018) Progress and future prospects of high-voltage and high-safety electrolytes in advanced lithium batteries: from liquid to solid electrolytes. J Mater Chem A 6:11631–11663
61. Forsyth M, Porcarelli L, Wang X, Goujon N, Mecerreyes D (2019) Innovative electrolytes based on ionic liquids and polymers for Next-generation solid-state batteries. Acc Chem Res 52(3):686–694
62. Azhari L, Bong S, Ma X, Wang Y (2020) Recycling for all solid-state lithium-ion batteries. Matter 3:1845–1861
63. Xiong Y, Zhang D, Ruan X, Jiang S, Zou X, Yuan W, Liu X, Zhang Y, Nie Z, Wei D, Zeng Y, Cao P, Zhang G (2024) Artificial intelligence in rechargeable battery: advancements and prospects. Energy Storage Mater 73:103860
64. Sharma SK, Sharma G, Gaur A, Arya A, Mirsafi FS, Abolhassani R, Rubahn H-G, Yu J-S, Mishra YK (2022) Progress in electrode and electrolyte materials: path to all-solid-state Li-ion batteries. Energy Adv 1:457–510
65. Go D, Shin JW, Lee S, Lee J, Yang BC, Won Y, Motoyama M, An J (2022) Atomic layer deposition for thin film solid-state battery and capacitor. Int J Precis Eng Manuf-Green Technol 10:851–873
66. Cao K, Ma J, Yue Z, Li H, Fan Y, Liu H (2025) Status and challenges of solid-state electrolytes for potassium batteries, Small 2500762
67. Wu X, Leonard DP, Ji X (2017) Emerging non-aqueous potassium-ion batteries: challenges and opportunities. Chem Mater 29:5031–5042
68. Feng X, Fang H, Wu N, Liu P, Jena P, Nanda J, Mitlin D (2022) Review of modification strategies in emerging inorganic solid-state electrolytes for lithium, sodium, and potassium batteries. Joule 6:543–587
69. Yao Q, Zhu C (2020) Advanced post-potassium-ion batteries as emerging potassium-based alternatives for energy storage. Adv Func Mater 30:2005209
70. Taghavi-Kahagh A, Roghani-Mamaqani H, Salami-Kalajahi M (2023) Powering the future: a comprehensive review on calcium-ion batteries. J Energy Chem 90:77–97
71. Zhang W, Lu J, Guo Z (2021) Challenges and future perspectives on sodium and potassium ion batteries for grid-scale energy storage. Mater Today 50:400–417
72. Gummow RJ, Vamvounis G, Kannan MB, He Y (2018) Calcium-ion batteries: current state-of-the-art and future perspectives. Adv Mater 30:1801702
73. Dompablo MEA-D, Ponrouch A, Johansson P, Palacín MR (2019) Achievements, challenges, and prospects of calcium batteries. Chem Rev 120:6331–6357
74. Wu X, Brinckerhoff A, Nguyen J, Pasternak M, Fang H (2025) Advances in developing inorganic calcium solid-state electrolytes: a minireview. Energy Fuels 39:5691–5756

Recent Approaches to Enhance Metal Oxide Performance

Apoorva Rai, H. S. Tewari, and Jai Singh

Abstract Metal oxide and its composites have recently emerged as a compelling area of research for efficient and sustainable energy storage systems. These materials are pivotal to next-generation energy technologies, offering significant improvements in electrical conductivity, reversible capacity, and cycling stability. This chapter explores the key properties of metal oxide and composites, with a focus on enhancing their conductivity through the integration of conductive additives such as graphene and carbon nanotubes. Addressing current challenges requires innovative synthesis techniques, precise nanostructuring, and cross-disciplinary collaboration. Techniques like doping, surface modification, and composite formation have shown promise in improving ion transport, enhancing mechanical stability, and reducing interfacial resistance. Ultimately, the chapter underscores the transformative potential of metal oxide composites in advancing energy storage solutions, emphasizing the importance of continued research to harness their full potential in a renewable energy future. Furthermore, by engineering these properties at the nanoscale, researchers aim to develop SSBs that meet the demands of next-generation energy storage systems.

Keywords Metal oxides · Nanostructures · Synthesis methods · Application · Energy storage

1 Introduction

Batteries are critical components of modern energy storage systems due to their ability to store large amounts of energy and deliver it quickly when required. They are widely used in applications ranging from smartphones to electric vehicles. Among various battery types, lithium-ion batteries (LIBs) are currently the most prevalent,

---

A. Rai · H. S. Tewari · J. Singh (✉)
Department of Pure and Applied Physics, Guru Ghasidas Vishwavidyalaya, Bilaspur, India
e-mail: jai@physics.du.ac.in

J. Singh
Department of Physics and Astrophysics, University of Delhi, New Delhi 110007, India

A. K. Rai et al. (eds.), *Electrochemistry of Next-Generation Solid-State Batteries*, Green Energy and Technology, https://doi.org/10.1007/978-981-95-7603-6_8

known for their high energy density, reliable performance, and efficiency. However, LIBs also pose several challenges, such as high production costs, safety and disposal concerns, limited lifespan, and the uneven global distribution of lithium resources [1, 2]. These limitations have sparked growing interest in alternative battery chemistries that are both cost-effective and sustainable.

As a result, metal-sulfur rechargeable batteries based on alkali metals—such as lithium, sodium, and potassium—have gained significant attention as promising alternatives. In particular, sodium-based systems are appealing due to the abundance and low cost of sodium. Conventional alkali metal-based batteries often utilize intercalation-type cathode materials, such as layered transition metal oxides (TMOs), which allow reversible insertion and extraction of alkali metal ions. Despite their advantages, TMOs suffer from limited energy density, high operating voltages, and structural degradation during cycling, especially under high-voltage conditions [3].

In contrast, sulfur has emerged as a highly attractive cathode material due to its high theoretical specific capacity (~ 1675 $mAh{\cdot}g^{-1}$), natural abundance, and low cost [4]. However, the practical application of sulfur cathodes is hindered by complex conversion reactions, which lead to the formation of soluble polysulfides. These polysulfides tend to migrate between the cathode and anode—a phenomenon known as the "shuttle effect"—which severely impacts cycling stability and capacity retention [5]. Sodium–sulfur (Na/S) batteries, especially in high-temperature configurations, offer a compelling solution by combining high energy density (~ 760 $Wh{\cdot}kg^{-1}$), long cycle life, and environmental friendliness. Despite successful use in space and large-scale energy storage, these systems still face challenges related to safety and maintenance [6, 7]. To address the drawbacks of sulfur cathodes and enhance the electrochemical performance of Na/S batteries, researchers have explored various host materials. Early efforts involving carbon-based and polymer-based materials have led to modest improvements, yet often fail to solve the core issues related to reaction kinetics and structural degradation over prolonged cycling. In response, transition metal oxide (TMO)-based composites have emerged as a promising solution. Their tunable properties—enabled by strategies such as atomic doping, defect engineering, heterojunction formation, and alloying—offer a pathway to improve the redox kinetics and suppress the shuttle effect [1]. However, the performance of TMO composites depends greatly on the specific type of compound used (e.g., sulfides, carbides, nitrides), and the mechanisms by which they facilitate polysulfide conversion remain underexplored. To further mitigate polysulfide shuttling, carbon-based materials like carbon nanotubes, graphene, and hollow carbon spheres have been investigated due to their porous structures, which can physically confine polysulfides. Nevertheless, the nonpolar nature of carbon results in weak chemical interactions with sodium polysulfides (NaPSs), limiting their adsorption capability and overall effectiveness [1].

Within the broader context of sodium-ion batteries (SIBs), various cathode materials have been studied, including polyanionic compounds, Prussian blue analogs, and organic materials. However, layered transition metal oxides, particularly O3-type compounds such as $NaNi_{1/3}Fe_{1/3}Mn_{1/3}O_2$, have attracted considerable interest due to their high specific capacity, straightforward synthesis, and favorable electrochemical properties [8]. These materials can be classified based on their morphology into polycrystalline and single-crystal types. Polycrystalline cathodes consist of secondary particles formed from nanocrystalline grains with random lattice orientations, resulting in numerous grain boundaries. While this structure can enhance ion diffusion and rate performance, it also introduces susceptibility to microcracking, which accelerates side reactions with the electrolyte and compromises long-term stability and safety. In contrast, single-crystal cathodes feature uniform particle sizes and a continuous, well-ordered crystal structure with consistent lattice orientation throughout. This reduces grain boundary-related degradation and improves structural integrity, offering better long-term performance. To further improve the electrochemical performance of layered oxide cathodes in SIBs, researchers have focused on two key strategies: doping and coating. Doping involves incorporating foreign elements such as Co, Ni, or Mn into the crystal lattice, enhancing electrochemical activity, ionic and electronic conductivity, and structural stability [9]. Coating techniques, using materials like metal oxides, carbon, or polymers, create protective layers on the cathode surface to prevent degradation and reduce side reactions with the electrolyte [9]. These combined strategies significantly enhance the cycling life, stability, and rate capability of cathode materials.

Recent advances in co-doping, for example, Ti and Mg have been particularly effective. These dopants help suppress phase transitions and reduce cation disorder, thereby reinforcing the layered structure and enhancing both rate performance and cycling stability [9]. In summary, the integration of sulfur based chemistry with advanced cathode engineering including TMO composites, carbon supports, doping, and coating represents a promising pathway to overcoming the limitations of conventional sodium-ion and sodium–sulfur batteries. These developments could pave the way for safer, more cost-effective, and environmentally sustainable energy storage solutions beyond lithium-ion technologies. Figure 1 illustrates the theoretical energy densities of various battery types, highlighting a comparison between commercial high-temperature sodium–sulfur batteries and Ni–Cd batteries. It also includes emerging battery technologies that demonstrate similarly enhanced energy densities, an essential attribute for their potential commercialization.

1.1 Role of Metal Oxides in SSBs

Metal oxides, particularly in nanomaterial form, have been extensively explored as electrode materials for lithium-ion batteries (LIBs) due to their ease of large-scale production and rich redox chemistry. This redox activity enables metal oxides to achieve high specific capacities and capacitances, often surpassing those of carbon-based materials [1]. Additionally, certain metal oxides like iron and manganese oxides are naturally abundant and cost-effective. The size, shape, and orientation of metal oxide nanostructures can be precisely controlled, allowing for systematic

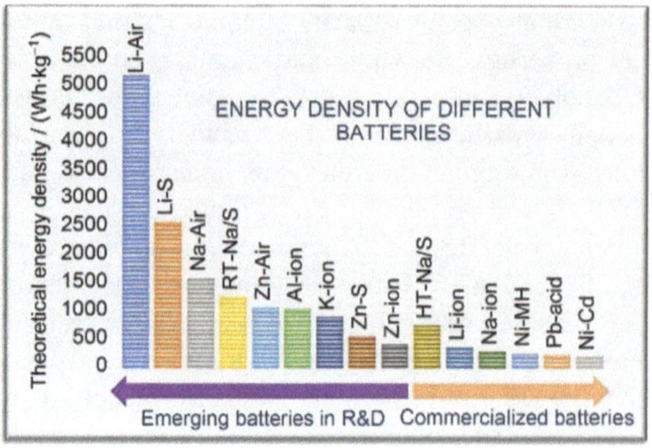

Fig. 1 Comparison of energy densities among various commercial and emerging battery technologies [1]

investigation of the relationships between structure and electrochemical properties [9].

1.2 Challenges and Limitations of Metal Oxides in SSBs

Metal oxides are increasingly utilized in solid-state batteries (SSBs) for their high ionic conductivity, stability, and environmental friendliness. However, several challenges and limitations hinder their widespread adoption.

(a) Many widely researched oxides, including MnO_2, Co_3O_4, CoO, NiO, and V_2O_5, are wide bandgap semiconductors or insulators, which typically results in poor electrical conductivity. This limitation is particularly problematic for electrode materials, as it can lead to safety issues due to excessive joule heat generation during repeated charging and discharging cycles [10].
(b) Furthermore, metal oxides typically exhibit sluggish ion transport kinetics [11]. Although reducing oxide particle sizes to the nanoscale and creating hierarchically porous structures can enhance ion diffusion by providing additional transport pathways, these modifications also inevitably increase the electrical resistance within the solid electrode due to the heightened presence of grain boundaries [12–14].
(c) Additionally, metal oxides used specially in lithium-ion batteries (LIBs) experience significant volume expansion and contraction during charge/discharge cycles, leading to electrode film pulverization [15, 16]. While innovative strategies, such as utilizing hollow structures, one-dimensional (1D) nanostructures [17–22] and "self-wound" nanomembranes [16] have been developed to mitigate volume changes, the cycling stability of single-phase oxides remains inadequate, with lifetimes typically limited to less than 150 cycles, far short of the commercial standard of 400–1200 cycles.

To overcome these challenges and unlock the full potential of high-efficiency electrochemical energy storage devices, further research and development are urgently required, focusing on the creation of innovative electrode materials, such as multifunctional hybrid materials, and the design of advanced electrode architectures.

2 Importance of Enhancing Metal Oxide Performance

Enhancing the performance of metal oxides is essential for advancing solid-state batteries (SSBs), as these materials serve as vital components in electrodes and electrolytes. To improve their efficacy, researchers focus on several strategies. One approach is optimizing the material's microstructure through nanoscale engineering, which can enhance ionic conductivity and reduce grain boundary resistance. Doping metal oxides with various elements can also create defect sites that facilitate ion transport, improving overall conductivity. Additionally, developing composite materials that integrate metal oxides with polymers or other conducting materials can help mitigate mechanical stress during cycling. Advanced characterization methods, such as in-situ electrochemical techniques, provide insights into the dynamic behavior of these materials, aiding in the design of more robust SSB architectures. Ultimately, addressing these enhancements will contribute to higher energy densities, longer cycle life, and improved safety, pushing the boundaries of solid-state battery technology.

2.1 Doping Strategies and Its Effects on Metal Oxide Properties

Doping strategies play a critical role in enhancing the performance of metal oxides in solid-state batteries (SSBs) by modifying their electrical, ionic, and structural properties. Here are some key doping strategies:

(a) **Nanostructuring**: Doping metal oxides while employing nanoscale fabrication techniques can result in high surface area and reduced diffusion paths for ions, further enhancing performance. Techniques like sol–gel synthesis and hydrothermal methods facilitate this approach [23].
(b) **Doping of Cations**: Introducing cations with varying charge states or sizes into a metal oxide lattice can generate defect sites, which can enhance ionic conduction. Specifically, doping lithium metal oxides with elements like sodium or magnesium can increase lithium-ion mobility, resulting in improved electrical conductivity [24].
(c) **Doping of Anions**: Replacing oxygen anions with alternative anions, such as sulfur or nitrogen, in metal oxides can modify their electronic structures

and create more favorable pathways for ion transport. This anion substitution strategy can significantly enhance both ionic and electronic conductivity [25].

(d) **Multiple Element Doping**: Co-doping metal oxides with multiple cations or anions can lead to synergistic improvements in their properties. By combining different dopants, it's possible to optimize defect concentrations and charge carrier mobility, resulting in enhanced electrochemical performance. This co-doping strategy can unlock superior properties, making metal oxides more suitable for various applications [26].

(e) **Surface Doping**: Modifying the surface of metal oxides through doping can improve electrochemical activity and stability. Surface dopants can enhance the interface between electrodes and solid electrolytes, reducing interfacial resistance and improving overall battery performance [26].

These doping strategies can lead to improved ionic conductivity, increased electrochemical activity, and enhanced structural stability, all of which are crucial for the development of efficient solid-state batteries, thereby paving the way for higher performance energy storage solutions.

2.1.1 Types of Dopants (e.g., Rare Earth Elements, Transition Metals)

Doping in metal oxide solid-state batteries refers to the intentional introduction of different atoms or ions into the crystal lattice of the oxide material. This process is used to modify the material's electronic, ionic, and structural properties, which can enhance the performance of solid-state batteries in terms of conductivity, stability, and capacity. Two common types of doping in these materials are rare earth doping and transition metal doping, and each plays a unique role.

(i) **Doping of Rare Earth Elements**

Rare earth elements (REEs) are a group of 15 chemical elements in the lanthanide series (e.g., Lanthanum, Cerium) along with Yttrium. These elements are known for their unique electronic properties, making them useful dopants in metal oxide materials [27].

Effects:

- Ionic Conductivity Enhancement: Rare earth dopants can help to stabilize the crystal structure of the metal oxide, making the ionic pathways more efficient, thus enhancing the ionic conductivity of the electrolyte in solid-state batteries.
- Electrochemical Performance: Doping with rare earth elements can improve the electrochemical stability and performance of the electrodes, potentially leading to higher charge/discharge capacities, longer cycle life, and better rate capabilities.
- Modification of Electronic Structure: Rare earth doping can influence the electronic structure of metal oxides, improving their electronic conductivity and facilitating better charge transport within the battery.

- Stabilization of Oxygen Vacancies: Rare earth elements like Cerium (Ce) have a unique ability to stabilize oxygen vacancies, which are essential for fast ionic conductivity in oxide materials.

Lanthanum (La) and Cerium (Ce) are commonly doped into oxide materials like $LiCoO_2$, Li_2TiO_3 or $LiFePO_4$ to increase the battery's stability and conductivity [28].

(ii) **Doping of Transition Metal Elements**

Transition metals, situated in the central block of the periodic table, include elements such as Chromium (Cr), Manganese (Mn), Cobalt (Co), and Nickel (Ni). A distinctive characteristic of these metals is the presence of partially filled d-orbitals, which confer unique electronic properties. These properties make transition metals particularly valuable for enhancing the performance of battery materials [29].

Effects:

- Modification of Redox Properties: Transition metals can introduce new oxidation states into the metal oxide lattice, which enhances the electrochemical activity of the material. This allows for better charge and discharge performance [27].
- Increased Electronic Conductivity: Transition metal dopants can improve the electronic conductivity of the oxide materials, which is particularly important in the case of cathodes in solid-state batteries.
- Structural Stability: Some transition metal dopants help to stabilize the crystal structure of metal oxides under cycling conditions. This can improve the mechanical integrity and longevity of the battery material.
- Enhanced Lithium or Sodium Ion Diffusion: Transition metals can improve the diffusion rate of lithium ions in cathode materials, which is essential for achieving high charge/discharge rates.

Titanium (Ti) and Chromium (Cr) are commonly doped into oxides like $LiCoO_2$, $LiMn_2O_4$, and $LiFePO_4$ to improve stability and charge storage capacity. The comparison, effect, and applications of dopant type in metal oxides are summarized in Table 1.

In summary, both rare earth and transition metal doping have been shown to significantly enhance the properties of metal oxide materials utilized in solid-state batteries. Rare earth doping primarily serves to improve ionic conductivity and stabilize oxygen vacancies, whereas transition metal doping is focused on tuning redox properties, boosting electronic conductivity, and ultimately upgrading the overall performance of battery materials.

2.1.2 Case Studies: NiO/Ni Composites, Al-Doped ZnO

(a) **NiO/Ni Nanocomposite for Li-Ion Storage**: A novel NiO/Ni nanocomposite embedded in a 3D porous carbon network was successfully synthesized through a facile two-step process as reported by Yang et al. [30] Initially, a precursor

Table 1 Comparison and summary of dopant type

Dopant type	Dopants	Impact of doping on metal oxides	Utilizations
Transition metal	Chromium (Cr), Titanium (Ti), Nickel (Ni), Manganese (Mn)	Modulates redox behavior, reinforces structural stability, amplifies electronic conductivity	Frequently used in popular cathode materials, including **$LiNiO_2$**, **$LiFePO_4$**, and **$LiCoO_2$** to improve electrochemical performance, rate capability, and lifespan
Rare earth	Cerium (Ce), Lanthanum (La), Neodymium (Nd)	Boosts ionic conductivity, enhanced electrochemical stability, stabilizes oxygen vacancies	Commonly employed in lithium-based oxides such as **Li_2TiO_3**, **$LiNiO_2$** and **$LiFePO_4$** to optimize performance and longevity

mixture of nickel carbonate and glucose was heat-treated at 700 °C, followed by annealing at 270 °C in air as displayed in Fig. 2a, b. When employed as a Li-storage electrode in Li-ion batteries, the NiO/Ni@C composite exhibited exceptional electrochemical properties, including a reversible specific capacity of 934 mAh/g after 200 cycles at 300 mA/g. Furthermore, it demonstrated superior rate performance, retaining a specific capacity of 505 mAh/g at 2 A/g and displaying remarkable stability with a capacity of 683 mAh/g after 1000 cycles at 1 A/g as shown in Fig. 3. This synthesis method is efficient, scalable, and suitable for large-scale production.

(b) **Al-Doped ZnO for Li–S Batteries**: Lithium-sulfur batteries have attracted a lot of interest because of its remarkable theoretical capacity, energy density, and affordability. Nevertheless, issues with sulfur cathodes include volume expansion, shuttle effects, and uncontrolled lithium sulfide deposition, which lead to low Coulombic efficiency and quick capacity loss. Metal-oxide nanostructures have been widely used in Li–S batteries in recent years to regulate the deposition of lithium sulfide and lessen the shuttle effect. However, the application of non-conductive metal oxides complicates electrochemical reaction kinetics by adding a new diffusion phase.

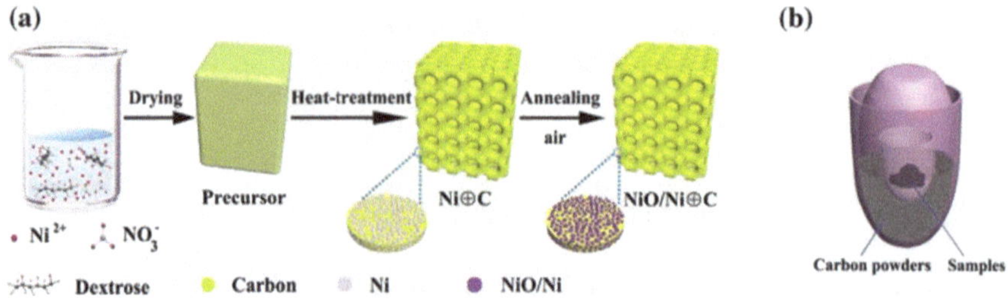

Fig. 2 **a** Synthesis of NiO/Ni composites, **b** NiO/Ni nanocomposite embedded in a 3D porous carbon network (NiO/Ni@C) [30]

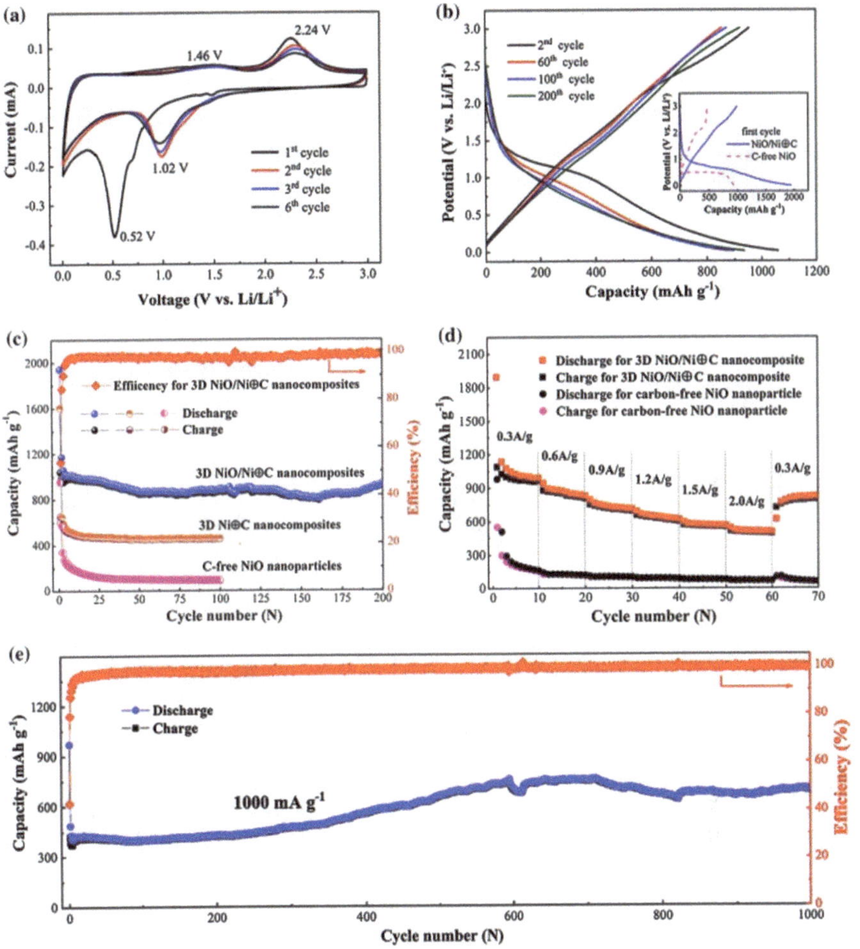

Fig. 3 Electrochemical performance of the studied electrodes is, **a** Cyclic voltammetry (CV) curves of the NiO/Ni@C electrode at a scan rate of 0.1 mV s^{-1}, **b** Charge/discharge curves of the NiO/Ni@C electrodes at a current density of 300 mA g^{-1}. The inset compares the first charge/discharge curves of NiO/Ni@C and carbon-free NiO electrodes, **c** A comparison of the cycling performance of NiO/Ni@C, Ni@C, and carbon-free NiO electrodes at 300 mA g− 1, **d** A comparison of the rate performance of NiO/Ni@C and carbon-free NiO electrodes at various current densities, **e** The long-term cycling performance of the NiO/Ni@C electrode at a high current density of 1000 mA g^{-1} for 1000 cycles [30]

Kong et al. [31] reported a simple biotemplating technique that uses kapok fibers as a carbon supply and template to create carbon nanoflakes embellished with conductive aluminum-doped zinc oxide (AZO@C) nanoparticles. The sulfur cathode based on the AZO@C nanocomposite performs better electrochemically than cathodes made of the less conducting ZnO and Al_2O_3. Notably, after 100 cycles at 0.1C, the AZO@C/S composite reaches a stable capacity of 927 mAh/g, and after 300 cycles at 0.5C, it reaches a reversible capacity of 544 mAh/g. After 100 cycles at 0.5C, the composite maintains a capacity of 466 mAh/g, even with a high sulfur loading of 6.96 mg·cm^{-2}. The outstanding electrochemical performance of the AZO@C/S composite can be attributed to the high conductivity of the polar AZO host, which suppresses the shuttle effect while enhancing redox kinetics during the reversible transformation of lithium sulfide species as shown in Fig. 4.

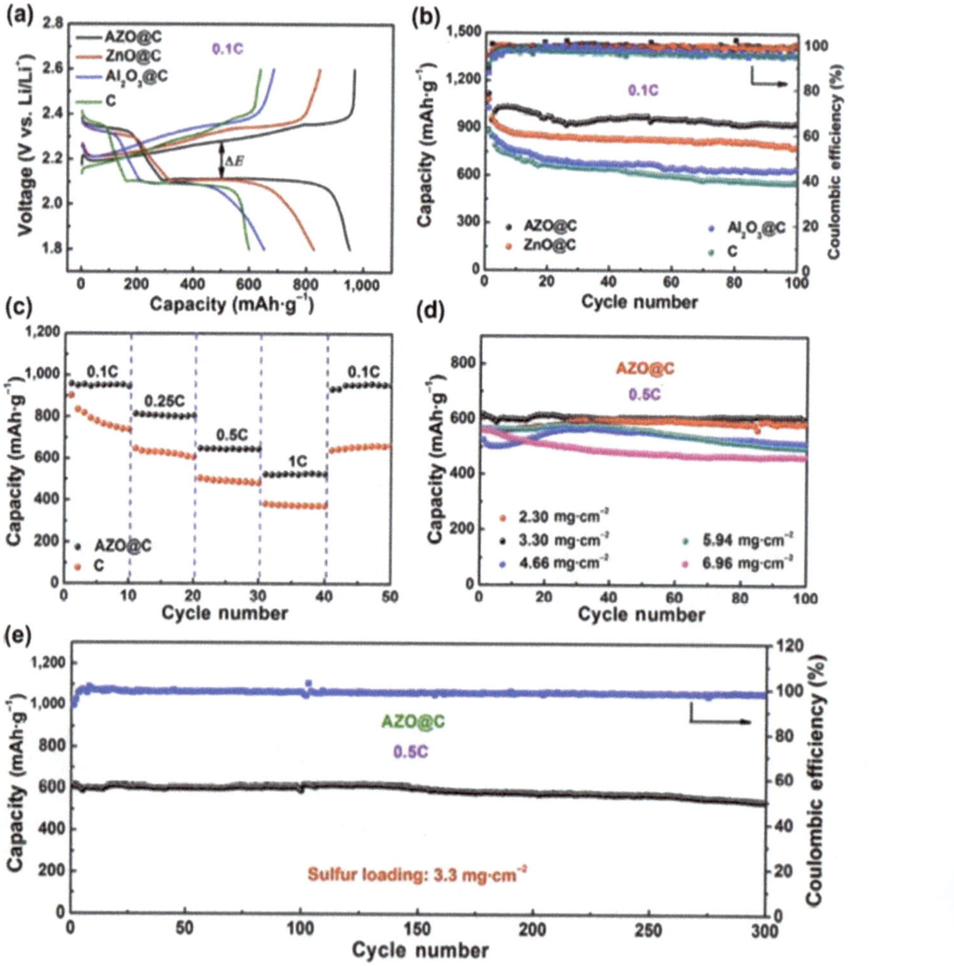

Fig. 4 **a** Charge–discharge curves of sulfur cathodes incorporating various metal oxide@carbon nanostructures (AZO@C, ZnO@C, Al_2O_3@C, and C) at a rate of 0.1C, **b** Cycling stability of sulfur cathodes at 0.1C, **c** Rate capability of the AZO@C/S composite cathode, **d** Influence of sulfur loading on the cycling performance of the AZO@C/S electrode, **e** Long-term cycling stability (1000 cycles) and Coulombic efficiency of the AZO@C/S electrode with high sulfur loading (3.3 mg·cm^{-2}) at 0.5C [31]

3 Enhancing Metal Oxide Performance by Nanostructuring

Enhancing the performance of metal oxide materials through nanostructuring is a promising approach to ameliorate the properties and make them more efficient for numerous applications, such as in energy storage, catalysis, and sensors. Nanostructuring refers to the design and fabrication of materials at the nanoscale, typically in the range of 1–100 nm. By tailoring the size, shape, and surface area of metal oxides, researchers can dramatically enhance their physical, chemical, and electronic characteristics, unlocking new possibilities for improved performance and functionality.

3.1 Effects of Nanostructuring on Metal Oxide Properties (e.g., Surface Area, Reactivity)

Since metal oxides are versatile materials with diverse applications in energy storage, catalysis, and electronics. Nevertheless, their full potential is frequently hindered by several key limitations, including: low surface area, poor electrical conductivity, limited mechanical strength. By engineering metal oxides at the nanoscale, researchers can unlock significant enhancements in their properties and functionality. This strategic manipulation of metal oxide materials can yield substantial benefits, including:

(a) **Enhanced Surface Area**: Nanostructuring dramatically boosts the surface area-to-volume ratio, a critical factor in catalytic applications. By increasing the surface area, the reactivity of metal oxides increases because of more active sites created by nanostructuring.
(b) **Enhanced Electrical Conductivity**: Nanostructuring can improve the electrical conductivity of metal oxides by reducing grain boundaries and increasing the density of charge carriers.
(c) **Enhanced Mechanical Strength**: Nanostructuring can improve the mechanical strength of metal oxides by reducing defects and increasing the density of the material.
(d) **Boosts Catalytic Efficiency**: The unique combination of high surface area and tailored electronic properties in nanostructured metal oxides significantly enhances their catalytic activity. Notably, metal oxide nanoparticles and nanowires have shown great results in applications such as catalytic converters and environmental remediation, where they efficiently break down pollutants and facilitate chemical reactions.

3.2 *Methods for Nanostructuring Metal Oxides*

To overcome the limitations of metal oxides, researchers have turned to nanostructuring techniques, which enable the precise control of metal oxide size, shape, and structure. Some of the techniques are listed below.

3.2.1 Sol–Gel Method

The sol–gel process is a versatile method for synthesizing nanomaterials, involving the transformation of a precursor solution into a gel-like network, which is then dried and annealed to produce the desired nanostructure [32]. By carefully controlling parameters such as precursor selection, catalyst addition, solvent choice, and drying conditions, this technique can be tailored to meet specific requirements. For instance, the synthesis of $NaCoO_2$ cathode via sol–gel method for Na-ion batteries is shown in Fig. 5. Moreover, the sol–gel process enables the incorporation of organic and temperature-sensitive materials, making it a valuable tool for a wide range of applications [33, 34]. Despite its versatility, the sol–gel method also has several drawbacks, including:

- Substantial shrinkage occurs during the drying and firing stages, which can cause cracking and compromise the material's integrity.
- Prolonged processing times, often requiring multiple days to complete.
- Difficulty in translating the process to large-scale industrial production, posing significant scalability challenges [34].

Despite these limitations, the sol–gel method remains a valuable and widely used technique for synthesizing high-purity materials, owing to its unique advantages, including:

- The sol–gel process is used to synthesize high-quality metal oxide nanosheets with optimal properties for battery applications, leveraging a citrate-assisted method involving evaporation, drying, and calcination [35].
- This process offers exceptional flexibility, enabling the synthesis of various metal oxides, including aluminum oxide, titanium dioxide, zinc oxide, and iron oxide, and facilitating the creation of customized composites for diverse applications [36].
- This method can strengthen the interfacial bond between carbon nanotubes (CNTs) and metal oxides, substantially enhancing the mechanical strength and electrical conductivity of the resulting composites [37].
- Recently, anisotropic ZnO/rGO and $LiFePO_4$/C nanocomposites have emerged as promising electrode materials, exhibiting high performance in supercapacitors (SCs) and batteries, respectively [34].

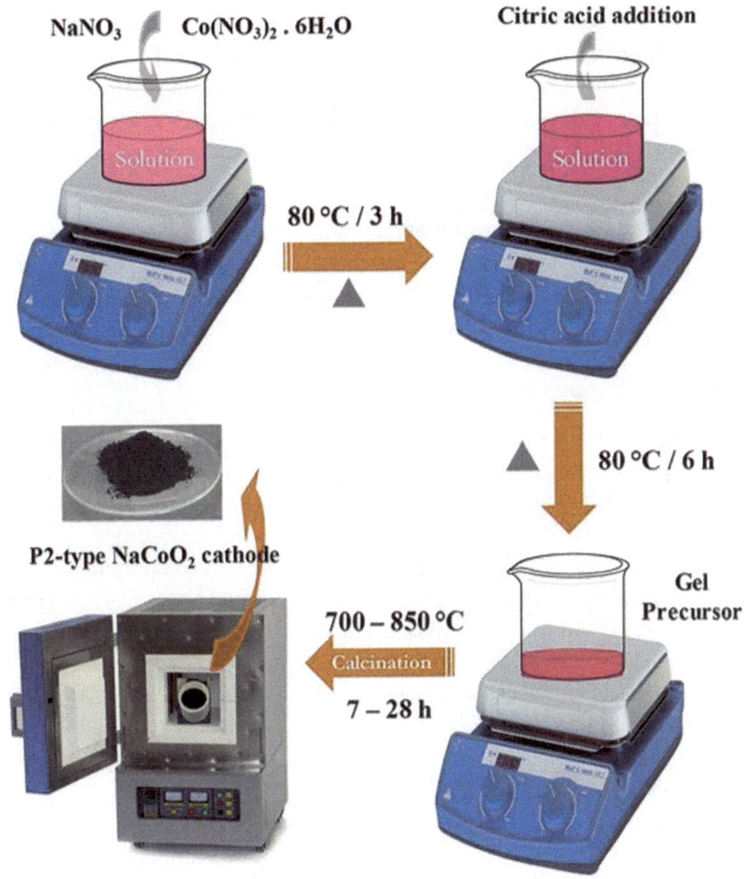

Fig. 5 Illustration of synthesis of NaCoO2 cathode via sol–gel method for Na-ion batteries [37]

3.2.2 Hydrothermal and Solvothermal Method

In hydrothermal and solvothermal synthesis, a precursor-solvent mixture is confined within a high-pressure vessel. The solvent plays a crucial role, serving as both a reaction medium and catalyst, and influencing the growth of one-dimensional (1D) or two-dimensional (2D) nanomaterials through heterogeneous reactions. Hydrothermal synthesis employs aqueous solutions, while solvothermal synthesis uses organic solvents or non-aqueous liquids. Both techniques involve subjecting metal precursors to precisely controlled high-temperature and high-pressure conditions, facilitating the formation of nanomaterials [34, 39]. Typical solvents used in solvothermal synthesis include alcohols, glycols, and supercritical fluids like carbon dioxide [40]. Hydrothermal and solvothermal synthesis methods offer unparalleled precision in crafting tailored zeolites and nanostructures with specific lattice arrangements and chemical compositions, rendering them optimally suited for high-performance applications, including batteries and supercapacitors [41]. Controlling the temperature, pressure, pH, and duration of the reaction is important, as these techniques permit detailed adjustment of these parameters, which in turn adjusts the particle size, shape, and degrees of crystallinity [41]. This method has distinct benefits, including the following advantages:

- By adjusting some conditions (temperature, pressure, pH), it is possible to fabricate complex nanostructures and composites, offering a wider range of structural configurations than other synthesis methods, such as sol–gel or exfoliation [41].
- This method produces less waste in comparison to other synthesis procedures.
- It enables the growth of crystalline phases that would otherwise be unstable at their melting temperatures.

These methods stand out for their ability to engineer a wide range of compositions and structural configurations. Notably, Pazhamalai et al. [34] developed innovative composite structures, such as a unique double sandwich-like Co_2SiO_4/derivatives with graphene oxide (GO) composite. This demonstrates the flexibility of these methods in layering and integrating materials. Furthermore, the ability to precisely control nanostructure morphology is exemplified by the various configurations of $MnCO_2O_4$, including nanorods, porous/nonporous nanospheres, nanoflakes, and nanosheets. Also, Su et al. [42] reported a hydro/solvothermal method, facilitated by urea, has been successfully employed to synthesize α-Fe_2O_3 and Fe_3O_4 with tunable particle sizes and morphologies. This versatile approach enables precise control over the structural and morphological properties of α-Fe_2O_3 and Fe_3O_4 nano- and microstructures as shown in Fig. 6.

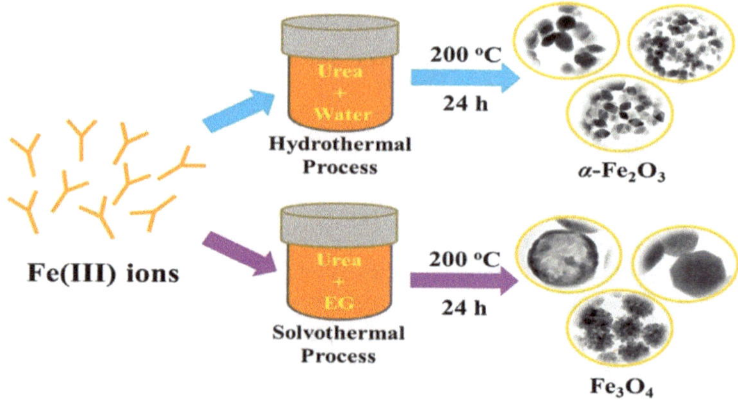

Fig. 6 α-Fe_2O_3 and Fe_3O_4 nano- and microstructures with different morphologies via hydrothermal and solvothermal synthesis [42]

3.2.3 Microwave Method

Microwave processing has emerged as a groundbreaking technology in materials processing, offering numerous benefits over traditional methods. Its widespread adoption in industry is evident, given its advantages. A key benefit of microwave processing is its ability to directly transfer energy to materials, outperforming conventional techniques. Furthermore, microwave processing surpasses traditional techniques, and its scalability far exceeds that of solid-state and wet chemical methods [43]. For instance, First-generation Li-ion batteries utilize lithium metal oxide (cathode), graphite/hard carbon (anode), and Li salt in nonaqueous solvents (electrolyte). The primary factor contributing to the battery's poor performance is the instability of its electrode materials. The instability of the electrode material could be due to given factors:

- Inhomogeneous compound formation [44, 45].
- The dissolution of the cathode material into the electrolyte [46].
- The volume changes during charging-discharging cycle [46].
- Jahn–Teller distortion effects [47, 48].

To optimize battery performance, the development of homogeneous and phase-pure materials is crucial. Selecting the optimal synthetic method plays an important role in the synthesis of uniform, phase-pure, and homogeneous materials, as it directly impacts the quality of the final product. For instance, $LiCoO_2$ is the most common cathode material used for commercial Li-ion batteries due to its stability, high working voltage and good cycle life [49]. Yan et al. [49] pioneered the rapid microwave synthesis of $LiCoO_2$, offering an alternative to traditional soft-chemistry methods. This approach yields uniform, fine, and dense microstructures. Notably, the synthesis temperature is reached within 3–5 min, resulting in grain sizes of 0.5–1 μm. This significant reduction in grain size, compared to conventional $LiCoO_2$, effectively prevents particle coarsening by minimizing grain growth time [50]. Also, in recent times Gouraha et al. [51] studied the effect of microwave treatment on Scheelite $CaMoO_4$ nanoparticles preparation, obtaining higher crystallinity for microwave treated sample. In this method, microwaves directly interact with the reaction mixture's molecules, causing a swift temperature increase. Unhindered by the vessel's thermal conductivity, this process generates instantaneous, localized superheating in substances that respond to microwave energy through dipole rotation or ionic conduction, the two primary mechanisms of microwave energy transfer [52]. The difference between the conventional and microwave heating is displayed in Fig. 7.

Clearly, microwave synthesis opens up new avenues for chemical reactions by enabling direct energy transfer to reactive species. This capability facilitates unprecedented transformations that are beyond the reach of conventional heating methods, thereby expanding the frontiers of synthetic organic chemistry [53]

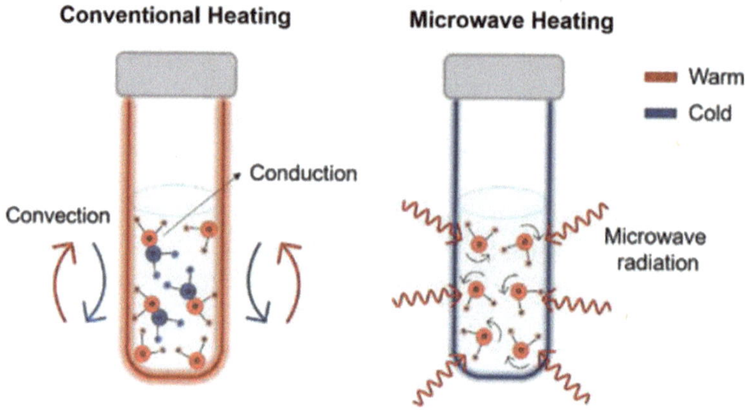

Fig. 7 Schematic difference between conventional and microwave heating [53]

3.2.4 Green Synthesis Methods

Green synthesis methods are revolutionizing battery technology and energy storage by offering a sustainable and environmentally conscious approach to manufacturing, enabling the creation of more eco-friendly products [54]. Although traditional methods have been employed for decades, research has demonstrated that green methods offer a more effective approach to generating nanoparticles (NPs) [55]. The advantages of green methods include reduced risk of failure, lower costs, and simplified characterization processes. Green synthesis methods offer an eco-friendly approach to producing metal oxide composites. The primary objective is to develop sustainable production processes that [56]:

- Eliminate or minimize the use of toxic chemicals.
- Reduce energy consumption.
- Leverage renewable resources.

To achieve this, scientists utilize plant extracts and microbial synthesis methods. These biological agents serve as:

- Reducing agents: facilitating the conversion of metal ions into metal oxide nanoparticles.
- Stabilizing agents: preventing the aggregation of metal oxide nanoparticles, ensuring uniform size and shape.

This innovative approach enables the production of metal oxide composites while promoting environmental sustainability. The integration of sustainable biomaterials, including cellulose, chitosan, and other naturally occurring polymers, as templates or supports enables the eco-friendly synthesis of metal oxide composites. Some notable systems utilized for the green synthesis of NPs are [56]:

- Green synthesis via enzymes [57, 58].

Fig. 8 Outline of preparation of silver NPs via green synthesis [65]

- Green synthesis via vitamins [59].
- Green synthesis by using actinomycetes and bacteria [60, 61].
- Green synthesis from fungi, algae, and yeasts [62–64].

Hence, Green synthesized nanoparticles (NPs) have a wide range of applications across various disciplines, including [65]:

- Dentistry.
- Pharmaceuticals.
- Bio-sensing.
- Energy storage.
- Several other fields (Fig. 8).

3.3 Case Studies: Al-Ion Batteries, Na-Ion SSBs

(a) **Recent Developments in Nanofiber Cathodes for Next-Generation Aluminum-Ion Batteries**

Rechargeable aluminum-ion batteries (AIBs) offer a promising, cost-effective alternative to lithium-ion batteries (LIBs), boasting higher theoretical volumetric

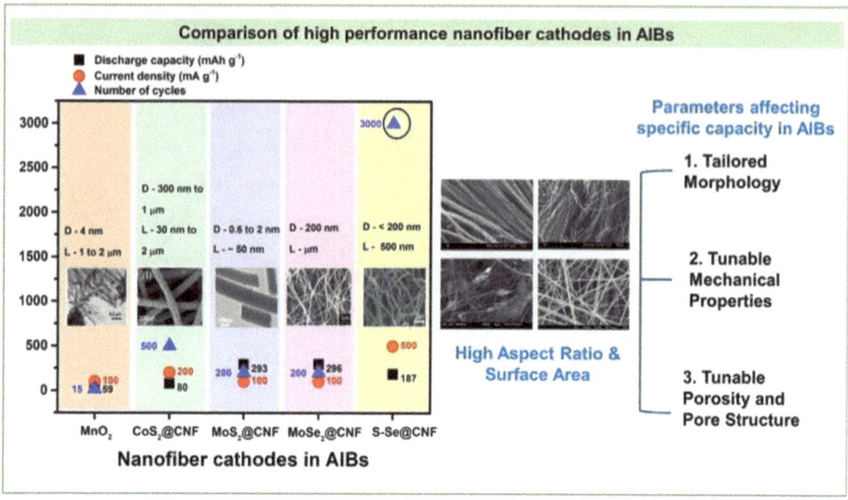

Fig. 9 Factors influencing performance enhancements in aluminum-ion batteries (AIBs) are shown for various nanofiber cathodes [68]

capacity. Figure 9 compares various AIB NF cathodes and their influential properties. However, their performance falls short of commercial expectations due to key challenges, including [66–69]:

- Volume changes caused by interfacial issues.
- Electrolyte-side reactions with electrodes.
- Limited cyclic stability.

These issues stem from the inadequacy of existing cathode materials, hindering optimal AIB performance. One-dimensional (1D) nanostructures, particularly nanofiber (NF) cathodes, present a viable solution to overcome these challenges. Their unique characteristics [67], including:

- Higher aspect ratios.
- Increased specific surface area.
- Enhanced flexibility.
- Quantum scale effects.

make NF cathodes an attractive option for improving performance. Ramasubramanian et al. [68] discussed in his review about aluminum-ion batteries (AIBs) that it is operated by utilizing ion complexes as charge carriers, with their efficiency and performance heavily influenced by the electrolyte type, categorizing AIBs into aqueous and non-aqueous systems. Recent research has focused on integrating cathode materials, including metal chalcogenides (e.g., VaS_2, MoS_2) [70, 71] and transition metal oxides (e.g., MnO_2, V_2O_5), [71] as well as carbon-based materials, to enhance rechargeable AIBs. Other critical components, such as separators, current collectors, and cell packaging, are also being redesigned to optimize overall battery performance. The primary objective is to develop high-capacity cathodes using metal chalcogenides and oxides. However, widespread commercialization of AIBs remains hindered by challenges in achieving desired electrochemical performance (100–400 mAh g^{-1}) and consistency. Nano fibers (NFs) and 1-dimensional (1D) nanomaterials offer advantages, including:

- Higher capacity.
- Expedited ion transport due to shorter diffusion pathways.
- Enhanced mechanical integrity.
- Prevention of electrode degradation.
- Reduced internal resistance.
- Improved cyclic stability through better electrolyte infiltration [72, 73].

(b) **Sodium-Ion SSBs**

The unique combination of high energy density and long lifespan makes all-solid-state sodium-ion batteries (ASIBs) an attractive solution for powering renewable energy systems and electric vehicles. While significant research in Sodium-Ion Batteries (NIBs) has focused on enhancing electrode materials, the success of this technology hinges on optimizing all components [73]. Notably, electrolytes play a crucial role, requiring exceptional stability within the broad electrochemical window defined by the anode's low reduction potential and the cathode material's high potential. The development of suitable anodes for Sodium-Ion Batteries (NIBs) has proven to be a significant hurdle. Traditional anode materials, such as graphite and silicon, demonstrate limited sodium storage capacity, rendering them ineffective. Furthermore, metallic sodium is not a viable alternative due to its propensity to form dendrites and its low melting point, which raises substantial safety concerns.

In response, researchers have investigated a diverse array of materials as potential NIB anodes, encompassing:

- Organic Compounds: Carbonyls, Schiff bases, and Quinone derivatives [74].
- Inorganic Oxides: Co_3O_4, CuO, Fe_2O_3, TiO_2, and $Na_2Ti_3O_7$.
- Alloying Elements: Tin, Phosphorus, Antimony, Bismuth, and Germanium.
- Disordered carbons.

These materials facilitate sodium storage through insertion, conversion, and alloying mechanisms, offering promising solutions for NIB anodes [75]. Upon examining, it becomes apparent that oxide-based compounds, such as $Na(Fe,Ti)O_4$ and $Na_2Ti_3O_7$ [76], exhibit limited specific capacity due to their high molar mass. However, their inorganic framework provides exceptional stability, resulting in impressive cycling performance. Titanium dioxide, in its various forms, including anatase [77], rutile [78] and β-TiO_2 [79] has also been investigated as a potential anode material. Each polymorph displays a unique degree of reversible sodium insertion, undergoing distinct structural transformations during electrochemical reactions [80–82]. Also the metal oxide conversion reactions have shown promise for Sodium-Ion Battery (SIB) anodes. The concept was first demonstrated using the spinel $NiCo_2O_4$, which achieved a reversible capacity of approximately 200 mAh/g after an initial discharge of 600 mAh/g [83]. Further analysis revealed that the conversion reaction to form sodium oxide (Na_2O) was the primary reaction mechanism. Notably, when $NiCo_2O_4$ was paired with excess $Na_{0.7}CoO_2$ in a full cell, the anode material's reversible capacity improved significantly. This highlights the importance of considering alternative counter electrodes, as sodium metal performs poorly in this role. A study by Sun et al. [84] explored the conversion reaction of Sb_2O_4 thin films, which involves the reduction of the metal oxide followed by alloying to form Na_3Sb. This process achieved an impressive reversible capacity of 800 mAh/g over 20 cycles, with an average potential of 0.5 V versus Na/Na^+ (Fig. 10).

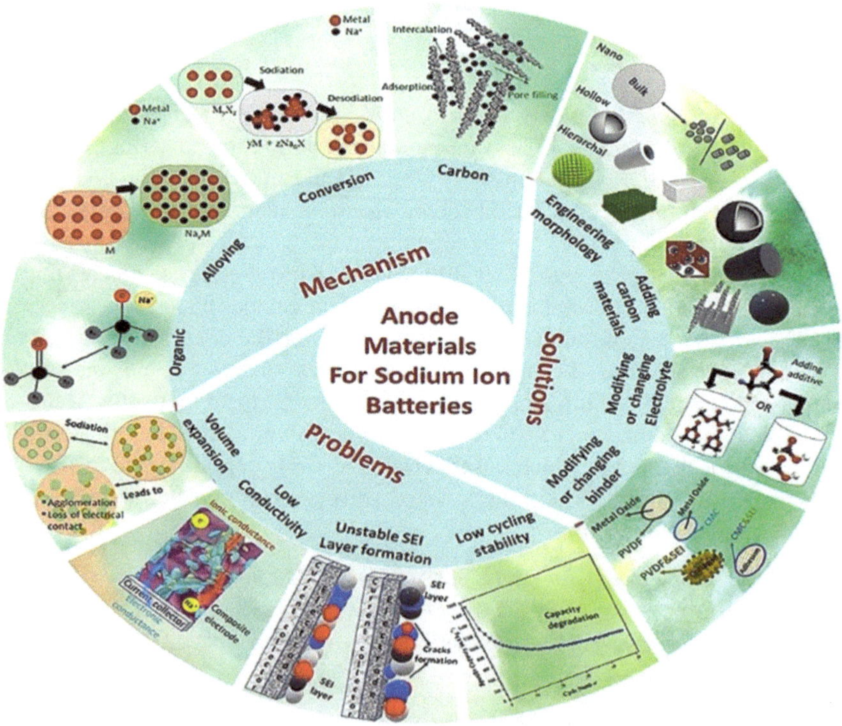

Fig. 10 Emerging opportunities in anode materials for sodium-ion batteries [85]

4 Enhancing Metal Oxide Performance by Surface Modification

Surface modification is a crucial strategy in materials science for improving the efficiency and functionality of metal oxide-based materials. This technique alters the surface properties of a material without changing its bulk properties, enhancing its performance in various applications such as:

- Catalysis.
- Energy storage.
- Sensors.
- Environmental remediation.

Metal oxides are attractive anode materials due to their exceptional theoretical capacities, stability, and natural abundance. Nevertheless, their performance is hindered by several key limitations:

- Intrinsic electrical conductivity issues, leading to poor charge transfer.
- Restricted ion diffusion, resulting in slow sodium-ion insertion and extraction.
- Insufficient surface area, limiting the reaction sites for sodium-ion intercalation.
- Incompatibility with electrolytes, causing instability and degradation.

4.1 Methods for Surface Modification

Addressing those challenges is crucial to unlocking the full potential of metal oxides as high-performance electrode materials. To overcome these limitations, researchers utilize several surface modification strategies out of which coating and functionalization techniques are explained:

(a) **Coating:** Metal oxide coatings can significantly benefit electrode materials in many ways, like electrolyte protection, stabilizing surface reactions, preventing electrolyte decomposition, and providing conductive pathway for efficient charge transfer. Research reveals that post-coating electrodes surpass loose active powders in performance. Notably, applying metal oxide coatings to pre-assembled $LiCoO_2$-based positive electrodes has yielded remarkable results and the electrode consisting of [86]:

    - $LiCoO_2$ (active material).
    - Carbon black (conductive additive).
    - N-methyl pyrrolidone (NMP) (solvent).
    - Polyvinylidene fluoride (PVDF) (binder).
    - Current collector.

exhibited significantly enhanced performance when coated with metal oxides, underscoring the potential of this method to boost electrode efficiency [86, 87]. One of the studies by Shim et al. [88] investigated the synthesis of magnesium oxide (MgO)-coated $LiCoO_2$ cathode materials at different annealing temperatures (750–810 °C). The results showed that the coated samples exhibited exceptional capacity retention of up to 85% and improved initial discharge capacity which is displayed in Table 2. Notably, the optimal MgO@$LiCoO_2$ material, annealed at 810 °C, demonstrated: 85% capacity retention after 60 cycles and a significant improvement over uncoated $LiCoO_2$, which retained only 72.8% capacity after 60 cycles.

(b) **Functionalization**: Functionalization of metal oxides helps in enhancing the interactions and stability of the material. Attaching functional groups (-OH, -COOH, or amines) to metal oxide surfaces can:

- Enhance molecular interactions.
- Improve stability.
- Boost catalytic performance.
- Enhance sensor sensitivity.
- Facilitate uniform dispersion in solvents or polymers.

By functionalizing metal oxides, researchers can tailor their surface properties to improve compatibility with electrolytes, stability, and overall performance.

Table 2 Impact of metal oxide coatings on the electrochemical behavior of electrode materials [87]

Coating	Thickness of coating	Technology used	Initial discharge capacity (mAh/g)	Capacity retained (mAh/g)	Rate capability (mAh/g)	References
ZnO	17 nm	Radio frequency magnetron sputtering	193–191	140–175.53	75–106	[87, 89]
CuO	2 wt%	Wet chemical method	162–166	110–158	96.5–120	[87, 90]
MgO		Ball milling	150–153	100–132.5	–	[87, 88]
TiO2	1.5 nm	Wet chemical method	120–133	58–110	16–98	[87, 91]
ZrO2	1.5 wt%	Wet chemical method	143–135	137–140	–	[87, 92]

4.2 *Effects of Surface Modification on Metal Oxide Properties*

(a) **Wettability: A Key Property of Material Surfaces**: Wettability is a crucial property in the preparation of electrodes using melt infusion techniques. Surface composition is often considered the primary factor governing wettability, as it determines the surface free energy. Surface energy refers to the strength of cohesive forces per unit area required to bring interior particles to the surface. Metals typically exhibit high surface energies, roughly ten times that of water, due to their robust internal bonds [93, 94]. For pure liquids, surface energy is equivalent to surface tension. The high surface tension of liquid metals causes them to contract into spherical shapes, making it challenging for them to spread onto substrates [95].

- **Enhancing Surface Wettability by Controlling the Temperature**

Weakening metallic bonding is key to improving surface wettability. At high temperatures, thermal motion increases, expanding interatomic distances and weakening interior interactions. This facilitates the migration of metal atoms from the interior to the surface, effectively reducing surface tension. Elevating the temperature offers a promising strategy for decreasing surface tension and enhancing the wettability of molten metals (e.g.; Li) [96]. The relationship between surface tension and temperature is described by Eq. (1).

$$\gamma_T = \gamma_m + \frac{d\gamma}{dT}(T - T_m) \tag{1}$$

where γ_T and γ_m denotes the surface tension at temperature T and surface tension at melting point respectively. T and T_m specifies the heating temperature and melting point respectively and $\frac{d\gamma}{dT} < 0$ [94].

(b) **Interfacial Engineering in Batteries**: The mesophase formed at the electrode–electrolyte interface plays a crucial role in metals, especially Li^+ and electron transport, influencing cycle stability. The interface's chemical stability depends on the alignment of intrinsic energy levels between the electrode and electrolyte materials. A thermodynamically stable interface forms when:

i. The electrolyte material's lowest unoccupied molecular orbital (LUMO) is lower than the negative material's chemical potential.
ii. The electrolyte material's highest occupied molecular orbital (HOMO) is higher than the positive material's chemical potential.

Conversely, an interface reaction occurs, forming an intermediate layer similar to the solid electrolyte interphase film [97, 98]. This process is driven by chemical reactions and doesn't require external voltage. From an electrochemical perspective, the growth of the interface layer can be inhibited if the reaction products have suitable ionic and electronic conductivity, allowing for passivation [99, 100].

4.3 Case Studies: Surface-Modified $LiCoO_2$, Al_2O_3-Coated $LiNiO_2$

(a) **Surface-Modified $LiCoO_2$**: A study by Cho et al. [101] demonstrated that a metal oxide layer on $LiCoO_2$, formed through coating, effectively suppresses phase transition during lithium deintercalation displayed in Fig. 11. This is attributed to the high fracture oughness of metal oxide coatings, such as ZrO_2 (8–12 MPa-m1/2) and Al_2O_3 (2.7–4.2 MPa-m1/2) [102]. In contrast, $LiCoO_2$ exhibits relatively low fracture toughness, with a mean value of 1.7 ± 0.4 MPa-m1/2 and a median value of 0.9 ± 0.4 MPa-m1/2 [101]. As a result, the metal oxide coating minimizes lattice distortion, allowing the hexagonal phase to be retained. Notably, metal oxides with low fracture toughness, such as SiO_2 (0.77 MPa-m1/2), would be ineffective in suppressing phase transition.

(b) **Al_2O_3-coated $LiNiO_2$:** Lee et al. [103] reported a novel approach to enhancing the performance of high-energy–density aqueous lithium-ion batteries (ALIBs) by introducing a bifunctional Al_2O_3 artificial layer on a $LiNiO_2$ (LNO) cathode shown in Fig. 12. Theoretical calculations reveal that Al_2O_3 exhibits high adsorption energy with H_2O molecules, effectively preventing direct contact between water and the LNO surface. Experimental results demonstrate that the Al_2O_3 coating significantly improves the electrochemical properties of LNO cathodes, particularly in terms of cyclability, with a capacity retention of approximately 77.8% after 50 cycles. The Al_2O_3 layer forms a stable cathode-electrolyte interphase (CEI) and maintains the structural integrity of the LNO bulk. Notably, the Al_2O_3 layer undergoes an in situ transformation into the more stable $Al(OH)_3$ in the aqueous electrolyte, further enhancing its protective properties.

5 Enhancing Metal Oxide Performance by Composite Formation

Metal oxides are crucial components in battery technology. Their performance and safety significantly impact the overall battery efficiency. Composite formation is a promising strategy to enhance metal oxide properties for battery applications such as ionic and electronic conductivity, increases the active sites for ion insertion/extraction during cycling, structural stability, and can mitigate the risk of thermal runaway of the metal oxides.

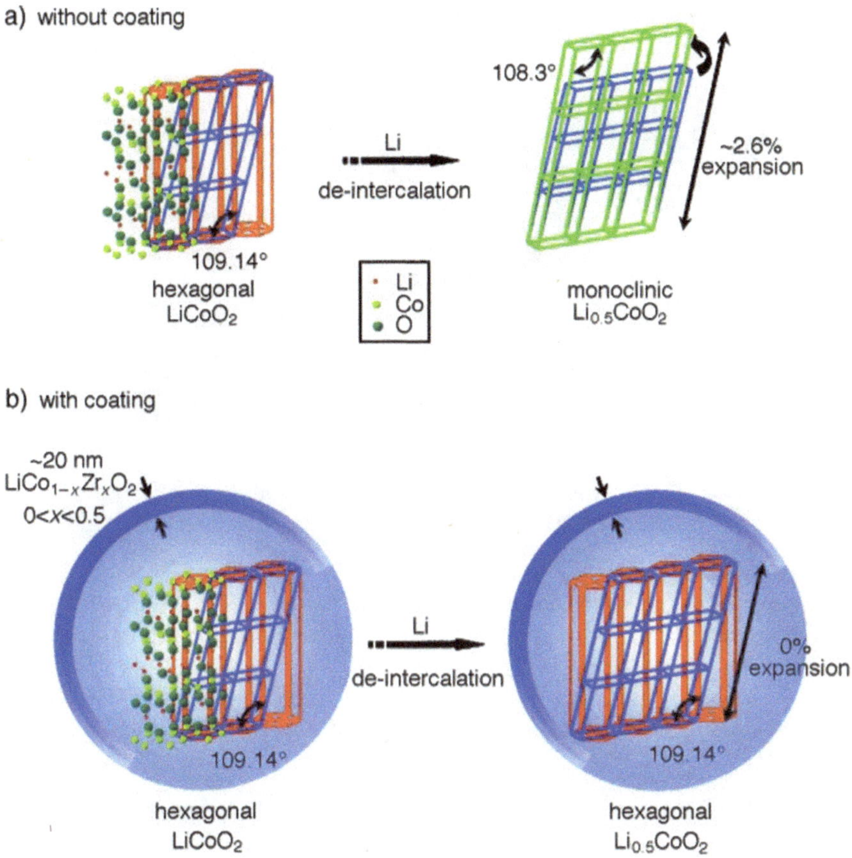

Fig. 11 **a** Bare $LiCoO_2$ undergoes a monoclinic phase transformation during charging (Li deintercalation), characterized by a nonuniform lattice expansion of 2.6%. and **b** A thin-film metal-oxide coating with enhanced fracture toughness effectively suppresses the undesirable phase transition from hexagonal to monoclinic in $LiCoO_2$ [102]

5.1 Types of Composites (e.g., Metal Oxide-Polymer, Metal Oxide–Carbon)

Here are various composite materials that can boost metal oxide performance in batteries:

- **Metal Oxide-Polymer Composite**: Conductive polymers, such as polypyrrole (PPy) and polyaniline (PANI), can be incorporated into metal oxides to enhance their electronic conductivity and provide additional stability during charge–discharge cycles.
- **Metal Oxide–Carbon Composite**: The incorporation of carbon materials into metal oxides is a common strategy to enhance their electrical conductivity and

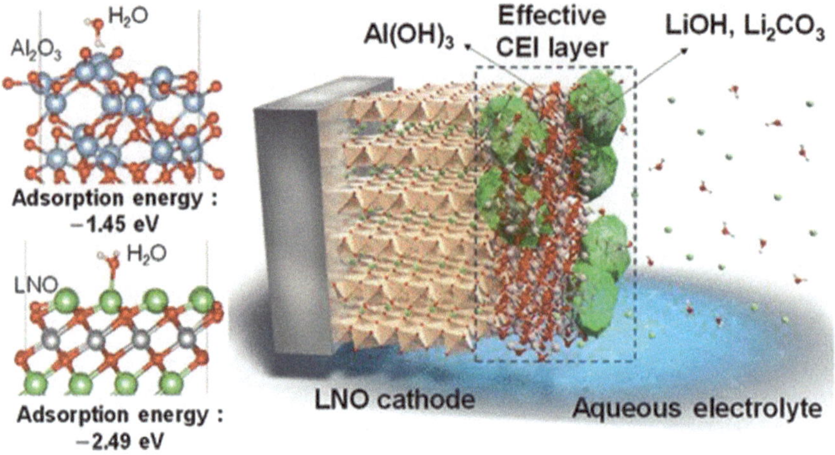

Fig. 12 Bifunctional Al_2O_3 artificial layer on a $LiNiO_2$ (LNO) cathode for high-energy–density aqueous Li-ion batteries (ALIBs) [103]

mechanical stability, as evident in lithium-ion battery anodes, such as NiO/N–C composite materials, where carbon additives improve overall performance.

5.2 *Case Study: TiO_2-Carbon Nanotube Composites*

TiO_2-Carbon Nanotube Composites: Titanium dioxide (TiO_2) is a promising anode material for next-generation lithium-ion batteries (LIBs) due to its safe operation and stable performance. However, its limited capacity hinders its practical application. To address this, researchers have developed TiO_2-based composites with other metal oxides (MOs) to enhance anode performance. A notable example is the TiO_2@Fe_2O_3 core–shell nanotube array synthesized by Zhong et al. [104, 105] This innovative structure leverages the high specific capacity of Fe_2O_3 and the stable backbone structure of TiO_2. Figure 13 illustrates the fabrication process of this core–shell nanotube array.

As a result, the hybrid electrode exhibits exceptional electrochemical performance, featuring:

- High Reversible Capacity: 520 mAh/g at 100 mA/g, surpassing bare TiO_2 electrodes by three times.
- Excellent Long-Term Stability: 87% capacity retention after 600 cycles.
- Outstanding Rate Capability: reversible operation even at high current densities of 10 A/g [104].

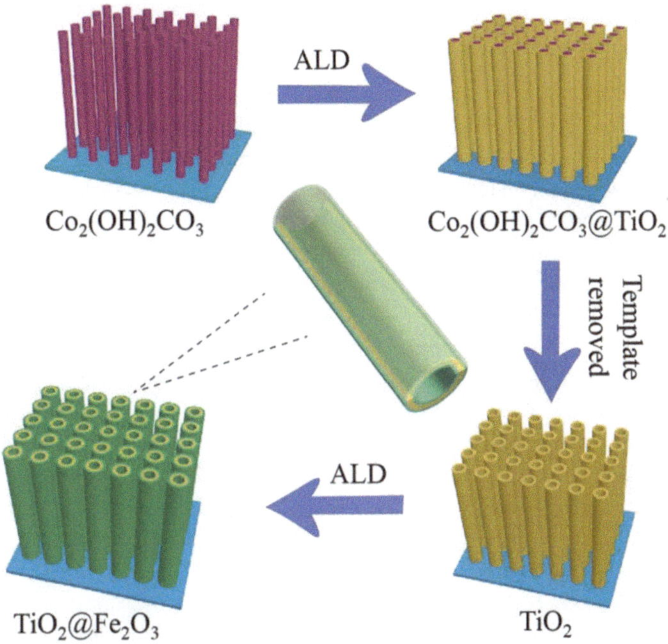

Fig. 13 Schematic representation of the fabrication process for double-sided, core–shell $TiO_2@Fe_2O_3$ nanotube arrays [104]

6 Conclusion

6.1 Summary of Recent Approaches to Enhance Metal Oxide Performance

Recent approaches to enhance the performance of metal oxides, especially for applications in catalysis, energy storage, and sensors, have focused on several key strategies which is explained above, including:

- Surface Modification
- Electrochemical Activation
- Synthesizing Novel Oxide Materials with Tailored Properties
- Hybridization with Conductive Materials
- Doping and Composite formation
- Coating and Interface Engineering.

6.2 Future Directions and Challenges

The growing demand for battery research has brought together a vast and diverse community of experts, fostering a multifaceted approach to innovation. Solid-state chemistry plays a crucial role in integrating experimental and computational methods, advanced characterization techniques, and operando analysis to tackle complex materials discovery and reactivity challenges. By converging these disciplines, we can accelerate the development of superior materials for rechargeable batteries, driving progress toward a more sustainable energy future [106]. Modern solid-state chemistry, empowered by advanced characterization techniques and computational tools, can systematically link composition, structure, and material properties. Computational methods, such as Density Functional Theory (DFT) and Molecular Dynamics (MD), provide valuable insights into the electrochemical behavior of battery materials. Although these methods require significant computational power and oversimplify complex chemical systems, they have successfully predicted several experimentally confirmed polyanion electrode materials. Examples include carbonophosphates ($A_3MPO_4CO_3$) (A = Li, Na, M = Fe, Mn) [107] pyrophosphates ($LiMoP_2O_7$) [108, 109] phosphate pyrophosphates ($Li_9V_3(P_2O_7)_3(PO_4)_2$) [110] and phosphates ($Li_3Cr_2(PO_4)_3$) [111, 112]. Despite these advances, combining solid-state chemistry approaches with computational tools remains essential for understanding complex battery materials. Recent discoveries of metal-ion battery cathodes, inspired by the concept of anionic redox and its crystal chemistry expansions, illustrate the potential of this integrated approach.

Beyond advancing materials research, it's crucial to investigate fundamental scientific issues. The primary challenge that warrants focus is improving the electrical and mechanical properties of cathode materials. This includes understanding the electrochemical mechanisms of emerging metal oxide-based hybrid films/arrays and exploring the physical/chemical properties of interfaces within these structures. Additionally, the impact of these properties on electrochemical performance must be examined [113].

Furthermore, engineering aspects, such as mechanical strain tolerance in electrodes/devices, require systematic research. Ensuring mechanical integrity and optimizing electrochemical kinetic properties are essential for developing high-performance energy storage devices [114].

Achieving breakthroughs in advanced electrode materials, electrode structure design, and device fabrication will be critical to overcoming the challenges in electrochemical energy storage. This presents a rich opportunity for materials scientists, who can make meaningful contributions to the field and potentially achieve significant breakthroughs in the near future [114, 115].

6.3 Potential Impact on Solid-State Battery Development

Solid-state batteries (SSBs) are poised to transform the energy storage landscape, offering enhanced safety, increased energy density, and faster charging speeds compared to conventional lithium-ion batteries. Breakthroughs in solid-state chemistry and materials science are expected to have a profound impact on SSB development, driving innovation and improvement in this promising technology. Lithium-ion batteries have become the cornerstone of electric vehicle technology, offering high voltage, energy density, and lifespan [116, 117]. However, conventional liquid lithium-ion batteries face significant challenges, including limited energy density and safety concerns. Recent incidents of spontaneous combustion in electric vehicles have underscored the need for safer, higher-performance batteries [118, 119]. All-solid-state batteries (ASSBs) have emerged as a promising alternative, offering enhanced safety and energy density. By replacing the liquid electrolyte and separator with a solid-state electrolyte, ASSBs minimize the risk of thermal runaway and enable the use of lithium metal anodes, significantly boosting energy density [120, 121]. While preliminary research has yielded promising results, further advancements are needed to optimize ASSB design and performance. Modeling studies can play a crucial role in this effort, providing powerful simulation tools to inform design decisions and accelerate development [122–124]. Investigating ASSB aging mechanisms and developing corresponding models can also provide valuable insights, guiding research toward the creation of longer-lasting ASSBs.

Current modeling studies on all-solid-state batteries (ASSBs) primarily focus on simulating their kinetic and mass transport properties. Commonly used battery models include black-box models, equivalent circuit models, and electrochemical models. Among these, electrochemical models are preferred due to their ability to capture the internal reaction mechanisms of batteries and facilitate the analysis of parameter changes on battery performance [125]. Notably, Danilov et al. [126] developed an electrochemical model for thin-film ASSBs, specifically $Li/Li_3PO_4/LiCoO_2$ batteries. This model accounts for charge transfer kinetics at the electrode–electrolyte interface, lithium diffusion in electrodes, and ion diffusion and migration in the electrolyte. The results demonstrated excellent agreement between model predictions and experimentally measured voltage profiles under constant current conditions. As a result, significant research efforts have focused on developing next-generation lithium-ion batteries (LIBs), Sodium-ion batteries by combining high-capacity, low-potential silicon (Si) and/or silicon-based/derivative (Si-B/Si-D) anodes with cathode materials [127–129]. For instance, the lithium-ion battery (LIB) has undergone significant evolution, transforming from the original "rocking-chair" battery concept to the advanced LIBs of today, and now paving the way for next-generation batteries featuring silicon (Si) and silicon-based/derivative (Si-B/Si-D) anodes paired with the various cathode materials as shown in Fig. 14 [130, 131].

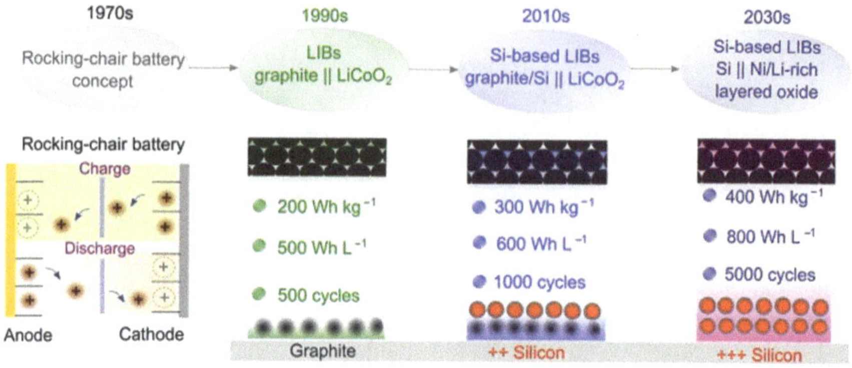

Fig. 14 Evolution of LIBs [131]

Acknowledgement The authors gratefully acknowledge the support provided by ANRF under the project grant no. ANRF/PAIR/2025/000003/PAIR.

References

1. Pugalendhi K, Natesan B (2025) Recent progress in tailoring transition metal cathodes for high-performance room temperature sodium–sulfur batteries. Tungsten 32
2. Skeete JP, Wells P, Dong X, Heidrich O, Harper G (2020) Beyond the event horizon: battery waste, recycling, and sustainability in the United Kingdom electric vehicle transition. Energy Res Soc Sci 69:101581
3. Wu LC, Fu HW, Lyu W, Cha LM, Rao AM, Guo KK, Zhou J, Wen SC, Lu BG (2024) Rational regulation of high-voltage stability in potassium layered oxide cathodes. ACS Nano 18:13415
4. Manthiram A, Fu YZ, Chung SH, Zu CX, Su YS (2014) Rechargeable lithium-sulfur batteries. Chem Rev 114:11751
5. Peng LL, Wei ZY, Wan CZ, Li J, Chen Z, Zhu D, Baumann D, Liu H, Allen CS, Xu X, Kirkland AI, Shakir I, Almutairi Z, Tolbert S, Dunn B, Huang Y, Sautet P, Duan XF (2020) A fundamental look at electrocatalytic sulfur reduction reaction. Nat Catal 3:762
6. Chang R, Minck R (1990) Sodium–sulfur battery flight experiment definition study. J Power Sources 29:555
7. Degruson JA (1992) Sodium sulfur batteries for space applications. In: NASA. Marshall Space Flight Center, the 1991 NASA Aerospace Battery Workshop
8. Huang Q, Wang X, Jin D, Yao Z, Hao Y, Liu T, Ji J, Tian Q, Zhang D (2025) Single crystal layered transition metal oxide cathode materials for sodium-ion batteries: potential and progress. MetalMat e70005
9. Shahzadi K, Zhao X, Liu Q, He W, Mu D, Li Y, Li L, Chen R, Wu F (2025) Advancing sodium-ion battery performance: innovative doping and coating strategies for layered oxide cathode materials. Adv Sustain Syst 9:2401045
10. Li N, Patrissi CJ, Che G, Martin CR (2000) Rate capabilities of nanostructured $LiMn_2O_4$ electrodes in aqueous electrolyte. J Electrochem Soc 47:2044
11. Bruce PG (2005) Energy materials. Solid State Sci 7:1463
12. Liu H, Su DW, Zhou RF, Sun B, Wang GX, Qiao SZ (2012) Highly ordered mesoporous MoS2 with expanded spacing of the (002) crystal plane for ultrafast lithium ion storage. Adv Energy Mater 2:970

13. Hwang H, Kim H, Cho J (2011) MoS_2 nanoplates consisting of disordered graphene-like layers for high rate lithium battery anode materials. Nano Lett 11:4830
14. Liu H, Du X, Xing X, Wang G, Qiao SZ (2012) Highly ordered mesoporous Cr_2O_3 materials with enhanced performance for gas sensors and lithium ion batteries. Chem Commun 48:867
15. Liu H, Wang G, Liu J, Qiao S, Ahn H (2011) Highly ordered mesoporous NiO anode material for lithium ion batteries with an excellent electrochemical performance. J Mater Chem 21:3052
16. Ji H, Mei Y, Schmidt OG (2010) Swiss roll nanomembranes with controlled proton diffusion as redox micro-supercapacitors. Chem Commun 46:3883
17. Arico AS, Bruce P, Scrosati B, Tarascon JM, Van Schalkwijk W (2005) Nanostructured materials for advanced energy conversion and storage devices. Nat Mater 4:377
18. Li Y, Tan B, Wu Y (2008) Mesoporous Co_3O_4 nanowire arrays for lithium ion batteries with high capacity and rate capability. Nano Lett 8:270
19. Wu ZS, Ren W, Wang DW, Li F, Liu B, Cheng HM (2010) High-energy MnO_2 nanowire/graphene and graphene asymmetric electrochemical capacitors. ACS Nano 4:5842
20. Wang Z, Su F, Madhavi S, Lou XW (2011) CuO nanostructures supported on Cu substrate as integrated electrodes for highly reversible lithium storage. Nanoscale 3:1623
21. Xiao W, Chen JS, Li CM, Xu R, Lou XW (2010) Synthesis, characterization, and lithium storage capability of $AMoO_4$ (A= Ni, Co) nanorods. Chem Mater 22:754
22. Ji HX, Wu XL, Fan LZ, Krien C, Fiering I, Guo YG, Mei Y, Schmidt OG (2010) Self-wound composite nanomembranes as electrode materials for lithium ion batteries. Adv Mater 22:4595
23. Mangaiyarkkarasi J (2025) Nanostructural innovations: powering the future of energy storage. In: Design, fabrication, and significance of advanced nanostructured materials, p 74
24. Chen Y, Chen X, Li J, Chen H, Su Y, Chen W, Yu C, Ling FC, Ru Q (2025) Cation W-doping motivated novel layered $K_{0.45}Rb_{0.05}Mn_{0.85-x}W_xMg_{0.15}O_2$ cathode for boosting potassium/sodium-ion storage. Chem Eng J 503:158632
25. Yin J, Huang S (2025) Progress of anion redox in Na-rich layered transition metal oxides (Na_2MO_3) as cathode materials for sodium-ion batteries. J Phys: Condens Matter 37:113002
26. Zhang H, Dong J, Chen G, Che H, Yan K, Wang X, Liu J, Liu D, Lu Y, Li N, Su Y (2025) Trifunctional surface engineering of Lithium-rich manganese-based oxides via Al^{3+}/PO_4 co-doping and oxygen vacancy regulation for High-performance lithium-ion batteries. Chem Eng J 506:159902
27. Jia Q, Yao Z, Xiang J, Shi J, Zhou Y, Huang J, Zhang H, Zhang X, Yang Y, Tu J (2025) Rare earth metal ion-doped halide solid electrolytes plus Ta5+ substitution for long cycling all-solid-state batteries. Adv Funct Mater 35:2426053
28. Yang P, Hu X, Lin W, Wu Y, Wang L, Shen C (2025) Revealing the lithium ion diffusion kinetics and cycling stability of rare earth doped $LiFePO_4$/C cathodes. J Energy Storage 109:115161
29. Cho E, Jeong S, Kim D, Choi W (2025) Sodium-doped $LiFe_{0.5}Mn_{0.5}PO_4$ using sodium gluconate as both reducing agent and a doping source in lithium-ion batteries. J Ind Eng Chem 143:729–39
30. Yang Z, Li Z, Li P, Gao C, Zhang H (2020) NiO/Ni nanocomposites embedded in 3D porous carbon with high performance for lithium-ion storage. J Mater Sci 55:1672
31. Kong Y, Luo J, Jin C, Yuan H, Sheng O, Zhang L, Fang C, Zhang W, Huang H, Xia Y, Liang C (2018) Enhanced sulfide chemisorption by conductive Al-doped ZnO decorated carbon nanoflakes for advanced Li–S batteries. Nano Res 11:489
32. Niederberger M (2007) Nonaqueous sol–gel routes to metal oxide nanoparticles. Acc Chem Res 40:800
33. Abid N, Khan AM, Shujait S, Chaudhary K, Ikram M, Imran M, Haider J, Khan M, Khan Q, Maqbool M (2022) Synthesis of nanomaterials using various top-down and bottom-up approaches, influencing factors, advantages, and disadvantages: a review. Adv Colloid Interface Sci 300:102597
34. Pazhamalai P, Krishnan V, Mohamed Saleem MS, Kim SJ, Seo HW (2024) Investigating composite electrode materials of metal oxides for advanced energy storage applications. Nano Converg 11:30

35. Heli H, Yadegari H, Jabbari A (2011) Investigation of the lithium intercalation behavior of nanosheets of LiV_3O_8 in an aqueous solution. J Phys Chem C 115:10897
36. Wu ZS, Wang DW, Ren W, Zhao J, Zhou G, Li F, Cheng HM (2010) Anchoring hydrous RuO_2 on graphene sheets for high-performance electrochemical capacitors. Adv Funct Mater 20:3602
37. Boddu VR, Palanisamy M, Sinha L, Yadav SC, Pol VG, Shirage PM (2021) Hysteresis abated P2-type $NaCoO_2$ cathode reveals highly reversible multiple phase transitions for high-rate sodium-ion batteries. Sustain Energy Fuels 5:3228
38. Nandagudi A, Nagarajarao SH, Santosh MS, Basavaraja BM, Malode SJ, Mascarenhas RJ, Shetti NP (2022) Hydrothermal synthesis of transition metal oxides, transition metal oxide/carbonaceous material nanocomposites for supercapacitor applications. Mater Today Sustain 19:100214
39. Gupta SP, Patil VB, Tarwal NL, Bhame SD, Gosavi SW, Mulla IS, Late DJ, Suryavanshi SS, Walke PS (2019) Enhanced energy density and stability of self-assembled cauliflower of Pd doped monoclinic WO_3 nanostructure supercapacitor. Mater Chem Phys 225:199
40. Desai RS, Jadhav VS, Patil PS, Dalavi DS (2024) Recent advances in hydrothermally and solvothermally grown Co_3O_4 nanostructures for electrochemical energy storage (EES) applications: a brief review. Mater Adv 5:960
41. Byrappa K, Adschiri T (2007) Hydrothermal technology for nanotechnology. Prog Cryst Growth Charact Mater 53:166
42. Su M, He C, Shih K (2016) Facile synthesis of morphology and size-controlled α-Fe_2O_3 and Fe_3O_4 nano-and microstructures by hydrothermal/solvothermal process: the roles of reaction medium and urea dose. Ceram Int 42:14804
43. Ishizawa N, du Boulay D, Hayatsu M, Kuze S, Matsushima Y, Ikuta H, Wakihara M, Tabira Y, Hesterd JR (2003) Structural disorder along the lithium diffusion pathway in cubically stabilized lithium manganese spinel I. Synchrotron X-ray studies. J Solid State Chem 174:174
44. Yoon WS, Iannopollo S, Grey CP, Carlier D, Gorman J, Reed J, Ceder G (2004) Local structure and cation ordering in O_3 lithium nickel manganese oxides with stoichiometry $Li[Ni_x Mn_{(2-x)/3}Li_{(1-2x)/3}]O_2$ NMR studies and first principles calculations. Electrochem Solid State Lett 7:A171
45. Gadjov H, Gorova M, Kotzeva V, Avdeev G, Uzunova S, Kovacheva D (2004) $LiMn_2O_4$ prepared by different methods at identical thermal treatment conditions: structural, morphological and electrochemical characteristics. J Power Sources 134:117
46. Carvajal JR, Rousse G, Masquelier C, Hervieu M (1998) Electronic crystallization in a lithium battery material: columnar ordering of electrons and holes in the spinel $LiMn_2O_4$. Phys Rev Lett 81:4660–4663
47. Gummow RJ, de Kock A, Thackeray MM (1994) Improved capacity retention in rechargeable 4 V lithium/lithium–manganese oxide (spinel) cells. Solid 69:67
48. Sony Corporation Battery Group and Sony Energytec, Inc. (1993) Development and commercialization of lithium ion secondary batteries. JEC Batt Newsletter 2:23
49. Yan H, Huang X, Lu Z, Huang H, Xue R, Chen L (1997) Microwave synthesis of $LiCoO_2$ cathode materials. J Power Sources 68:532
50. Gouraha S, Jain N, Anchal K, Patel G, Banerjee S, Singh J (2024) Microwave assisted synthesis of $Er_xYb_yCa_{1-xy}MoO_4$ nano-phosphor for efficient temperature sensing and catalytic applications. Sci Rep 14:27385
51. Tompsett GA, Conner WC, Yngvesson KS (2006) Microwave synthesis of nanoporous materials. ChemPhysChem: Eur J Chem Phys Phys Chem 7:319
52. Hayes BL (2006) Recent advances in microwave-assisted synthesis. Aldrichimica Acta 37:77
53. Ferreira SH, Rovisco A, dos Santos A, Águas H, Igreja R, Barquinha P, Fortunato E, Martins R (2021) Porous ZnO nanostructures synthesized by microwave hydrothermal method for energy harvesting applications. Nanopores 172
54. Nwachukwu AC, Nwanya A, Alshoaibi C, Awada AB, Ekwealor FI (2023) Ezema recent progress in green synthesized transition metal-based oxides in lithium-ion batteries as energy storage devices. Curr Opin Electrochem 39:101250

55. Abdelghany TM, Al-Rajhi AMH, Al Abboud MA (2018) Recent advances in green synthesis of silver nanoparticles and their applications: about future directions. A review. Bionanoscience 8:16
56. Gour A, Jain NK (2019) Advances in green synthesis of nanoparticles. Artif Cells, Nanomedicine, Biotechnol 47:851
57. Smuleac V, Varma R, Baruwati B (2011) Nanostructured mem branes for enzyme catalysis and green synthesis of nanoparticles. ChemSusChem 4:1777
58. Manivasagan P, Venkatesan J, Kang K-H (2015) Production of a-amylase for the biosynthesis of gold nanoparticles using Streptomyces sp. MBRC-82. Int J Biol Macromol 72:78
59. Nadagouda MN, Varma RS (2006) Green and controlled synthesis of gold and platinum nanomaterials using vitamin B2: density assisted self-assembly of nanospheres, wires and rods. Green Chem 8:518
60. Shivaji S, Madhu S, Singh S (2011) Extracellular synthesis of antibacterial silver nanoparticles using psychrophilic bacteria. Process Biochem 46:1807
61. Silver S (2003) Bacterial silver resistance: molecular biology and uses and misuses of silver compounds. FEMS Microbiol Rev 27:353
62. Ahmed A-A, Hamzah H, Maaroof M (2018) Analyzing formation of silver nanoparticles from the filamentous fungus Fusarium oxysporum and their antimicrobial activity. Turk J Biol 42:62
63. Lara HH, Guisbiers G, Mendoza J (2018) Synergistic antifungal effect of chitosan-stabilized selenium nanoparticles synthesized by pulsed laser ablation in liquids against Candida albicans biofilms. IJN 13:2697
64. Bakir EM, Younis NS, Mohamed ME (2018) Cyanobacteria as nano gold factories: chemical and anti-myocardial infarction properties of gold nanoparticles synthesized by Lyngbya majuscula. Mar Drugs 16:217
65. Carmona ER, Benito N, Plaza T, Recio-Sánchez G (2017) Green synthesis of silver nanoparticles by using leaf extracts from the endemic Buddleja globosa hope. Green Chem Lett Rev 10:256
66. Gopakumar G, Sujith KV, Jayadevan S, Anas S (2023) Portable electronics and micro supercapacitors. In: Supercapacitors and their applications. CRC Press, Boca Raton, p 146
67. Fan X, Liu B, Liu J, Ding J, Han X, Deng Y, Lv X, Xie Y, Chen B, Hu W, Zhong C (2020) Battery technologies for grid-level large-scale electrical energy storage. Trans Tianjin Univ 26:103
68. Ramasubramanian B, Tipparaju SK, Vincent S, Koperski M, Chellappan V, Ramakrishna S (2025) Advances in nanofiber cathodes for aluminum-ion batteries. Adv Fiber Mater 29
69. Ramasubramanian B, Sundarrajan Subramanian PR, Prasada Rayavarapu MV, Reddy VC, Ramakrishna S (2022) Novel low-carbon energy solutions for powering emerging wearables, smart textiles, and medical devices. Energy Environ Sci
70. Zhang Z, Chen Y, Guo J (2019) ZnO nanorods patterned-textile using a novel hydrothermal method for sandwich structured-piezoelectric nanogenerator for human energy harvesting. Phys E Low Dimens Syst Nanostruct 105:218
71. Almodóvar P, Giraldo DA, Chancón J, Álvarez-Serrano I, López ML (2020) A C/V_2O_5 core-sheath nanofibrous cathode with mixed-ion intercalation for aluminium-ion batteries. Nano Express 1:010016
72. Mckerracher RD, Holland A, Cruden A, Wills RGA (2020) δ-MnO_2 nanofibers: a promising cathode material for new aluminum-ion batteries. Chem Electro Chem 7:2106
73. Mckerracher RD, Holland A, Cruden A, Wills RGA (2019) Comparison of carbon materials as cathodes for the aluminium-ion battery. Carbon NY 144:341
74. Shi F, Chen C, Xu ZL (2021) Recent advances on electrospun nanofiber materials for post-lithium ion batteries. Adv Fiber Mater 3:301
75. López Herraiz M, Castillo Martínez E, Carretero González J, Carrasco J, Rojo T, Armand M (2015) Oligomeric-Schiff bases as negative electrodes for sodium ion batteries: unveiling the nature of their active redox centers. Energy Environ Sci 8:3241

76. Cui J, Yao S, Kim JK (2017) Recent progress in rational design of anode materials for high-performance Na-ion batteries. Energy Storage Mater 7:114
77. Ni J, Fu S, Wu C, Zhao Y, Maier J, Yu Y, Li L (2016) Superior sodium storage in $Na_2Ti_3O_7$ nanotube arrays through surface engineering. Adv Energy Mater 6:1502568
78. Liu Z, Zhang W, Zhou Z, Liu X, Zhang H, Wei M (2020) Hierarchical porous anatase TiO_2 microspheres with high-rate and long-term cycling stability for sodium storage in ether based electrolyte. ACS Appl Energy Mater 3:3627
79. Hong Z, Hong J, Xie C, Huang Z, Wei M (2016) Hierarchical rutile TiO_2 with mesocrystalline structure for Li-ion and Na-ion storage. Electrochimica Acta 202:208
80. Lee J, Lee JK, Chung KY, Jung HG, Kim H, Mun J, Choi W (2016) Electrochemical investigations on TiO_2-B nanowires as a promising high capacity anode for sodium-ion batteries. Electrochimica Acta 200:28
81. Usui H, Yoshioka S, Wasada K, Shimizu M, Sakaguchi H (2015) Nb-doped rutile TiO_2: a potential anode material for Na-ion battery. ACS Appl Mater Interfaces 7:6573
82. Wu L, Bresser D, Buchholz D, Passerini S (2015) Nanocrystalline TiO_2 (B) as anode material for sodium-ion batteries. J Electrochem Soc 162:3052
83. Greco G, Mazzio KA, Dou X, Gericke E, Wendt R, Krumrey M, Passerini S (2019) Structural study of carbon-coated TiO_2 anatase nanoparticles as high-performance anode materials for Na-Ion batteries. ACS Appl Energy Mater 2:7151
84. Sun-Q, Ren QQ, Li H, Fu Z-W (2011) Electrochem Commun 13:1464
85. Perveen T, Siddiq M, Shahzad N, Ihsan R, Ahmad A, Shahzad MI (2020) Prospects in anode materials for sodium ion batteries—a review. Renew Sustain Energy Rev 119:109549
86. Slater MD, Kim D, Lee E, Johnson CS (2013) Sodium-ion batteries. Adv Funct Mater 23:958
87. Zhou A, Dai X, Lu Y, Wang Q, Fu M, Li J (2016) Enhanced interfacial kinetics and high-voltage/high-rate performance of $LiCoO_2$ cathode by controlled sputter-coating with a nanoscale $Li_4Ti_5O_{12}$ ionic conductor. ACS Appl Mater Interfaces 8:34131
88. Somo TR, Mabokela TE, Teffu DM, Sekgobela TK, Ramogayana B, Hato MJ, Modibane KD (2021) A comparative review of metal oxide surface coatings on three families of cathode materials for lithium ion batteries. Coatings 11:744
89. Shim JH, Lee S, Park SS (2014) Effects of MgO coating on the structural and electrochemical characteristics of $LiCoO_2$ as cathode materials for lithium ion battery. Chem Mater 26:2543
90. Dai X, Wang L, Xu J, Wang Y, Zhou A, Li J (2014) Improved electrochemical performance of $LiCoO_2$ electrodes with ZnO coating by radio frequency magnetron sputtering. ACS Appl Mater Interfaces 6:15859
91. Hao Q, Ma H, Ju Z, Li G, Li X, Xu L, Qian Y (2011) Nano-CuO coated $LiCoO_2$: synthesis, improved cycling stability and good performance at high rates. Electrochimica acta 56:9031
92. Jayasree SS, Nair S, Santhanagopalan D (2018) Ultrathin TiO_2 coating on $LiCoO_2$ for improved electrochemical performance as Li–ion battery cathode. Chem Sel 3:Mar 2766
93. Liu GQ, Kuo HT, Liu RS, Shen CH, Shy DS, Xing XK, Chen JM (2010) Study of electrochemical properties of coating ZrO_2 on $LiCoO_2$. J Alloy Compd 496:516
94. Morley NB, Burris J, Cadwallader LC, Nornberg MD (2008) GaInSn usage in the research laboratory. Rev Sci Instrum 79:056107
95. Zheng ZJ, Ye H, Guo ZP (2020) Recent progress in designing stable composite lithium anodes with improved wettability. Adv Sci. 7:2002212
96. Aqra F, Ayyad A (2012) Surface tension (γLV), surface energy (γSV) and crystal-melt interfacial energy (γSL) of metals. Curr Appl Phys 12:35
97. Zhao X, Xu S, Liu J (2017) Surface tension of liquid metal: role, mechanism and application. Front Energy 11:567
98. Goodenough JB, Park KS (2013) The Li-ion rechargeable battery: a perspective. J Am Chem Soc 135:1176
99. Peryez SA, Cambaz MA, Thangadurai V, Fichtnert M (2019) Interface in solid-state lithium battery: challenges, progress, and outlook. ACS Appl Mater Inter 11:22050
100. Famprikis T, Canepa P, Dawson JA, Islam MS, Masquelier C (2019) Fundamentals of inorganic solid-state electrolytes for batteries. Nat Mater 18:1291

101. Cho J, Kim YJ, Kim TJ, Park B (2001) Zero-strain intercalation cathode for rechargeable Li-Ion cell. Angew Chem Int Ed 40:3369
102. Seo JY, Oh D, Kim DJ, Kim KM, Kwon JS (2020) Enhanced mechanical properties of ZrO_2-Al_2O_3 dental ceramic composites by altering Al_2O_3 form. Dental Mater 36:125
103. Lee C, Choi JM, Miyahara Y, Jeon I, Miyazaki K, Abe T (2023) Bifunctional Al_2O_3 based artificial layers on $LiNiO_2$ cathodes for high-energy-density aqueous Li-Ion batteries. Chem Mater 36:869
104. Zhong Y, Ma Y, Guo Q, Liu J, Wang Y, Yang M, Xia H (2017) Controllable synthesis of TiO_2@Fe_2O_3 core-shell nanotube arrays with double-wall coating as superb lithium-ion battery anodes. Sci Rep 7:40927
105. Chen Y, Chen X, Zhang Y (2021) A comprehensive review on metal-oxide nanocomposites for high-performance lithium-ion battery anodes. Energy Fuels 35:6442
106. Abakumov AM, Fedotov SS, Antipov EV, Tarascon JM (2020) Solid state chemistry for developing better metal-ion batteries. Nat Commun 11:4976
107. Chen H, Hautier G, Jain A, Moore C, Kang B, Doe R, Wu L, Zhu Y, Tang Y, Ceder G (2012) Carbonophosphates: a new family of cathode materials for Li-ion batteries identified computationally. Chem Mater 24:2016
108. Hautier G, Jain A, Mueller T, Moore C, Ong SP, Ceder G (2013) Designing multielectron lithium-ion phosphate cathodes by mixing transition metals. Chem Mater 25:2074
109. Panin RV, Drozhzhin OA, Fedotov SS, Suchkova EA, Khasanova NR, Antipov EV (2018) Pyrophosphates $AMoP_2O_7$ (A= Li and Na): synthesis, structure and electrochemical properties. Mater Res Bull 106:175
110. Jain A, Hautier G, Moore C, Kang B, Lee J, Chen H, Twu N, Ceder G (2012) A computational investigation of $Li_9M_3(P_2O_7)_3(PO_4)_2$ (M= V, Mo) as cathodes for Li ion batteries. J Electrochem Soc 159:622
111. Hautier G, Jain A, Ong SP, Kang B, Moore C, Doe R, Ceder G (2011) Phosphates as lithium-ion battery cathodes: an evaluation based on high-throughput ab initio calculations. Chem Mater 23:3508
112. Herklotz M, Scheiba F, Glaum R, Mosymow E, Oswald S, Eckert J, Ehrenberg H (2014) Electrochemical oxidation of trivalent chromium in a phosphate matrix: $Li_3Cr_2(PO_4)_3$ as cathode material for lithium ion batteries. Electrochimica Acta 139:364
113. Jiang J, Li Y, Liu J, Huang X, Yuan C, Lou XW (2012) Recent advances in metal oxide-based electrode architecture design for electrochemical energy storage. Adv Mater 24:5180
114. Naseer M, Siyal SH, Najam T, Afzal S, Iqbal R, Ismail MA, Rauf A, Shah SS, Nazir MA (2025) Engineering of metal oxide integrated metal organic frameworks (MO@ MOF) composites for energy and environment sector. Mater Sci Eng: B 313:117909
115. Fan Y, Liu X, Li K, Zhang X, Xu L, Li B (2025) Defect engineering of transition metal oxides and synergistic interaction of two-dimensional conducting metal–organic frameworks for efficient photocatalytic hydrogen evolution. Sep Purif Technol 354:128832
116. Huo D, Li G, Fan G, Zhang X, Han J, Wang Y, Zhou B, Chen S, Jia L (2025) Characterization of cathode degradation and development of a coupled electrochemical-aging model for sulfide-based all-solid-state batteries. J Power Sources 627:235830
117. Rajaeifar MA, Ghadimi P, Raugei M, Wu Y, Heidrich O (2022) Challenges and recent developments in supply and value chains of electric vehicle batteries: a sustainability perspective. Resour Conserv Recycl 180:106144
118. Guo Y, Wu S, He YB, Kang F, Chen L, Li H, Yang QH (2022) Solid-state lithium batteries: safety and prospects. EScience 2:163
119. Zhang Y, Zhang L, Zhao Z, Zhang Y, Cui J, Liu C, Mu D, Su Y, Wu B, Wu F (2024) Lithiophilic Li-Si alloy-solid electrolyte interface enabled by high-concentration dual salt-reinforced quasi-solid-state electrolyte. J Energy Chem 95:230
120. Liu X, Cheng Y, Su Y, Ren F, Zhao J, Liang Z, Zheng B, Shi J, Zhou K, Xiang Y, Zheng J (2023) Revealing the surface-to-bulk degradation mechanism of nickel-rich cathode in sulfide all-solid-state batteries. Energy Storage Mater 54:723

121. Yu CY, Choi J, Dunham J, Ghahremani R, Liu K, Lindemann P, Garver Z, Barchiesi D, Farahati R, Kim JH (2024) Time-resolved impedance spectroscopy analysis of aging in sulfide-based all-solid-state battery full-cells using distribution of relaxation times technique. J Power Sources 597:234116
122. Kazemi N, Danilov DL, Haverkate L, Dudney NJ, Unnikrishnan S, Notten PH (2019) Modeling of all-solid-state thin-film Li-ion batteries: accuracy improvement. Solid State Ionics 334:116
123. Raijmakers LH, Danilov DL, Eichel RA, Notten PH (20200 An advanced all-solid-state Li-ion battery model. Electrochimica acta 330:135147
124. Ren D, Lu L, Hua R, Zhu G, Liu X, Mao Y, Rui X, Wang S, Zhao B, Cui H, Yang M (2023) Challenges and opportunities of practical sulfide-based all-solid-state batteries. eTransportation 18:100272
125. Adaikkappan M, Sathiyamoorthy N (2022) Modeling, state of charge estimation, and charging of lithium-ion battery in electric vehicle: a review. Int J Energy Res 46:2165
126. Danilov D, Niessen RA, Notten PH (2010) Modeling all-solid-state Li-ion batteries. J Electrochem Soc 158:215
127. Nitta N, Wu F, Lee JT, Yushin G (2015) Li-ion battery materials: present and future. Mater Today 18:264
128. Placke T, Kloepsch R, Dühnen S, Winter M (2017) Lithium ion, lithium metal, and alternative rechargeable battery technologies: the odyssey for high energy density. J Solid State Electrochem 21:1964
129. Bresser D, Hosoi K, Howell D, Li H, Zeisel H, Amine K, Passerini S (2018) Perspectives of automotive battery R&D in China, Germany, Japan, and the USA. J Power Sources 382:178
130. Judez X, Eshetu GG, Li C, Martinez RLM, Zhang H, Armand M (2018) Opportunities for rechargeable solid-state batteries based on Li-intercalation cathodes. Joule 2:2224
131. Eshetu GG, Zhang H, Judez X, Adenusi H, Armand M, Passerini S, Figgemeier E (2021) Production of high-energy Li-ion batteries comprising silicon-containing anodes and insertion-type cathodes. Nat Commun 12:5459

Recent Advances, Challenges, and Future Perspectives of Solid-State Batteries

Tejram Dewangan, Nishi Keshari, Tilak Gandharv, Kriteemoy Basu, Naveen Kumar, and Katchala Nanaji

Abstract Solid-state batteries are drawing attention due to their high-energy density and safety concerns, which are a requirement for portable electronic devices and electric vehicles. Lithium-ion liquid electrolyte batteries have the problem of fire hazards and leakage. Despite this, most current technologies adopt lithium-ion liquid electrolyte batteries as they do not face the problem of dendrite formation, which is common in solid-state batteries. For better contact between the electrode and the solid-state electrolyte, some liquid is added to the solid electrolyte. Solid-state batteries are safer as they don't have flammable organic electrolytes. In this chapter, we have discussed the various electrolytes for solid-state batteries. Further, we also discuss the challenges associated with the electrolyte–electrode interface and the potential strategies to mitigate the challenges. In addition, we have mentioned the recent advances and future perspectives of solid-state batteries.

Keywords All solid-state battery (ASSB) · Li-ion battery (LIB) · Solid-state electrolyte (SSE) · Challenges

1 Introduction

With the accelerating development of electric vehicles, mobile electronics, and grid energy storage, there is an urgent demand for batteries that offer both high-energy density and exceptional safety. The limitations of existing energy storage technologies have driven the development of solid-state batteries. In conventional systems like lithium-ion batteries, a liquid electrolyte is typically employed to enable the transport of Li-ions between the cathode and the anode. This conventional organic liquid electrolyte exhibits high lithium-ion conductivity, but it is flammable and poses

T. Dewangan · N. Keshari · N. Kumar · K. Nanaji (✉)
Department of Chemistry, Indian Institute of Technology (IIT) Bhilai, Durg, Chhattisgarh, India
e-mail: nanaji@iitbhilai.ac.in

T. Gandharv · K. Basu · K. Nanaji
Department of Electrical Engineering, Indian Institute of Technology (IIT) Bhilai, Durg, Chhattisgarh, India

A. K. Rai et al. (eds.), *Electrochemistry of Next-Generation Solid-State Batteries*, Green Energy and Technology, https://doi.org/10.1007/978-981-95-7603-6_9

a high risk of leakage. To overcome this problem, a solid electrolyte was introduced, which is capable of replacing the liquid electrolyte. The basic difference is that, in all-solid-state batteries (ASSBs), the organic liquid electrolytes and separator of conventional batteries are substituted with an inorganic solid electrolyte [1]. This distinctive feature provides several benefits, including the potential to reduce cell thickness (Fig. 1a and b), which enables downsizing of the battery, increasing volumetric energy density by packing more energy into a smaller space compared to conventional LIBs. This is particularly advantageous for space-limited applications such as EVs, as it allows greater energy storage in a compact design, making ASSBs highly attractive for high-energy–density requirements [2]. Moreover, the integration of solid-state electrolytes (SSEs) allows the use of higher-energy–density electrode materials, which potentially enhances energy storage capabilities. Driven by continuous R&D initiatives, ASSBs hold great promise for transforming electrochemical energy storage and emerging as strong alternatives to traditional lithium-ion batteries (LIBs) [2]. SSBs offer several advantages, including high-energy density, superior mechanical stability, reduced weight, and low self-discharge.

Despite having the potential to meet the energy storage demand of the world, it possesses many challenges, such as high ionic resistance at room temperature, low power density, electrode–electrolyte interface instability, and a costly fabrication process [3]. In solid-state batteries, different types of solid electrolytes are utilized, including polymer-based ones such as polyethylene oxide (PEO), polyacrylonitrile (PAN), polyvinylidene fluoride (PVDF), polymethyl methacrylate (PMMA), etc., and inorganic oxide-based solid-state electrolytes including perovskites, NASICONs, LISICONs, garnets, etc. [4]. Oxide-based electrolytes offer good ionic conductivity, mechanical strength, Li-metal compatibility, and high-voltage stability. However, their brittleness leads to stress, cracks, and poor electrode/electrolyte contact, increasing interfacial impedance, promoting Li dendrite growth, and reducing battery performance. High-temperature treatment or co-sintering is needed to improve such interfacial contact [5].

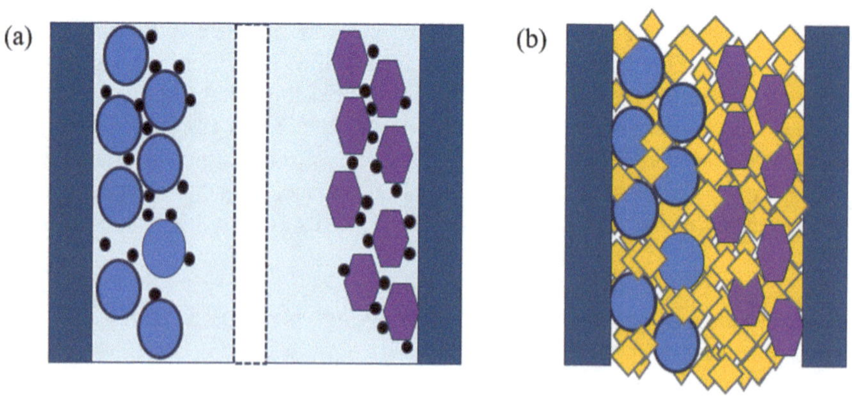

Fig. 1 Schematic representation of **a** conventional LIB with liquid electrolyte and **b** ASSB

2 Solid-State Battery *vs.* Conventional Li-Ion Battery—Advantages and Disadvantages

Solid-state battery (SSB) technology is advancing in applications from small devices to EVs and grid storage. Replacing the flammable liquid electrolytes (LEs) in conventional LIBs with solid electrolytes (SEs) is expected to significantly enhance safety. SEs could also enable Li-metal anodes, increasing energy density, but dendritic growth remains a challenge, potentially leading to short circuits.

Polymer SEs may block dendrites, but ceramic SEs remain vulnerable due to grain boundaries and fractures. High interfacial resistance between SEs and cathodes further limits performance, often addressed by adding LE, coatings, or structural modifications, though these increase costs. The addition of LE lowers the interfacial resistance, but it raises safety concerns, leading to greater emphasis on ASSBs, which contain no liquid. Table 1 compares solid-state batteries and conventional Li-ion batteries.

Table 1 Comparison between conventional lithium-ion batteries and solid-state batteries [2, 4, 6]

Conventional Li-ion battery versus solid-state battery

	Conventional Li-ion battery	Solid-state battery
1	Chance of electrolyte leakage if any holes are present	No leakage problem
2	It contains liquid electrolytes that facilitate the transport of Li-ions between the cathode and anode in both directions	Consists of a solid electrolyte
3	Dendrite formation, which makes it prone to explosion	Dendrite formation, but the solid electrolyte blocks the dendrite from reaching the other end of the battery
4	More weight and volume	Less weight and volume of the battery
5	High self-discharge	Low self-discharge
6	Low ionic resistance	High ionic resistance at room temperature
7	More hazardous due to flammable liquid electrolyte	Safer due to it's less flammable electrolyte–electrode system, offering excellent safety
8	High power density	Low power density
9	Low energy density	High-energy density
10	Operates between a temperature range of -20 to 60 °C	Capable of operating a wider temperature range, reaching up to approximately 200 °C
11	Poor thermal stability	High thermal stability

3 Advancements in Electrolytes for Solid-State Batteries

The performance of solid-state batteries is governed by the diffusion of ions through the electrolyte. Solid electrolytes are safer as there are no leakage problems with the electrolyte, and the electrolyte–electrode system is less flammable. Solid-state electrolytes are broadly classified into two categories: solid polymer and inorganic types. Polymer electrolytes outperform inorganic SSEs in terms of interface contact, electrochemical stability, manufacturing ease, flexibility, and cost-effective availability.

3.1 Solid Polymer Electrolytes (SPEs)

3.1.1 Poly (Ethylene Oxide) (PEO)-Based Solid Polymer Electrolyte

Among solid polymer electrolytes, poly(ethylene oxide) (PEO) is the most extensively studied and was one of the SPEs. PEO exhibits a molecular weight of almost 20,000 g/mol, with a repeating unit of $-(O-CH_2-CH_2)_n-$. It exhibits an ionic conductivity of about $10^{-7} - 10^{-8}$ S cm^{-1} at room temperature, whereas the PEO-Li-salt electrolyte shows an improved conductivity of 10^{-4} S cm^{-1} at a temperature of 60–80 °C, which is considered a practically useful value [7]. However, this conductivity remains notably lower than that of liquid electrolytes, which is 10^{-2} S cm^{-1} [8]. The PEO-based electrolyte faces some challenges, including low ionic conductivity, limited electrochemical window range, and dendrite formation when connected to the lithium anode. The lone electron pairs on the oxygen atoms of the PEO chain segments can coordinate with cations [9]. Dissociation and formation of the bond between the cation and oxygen, which causes the segmental rearrangement in the PEO chain, result in the cation (lithium) transport (Fig. 2) [10]. Only the amorphous regions can transport the lithium ions in the PEO. Above 60 °C, the PEO resides in an amorphous phase, and below 60 °C, the PEO crystallizes, in which the lithium ions cannot be transported. This temperature is called the glass transition temperature [11]. Shortening the PEO molecular segments can enhance lithium-ion conductivity by reducing resistance to segmental motion. Optimizing PEO production processes for this purpose is a key area for large-scale industrial development. Experiments show that prestretching PEO-based SPE (PL) under optimal stress aligns molecular chains, reducing lithium-ion transmission resistance and enhancing ionic conductivity [12].

The second challenge lies in the narrow electrochemical window within which the electrolyte remains stable. Relative to the standard hydrogen electrode (SHE), the electrochemical stability window of PEO is 9.12 V. Although PEO has the widest electrochemical window range compared to other solid polymer electrolytes, for instance, polyacrylonitrile (PAN) has 8.11 V, while polyethyleneimine (PEI) has a very low value of 6.32 V (Fig. 3). Complexation of PEO with lithium salt shows good lithium-ion conductivity, but the range of the electrochemical stability window

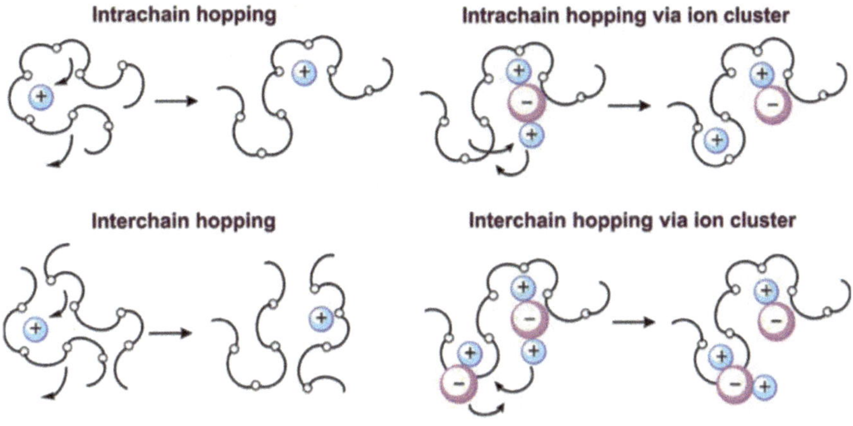

Fig. 2 Schematic illustration of lithium-ion conduction within the amorphous region of the polymer [13]. Reproduced with permission from ref. [13] Copyright 2023 Wiley–VCH

decreases. ESW for LiFSI, LiTFSI, and $LiCF_3SO_3$ are 4.77 V, 5.04 V, and 6.23 V, respectively. When complexed with PEO, the ESWs are reduced to 4.33 V for $LiCF_3SO_3$-PEO, 3.57 V for LiTFSI-PEO, and 2.02 V for LiFSI-PEO relative to SHE, respectively.

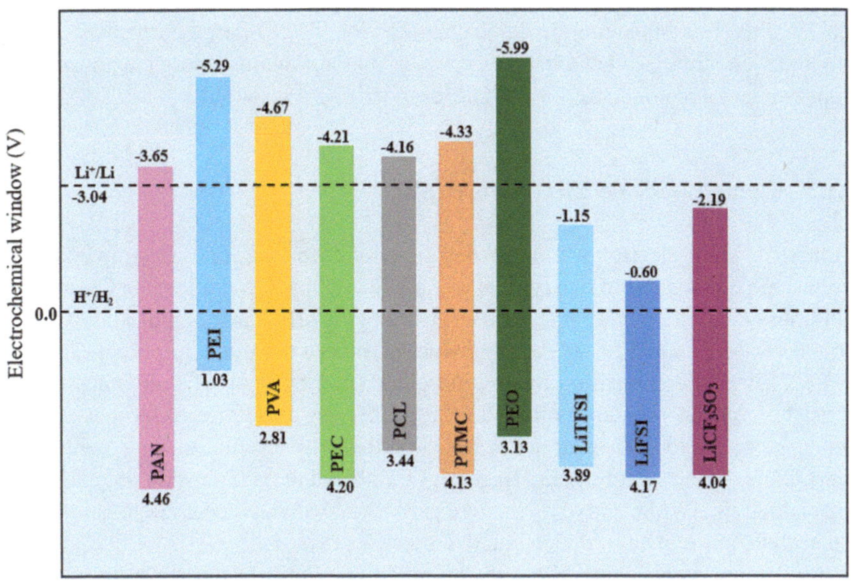

Fig. 3 Redox potentials of unmodified polymers and salts. The dashed lines at − 3.04 and 0 V correspond to the Li^+/Li reduction potential and the H^+/H redox potential, respectively

The third challenge involves the formation of lithium dendrites on the surface of the lithium anode. During the charge–discharge process, these dendrites continue to grow through the PEO electrolyte, leading to a rapid decline in capacity. The LiTFSI-based LAGP-PEO composite solid electrolyte can effectively inhibit the formation of dendrites on the lithium anode [14]. During cycling, LiI facilitates the formation of a uniform and dense SEI layer on the lithium anode, optimizing ion deposition, increasing critical current density, and preventing dendrite formation [15].

3.1.2 Polyacrylonitrile (PAN)-Based Solid Polymer Electrolyte

PAN is also a better candidate as a solid polymer electrolyte. Unlike PEO, where ion transport is caused by segmental rearrangement, PAN exhibits decoupled ion transport, allowing ion movement independent of the polymer matrix [16]. The ionic conductivity of the PAN-based electrolyte depends on the salt concentration, where the conductivity increases until the salt concentration reaches 70 wt.%. At this concentration, the ionic conductivity was 1.32×10^{-3} S cm^{-1}. High conductivity was achieved by compositing PAN with salt $LiClO_4$ and α-Al_2O_4. The composite (PAN-$LiClO_4$- α-Al_2O_4) polymer electrolyte (CPE) has achieved the conductivity of 5.7×10^{-4} S cm^{-1} at ambient temperature [7].

Fang Yuan et al. [17] synthesize copolymer polyacrylonitrile-polyethylene oxide (PAN-PEO) and then prepare PAN-PEO-$LiClO_4$ solid polymer electrolyte, which demonstrated an ionic conductivity of 6.79×10^{-4} S cm^{-1} with good mechanical stability. For enhancing the ionic conductivity of PEO-based electrolytes, many modification strategies have been developed. Among them, adding plasticizers can improve the ionic conductivity, but causes mechanical instability.

3.1.3 Composite Solid Electrolyte (CSE)

Composite solid electrolytes, consisting of a solid polymer and an inorganic electrolyte, are classified into two types based on their component ratios. (1) **Ceramic in polymers**—in which inorganic solid electrolytes (as filler) are composed into a solid polymer/lithium salt electrolyte. (2) **Polymer in ceramic**—where polymer/lithium salt (as filler) is compositing into an inorganic solid electrolyte. Polymer filler can consist of various substances like PEO, PAN, PVDF, and PMMA. Inorganic fillers can be categorized as inert or active based on their role in lithium-ion conductivity. Inert fillers, such as ZrO_2, ZnO, TiO_2, MgO, BN, and natural clay, do not contribute to conductivity. While active fillers, like garnets, NASICON, perovskite, and sulfide electrolyte, participate in the conductive process [18].

Adding inorganic inert fillers in the polymer matrix can participate in Lewis acid–base interactions, potentially influencing the composite's properties, which can increase the amorphous phase ratio by decreasing the crystallinity of the polymer. Incorporating inorganic inert fillers like Al_2O_3 and TiO_2 (5.8–13 nm) into PEO/$LiClO_4$ matrix can lower the polymer's glass transition temperature (Tg), enhance lithium-ion conductivity, and improve mechanical strength. Croce et al. [19] reported conductivities of 10^{-4} S cm^{-1} at 50 °C and 10^{-5} S cm^{-1} at 30 °C.

Adding inorganic active fillers into a polymer matrix can create new channels for lithium-ion transfer, which increases lithium-ion conductivity and also enhances the electrochemical stability window. Active fillers can also undergo Lewis acid–base interaction with the polymer. Apart from these, active fillers can also interact with polymers, such as hydrogen bonds, ionic dipole, π-π bond, and so on [18]. The optimal inorganic active filler content in polymer is typically between 10 wt.% and 30 wt.%. Excessive filler can cause aggregation, negatively affecting stress distribution and electrochemical performance. On the other hand, low filler amounts may not create a continuous lithium-ion network, with energy barriers between particles hindering ions [18]. Active fillers can also decrease crystallinity while enhancing the amorphous phase within the polymer matrix. Zhang et al. [20] incorporated $Li_{6.75}La_3Zr_{1.75}Ta_{0.25}O_{12}$ (LLZTO) into a PVDF/$LiClO_4$ matrix. The interaction between La in LLZTO and the solvent's N and C = O groups induced partial PVDF defluoridation, reducing crystallinity. This enhanced lithium salt complexation and dissociation, achieving a solid composite electrolyte conductivity of 5×10^{-4} S cm^{-1}. Zheng et al. [21] introduced $Li_7La_3Zr_2O_{12}$ nanofibers (LLZO NFs) into polyethylene oxide (PEO) to develop a composite solid electrolyte, which was further treated by quenching with liquid nitrogen to enhance its properties. Incorporating LLZO NFs in polymers can suppress dendrite growth, reduce crystallinity, and enhance both ionic conductivity and mechanical strength.

3.2 *Inorganic Solid-State Electrolytes*

3.2.1 NASICON

NASICON (Sodium Super Ionic Conductor) refers to a family of solid electrolytes with the general formula $NaM_2(PO_4)_3$, where M represents a cation species. A well-known example is $Na_{1+x}Zr_2Si_xP_{3-x}O_{12}$ ($0 \leq x \leq 3$), which typically exhibits a rhombohedral structure but undergoes a phase transition to a monoclinic structure when $1.8 \leq x \leq 2.2$. This structure features interconnected XO_4 tetrahedra and MO_6 octahedra, forming a 3D ion-conducting network. Increasing the value of x enhances conductivity by lowering the activation energy [22] $Na_3Zr_2PSi_2O_{12}$ achieves high ionic conductivity (2×10^{-1} S cm^{-1} at 300 °C), similar to β″-alumina. Li-substituted NASICON ($LiZr_2(PO_4)_3$) exhibits lower conductivity with phase transitions affecting ionic mobility [23].

Substituting Ti^{4+} in $LiTi_2(PO_4)_3$ with trivalent elements (e.g., Al, Fe) improves conductivity. $Li_{1+x}Al_xTi_{2-x}(PO_4)_3$ (LATP) exhibits high ionic conductivity but faces redox issues with some anodes, while $Li_{1.3}Al_{0.3}Ti_{1.7}(PO_4)_3$ achieves 7×10^{-4} S cm^{-1} at 25 °C. $Li_{1+x}Al_xGe_{2-x}(PO_4)_3$ (LAGP) also offers improved conductivity with good electrochemical stability. Both LAGP and LATP show stability in a wide voltage window, with reversible lithium deposition and dissolution.

$Li_{1.3}Al_{0.4}Cr_{0.1}Ge_{1.5}(PO_4)_3$ achieves the maximum conductivity of 6.65×10^{-3} S cm^{-1} at 26 °C. These materials enhance ionic conductivity via structural modifications and doping, with LAGP glass–ceramics offering the highest ionic conductivity [24].

3.2.2 Garnets

Lithium-containing garnets, first discovered in 1968 with the composition $Li_3Ln_3M_2O_{12}$ (M = Te, W; Ln = Y, Nd, Sm, Pr, etc.), exhibit varying lattice constants influenced by the ionic radius of the Ln element. $Li_3Ln_3Te_2O_{12}$ are capable of forming with a broader range of Ln elements than $Li_3Ln_3W_2O_{12}$ due to the stability of TeO_6 octahedra compared to the distorted WO_6 octahedra in the latter [25].

Thangadurai et al. [26] reported on $Li_5La_3M_2O_{12}$ (M = Ta, Nb), which exhibited ionic conductivities on the order of ~ 10^{-6} S cm^{-1} at 25 °C, with $Li_5La_3Nb_2O_{12}$ demonstrating comparatively higher conductivity along with a lower activation energy. Ba-doped $Li_6BaLa_2Ta_2O_{12}$ demonstrated significantly enhanced conductivity (4.0×10^{-5} S cm^{-1} at 22 °C) and improved stability against molten lithium [27]. Murugan et al. [28] introduced $Li_7La_3Zr_2O_{12}$ (LLZO), featuring a cubic crystal structure and high conductivity (3×10^{-4} S cm^{-1} at 25 °C). Further Nb doping, as in $Li_{6.75}La_3(Zr_{1.75}Nb_{0.25})\,O_{12}$, increased conductivity to 8×10^{-4} S cm^{-1} at 25 °C and extended the electrochemical stability window to 0–9 V vs. Li/Li^+.

Further research, such as $Li_{6.4}La_3Zr_{1.4}Ta_{0.6}O_{12}$ and Al/Ga co-substituted $Li_7La_3Zr_2O_{12}$, showed increased ionic conductivity, accompanied by phase transitions and lowered activation energy. Sc and Ga co-doping have also been shown to enhance conductivity, with $Li_{6.65}Ga_{0.15}La_3Zr_{1.90}Sc_{0.10}O_{12}$ achieving 1.8×10^{-3} S cm^{-1} [29]. Similarly, Rb doping in $Li_{6.20}Ga_{0.30}La_{2.95}Rb_{0.05}Zr_2O_{12}$ resulted in 1.62×10^{-3} S cm^{-1}, attributed to the increased Li-ion concentration [30]. Although garnet-type Li-ion conductors show great potential for ASSBs, their susceptibility to degradation in ambient environments, particularly in the presence of moisture and CO_2, continues to pose a significant challenge.

3.2.3 Perovskite

Inaguma et al. [31] first developed perovskite-type solid electrolytes, $Li_{3x}La_{2/3-x}TiO_3$ (LLTO), which exhibited a bulk ionic conductivity of 1×10^{-3} S cm^{-1} and a total conductivity exceeding 2×10^{-5} S cm^{-1} at ambient temperature for $Li_{0.34}La_{0.51}TiO_{2.94}$. The crystal structure of LLTO, where Li and La ions occupy the A sites and Ti resides at the B sites, contains bottlenecks at the oxygen ion positions, which hinder efficient Li-ion migration. Doping LLTO with Sr^{2+} raised bulk conductivity to 1.5×10^{-3} S cm^{-1}, though grain boundary conductivity remained low (10^{-5} S cm^{-1}). However, its instability below 1.8 V vs. Li/Li^+ restricts compatibility with many anodes.

Chung et al. [32] explored Ti^{4+} substitutions with Zr^{4+}, Sn^{4+}, Ge^{4+}, and Mn^{4+}, finding that Mn^{4+} and Ge^{4+} enhanced ionic conductivity. However, full substitution caused secondary phases and Ti^{4+} reduction. Thangadurai et al. introduced $LiSr_{1.65}Zr_{1.3}Ta_{1.7}O_9$, stable with lithium but low conductivity (1.3×10^{-5} S cm^{-1}). Chen et al. reported $Li_{3/8}Sr_{7/16}Ta_{3/4}Zr_{1/4}O_3$ (LSTZO), with higher conductivities (2×10^{-4} S cm^{-1} bulk, 1.33×10^{-4} S cm^{-1} grain boundary), and stability above 1.0 V vs. Li/Li^+. Sintering further improved LSTZ's conductivity to 3.5×10^{-4} S cm^{-1} (bulk) and 2.7×10^{-4} S cm^{-1} (total).

Subsequent studies on LSTZ reported enhanced conductivity of 2.8×10^{-4} S cm^{-1} (bulk) and 2×10^{-4} S cm^{-1} (total) at 27 °C for $Li_{3/8}Sr_{7/16}Ta_{3/4}Zr_{1/4}O_3$. Substituting Zr with Hf yielded $Li_{3/8}Sr_{7/16}Ta_{3/4}Hf_{1/4}O_3$ (LSTH), which exhibited an ionic conductivity of 3.8×10^{-4} S cm^{-1} at 25 °C along with a broad electrochemical stability window (1.4–4.5 V vs. Li/Li^+). Despite these promising characteristics, both LSTZ and LSTH demand very high sintering temperatures (~1300 °C), making them unsuitable for co-sintering with electrode materials. Chen et al. [33] found that $Li_{3/8}Sr_{7/16}Nb_{3/4}Zr_{1/4}O_3$ (LSNZ) had optimal conductivity at a lower sintering temperature (1200 °C) but lower ionic conductivity (2×10^{-5} S cm^{-1}). Conductivity improved with excess lithium and B_2O_3, while La doping at Sr sites raised it to 3.33×10^{-5} S cm^{-1} for $Li_{3/8}Sr_{7/16-3x/2}La_xZr_{1/4}Nb_{3/4}O_3$ ($x = 0.05$). Despite these improvements, LSNZ's conductivity remains lower than LSTZ/LSTH. Interfacial issues with electrode materials still need further study to enhance the performance of these solid electrolyte ASSBs.

3.2.4 Lisicon

Hong et al. [34] first reported LISICON (Lithium Super Ionic Conductor) with $Li_{14}Zn(GeO_4)_4$, showing an ionic conductivity of 1.25×10^{-1} S cm^{-1} at 300 °C. Rapid Li-ion conduction is mainly attributed to large bottlenecks between Li-ion sites and weak Li-O^{2-} bonding. However, the ionic conductivity at lower temperatures (2×10^{-6} S cm^{-1} at 50 °C) remained too low for practical use. Substituting P^{5+} with Ge^{4+} and Li^+ with Zn^{2+} in γ-Li_3PO_4 improved ionic conductivity by enabling both vacancy and interstitial diffusion.

Kuwano et al. [35] developed $Li_{3+x}Ge_xV_{1-x}O_4$ solid electrolytes, with $Li_{3.5}Ge_{0.5}VO_4$ achieving high ionic conductivity of 4×10^{-5} S cm^{-1} at 18 °C. Song et al. enhanced the conductivity of LISICON electrolytes by substituting O^{2-} with Cl^- ($Li_{10.42}Ge_{1.5}P_{1.5}Cl_{0.08}O_{11.92}$), achieving conductivities up to 3.7×10^{-5} S cm^{-1} from larger bottlenecks and weaker bonding. Hu et al. [36] synthesized solid solutions of Li_4SiO_4 and Li_3PO_4 ($Li_{4-x}Si_{1-x}P_xO_4$), with the highest conductivity (~10^{-6} S cm^{-1}) observed for $Li_{3.5}Si_{0.5}P_{0.5}O_4$. Despite these improvements, the conductivity was still insufficient for ASSBs.

Kamaya et al. [37] enhanced conductivity by substituting O^{2-} with S^{2-} in thio-LISICON $Li_{10}GeP_2S_{12}$, achieving 1.2×10^{-2} S cm^{-1} at room temperature, comparable to organic liquid electrolytes.

3.2.5 LiPON

LiPON (Lithium Phosphorus OxyNitride) is an amorphous Li-ion solid electrolyte with ~ 2 × 10^{-6} S cm^{-1} conductivity at 25 °C, and good stability against lithium. However, increasing Li-ion content in the mixture reduces conductivity, likely due to a lower proportion of phosphate chains that aid Li-ion transport. Suzuki et al. [38] improved the conductivity by modifying the composition of the target material to a mixture of Li_3PO_4 and Li_2O in a 1:2 molar ratio, resulting in an ionic conductivity of 6.4 × 10^{-6} S cm^{-1} at 25 °C—three times higher than that of conventional LiPON made from pure Li_3PO_4.

Non-standard LiPON containing additional phosphate groups (PO_4^{3-}, PO_3^{2-}, PO^{2-}) showed ionic conductivity (6.7 × 10^{-6} S cm^{-1} at 24 °C) higher than standard LiPON (1.4 × 10^{-6} S cm^{-1}), as the extra groups induce structural disorder and enhance Li-ion mobility. Its thin, low-resistance films are ideal for thin-film ASSBs, where all components are vapor-deposited with a protective coating to prevent air-Li reactions. LiPON electrolytes are stable between 0–5.5 V vs. Li/Li^+, ensuring compatibility with various electrodes [39].

As LiPON is promising for thin-film ASSBs, its low ionic conductivity (10^{-6} to 10^{-5} S cm^{-1} at room temperature) makes it unsuitable for bulk ASSBs.

3.2.6 Sulfides

Sulfide-based solid electrolytes, obtained from LISICON-type γ-Li_3PO_4 with S^{2-} replacing O^{2-}, displayed significantly greater ionic conductivities and accelerated Li-ion movement than oxide-based SSEs. They are categorized as ceramics, glasses, and glass–ceramics, among which Li_2S-P_2S_5 and thio-LISICON $Li_{4-x}Ge_{1-x}P_xS_4$ (0 < x < 1) are the most promising. Thio-LISICON $Li_{4-x}Ge_{1-x}P_xS_4$ has gained attention for its superior properties. Kanno et al. [40] developed six new thio-LISICON electrolytes, enhancing conductivity through vacancy-doping ($Li_{4-x}Ge_{1-x}P_xS_4$) and interstitial-doping ($Li_{4+x}Ge_{1-x}Ga_xS_4$). The highest ionic conductivity was observed in vacancy-doped $Li_{4-x}Ge_{1-x}P_xS_4$. $Li_{10}GeP_2S_{12}$ stands out with ionic conductivity comparable to that of organic liquid electrolytes, while retaining performance even at − 30 °C and − 45 °C. It shows compatibility with lithium up to 5 V vs. Li/Li^+, and a $LiCoO_2$/$Li_{10}GeP_2S_{12}$ In battery exhibited sustained cycling stability with a capacity exceeding 120 mAh g^{-1}. Oxygen-substituted $Li_{10}MP_2O_{12}$ (M = Ge, Si, Sn, Al, P) showed lower conductivity and stability, while selenium-substituted variants had slightly larger cell volumes with minimal conductivity gains. Studies confirm that S^{2-} is well-suited to facilitate Li-ion conduction [41].

Kato et al. [42] reported Cl-doped $Li_{9.54}Si_{1.74}P_{1.44}S_{11.7}Cl_{0.3}$ with a record conductivity of 2.5 × 10^{-2} Scm^{-1}. Mizuno et al. [43] developed 70Li_2S-30P_2S_5 glass–ceramic electrolytes with conductivities of 3.2 × 10^{-3} S cm^{-1}. However, sulfide electrolytes react with moisture, generating H_2S. 67Li_2S-33P_2S_5 showed significant structural changes and H_2S release, whereas 75Li_2S-25P_2S_5 demonstrated better stability due to the PS_4-3 group's resistance to hydrolysis. To integrate sulfide electrolytes into ASSBs, selecting stable compositions and preventing atmospheric exposure is crucial.

3.2.7 Argyrodite

Argyrodite-type Li-ion solid electrolytes, Li_6PS_5X (X = Cl, Br, I), were first synthesized by Deiseroth et al. [44] with structures similar to Ag^+ and Cu^+ argyrodites. The lattice has 136 tetrahedral sites, four of which are occupied by P atoms, and the remaining 132, made of halogens and S_2, are partially filled by Li atoms. Schneider et al. [45] developed halogen-free, Si-containing argyrodites to overcome halogen-related issues, stabilizing the high-conductivity cubic phase at elevated temperatures by partially replacing P with Si. Compositions $Li_{2x}SiP_2S_{7+x}$ (10 < x < 12) achieved enhanced ionic conductivities of ~ 3 × 10^{-3} S cm^{-1}, comparable to halogen-containing counterparts.

Li_6PS_5Br solid electrolytes have been used in various ASSBs, such as Cu-Li_2S/Li_6PS_5Br/In, CuS/Li_6PS_5Br/In-Li, MoS_2/Li_6PS_5Br/In-Li, and S/Li_6PS_5Br/In-Li. The Cu-Li_2S/Li_6PS_5Br/In cell showed high initial capacities of 500 and 445 mAh g^{-1} (0.1C, theoretical capacity 490 mAh g^{-1}), but dropped to ~ 70 mAh g^{-1} after 20 cycles due to cathode volume changes [46]. Two discharge plateaus at 1.4 V and 0.9 V correspond to a two-step CuS reduction: initial intercalation with minimal volume change, followed by a conversion reaction causing significant expansion. Similarly, CuS/Li_6PS_5Br/In-Li delivered 650 mAh g^{-1} initially (0.02 C, theoretical capacity 560 mAh g^{-1}), but declined to 90 mAh g^{-1} after 20 cycles, as Li_6PS_5Br accommodated reduced Cu^+ forming $Cu_yLi_{6-y}PS_5Br$ with low conductivity, leading to increased resistance, active materials detachment, cracks, and rapid capacity fading.

The S/Li_6PS_5Br/In-Li battery exhibited an initial discharge capacity of 1355 mAh g^{-1} (0.1 C, theoretical capacity of 1672 mAh g^{-1}) and retained 1080 mAh g^{-1} after 50 cycles [47]. While argyrodite-type solid electrolytes show great potential for ASSBs, their sensitivity to analogous to sulfide-based electrolytes for practical large-scale deployment. To overcome stability challenges, reliable and scalable battery manufacturing methods must be developed.

3.2.8 Anti-perovskite

Anti-perovskite solid electrolytes are designed by inverting the conventional perovskite lattice from $A^+B^{2+}X_3$ to $A^-B^{2-}X^+{}_3$, where halogens (F, Cl, Br, I) or their mixture occupy the A-site, oxygen atom the B-site, and lithium the X-site [48]. Li_3OCl and Li_3OBr exhibited classic perovskite structures, and a compositional solid solution, $Li_3O(Cl_{1-z}Br_z)$, can be created across the two terminal phases. The room temperature ionic conductivities were 8.5 × 10^{-4} S cm^{-1} for Li_3OCl and 1.94 × 10^{-3} S cm^{-1} for $Li_3OC_{l0.5}Br_{0.5}$. This increased conductivity of $Li_3OCl_{0.5}Br_{0.5}$ can be ascribed to its capability to circumvent the small Li-ion hopping interstitial space found in Li_3OBr and the structural distortions observed in Li_3OCl. Further improvements in ionic conductivity were proposed through metal doping, A-site mixing, and Li-site depletion, paving the way for $Li_{3-x-\delta}M_{x/2}O(A_{1-z}A'_z)_{1-\delta}$ compositions exhibiting enhanced functional characteristics [22].

However, Zhang et al. [49] highlighted that an ideal anti-perovskite structure does not inherently demonstrate superionic conductivity due to the lack of a viable Li-ion migration pathway. Instead, structural anomalies introduced during synthesis played a crucial role in enhancing ionic mobility. By incorporating vacancies into the anti-perovskite framework, simulations revealed significantly improved Li-ion diffusion. Additionally, introducing Cl–O exchange disorder within the Li_3OCl structure further boosted Li-ion mobility. Building on this work, Braga et al. synthesized glass electrolytes with the formula $Li_{3-2x}M_xHalO$ (M = Ba, Mg, Ca, Hal = Cl or mixed halogens), achieving an ionic conductivity of 2.5×10^{-2} S cm^{-1} at 25 °C for $Li_{3-2x}Ba_xClO$ (x = 0.005), the maximum value recorded for solid electrolytes at the time.

Expanding on the Li_3OA (A = halogen) family, Fang et al. [50] introduced novel anti-perovskites, $Li_3O(BH_4)$ and $Li_3O(BH_4)_{0.5}Cl_{0.5}$, through halogen substitution with BH_4 super-halogens. Using molecular dynamics simulations, their ionic conductivities at room temperature were estimated to be 1.0×10^{-4} S cm^{-1} and 2.1×10^{-4} S cm^{-1}, respectively. A thin-film ASSB featuring a $LiCoO_2/Li_3OCl$/graphite architecture was fabricated using these solid electrolytes through layer-by-layer deposition techniques. The battery exhibited an initial discharge capacity of 120 mAh g^{-1} (10 mA g^{-1}) within a voltage range of 2.2–4.2 V, though the capacity declined to approximately 66 mAh g^{-1} after 20 cycles, likely attributable to irreversible side reactions and inadequate battery sealing. Additionally, the initial cycle exhibited a Coulombic efficiency of approximately 95%, revealing that the electrolyte still lacked adequate ionic conductivity to deliver the best battery performance [51]. Despite their enhanced ionic transport properties and suitability for bulk as well as thin-film ASSB configuration, anti-perovskite-type solid electrolytes suffer from strong moisture sensitivity and require handling in an inert atmosphere, hindering the transition of ASSBs from lab to application.

4 Challenges and Their Potential Solutions in the Commercialization of Solid-State Batteries

4.1 *Chemical Instability*

Chemical stability in batteries is vital for maintaining durability and safety, ensuring electrode/electrolyte materials maintain their composition and structure during fabrication, storage, and operation. Degradation at the electrode/electrolyte interface forms high-resistance layers, hindering ion transport and accelerating capacity fading—a key challenge for large-scale deployment. Moisture sensitivity in SSEs dictates synthesis and handling, while interfacial instability shortens shelf life (Calendar Failure). Environmental factors like air exposure, temperature fluctuation, and side reactions further impact reliability and safety. Strategies such as advanced protective coatings, stable electrolytes, and interface engineering are crucial for strengthening stability and commercialization.

Chemical stability of SSEs in which sulfide-based SSEs (e.g., $xLi_2S.P_2S_5$) are highly moisture-sensitive, releasing toxic H_2S gas upon exposure, causing severe structural degradation [52]. Stability enhancements, like As-substituted Li_4SnS_4, enhance sulfur bonding over oxygen bonding, mitigating degradation [53]. Oxide-based SSEs (e.g., LLZO) suffer Li^+/H^+ exchange, forming resistive Li_2CO_3/LiOH layers, increasing interfacial resistance and reducing Li-ion wettability [54, 55]. Humidity accelerates Li_2CO_3 formation, weakening garnet and perovskite electrolytes [56, 57]. Polymer electrolytes, though more stable, require inert handling, while LiPON decomposes in humid conditions, releasing PH_3/NH_3 gas [58, 59]. Compared to sulfides, oxide and NASICON-based materials offer better stability for large-scale adoption [60–62]. On the other hand, the degradation of the Electrode/Electrolyte interface is dictated by interdiffusion and electrochemical reactions. Interdiffusion forms high-resistance layers (e.g., La_2CoO_4 at $LiCoO_2$/LLZO interfaces), hindering ion transport performance [63]. High-temperature processing exacerbates decomposition, creating Li-free blocking layers (e.g., Li_3PO_4 at LNMO/LLZTO interfaces) [64]. The space charge layer effect depletes lithium at sulfide cathode interfaces, increasing interfacial resistance [65, 66]. Lithium metal anodes face interphase growth, as observed in Li/LAGP interfaces, where layers up to 20 μm increase resistance and hinder cycling stability [67]. Some sulfide SSEs (e.g., β-Li_3PS_4) form stable passivation layers, while others (e.g., LGPS) undergo continuous reduction, accelerating degradation. Garnet-type SSEs like LLZO exhibit relative lithium stability, though dopant-dependent reduction effects require further optimization [68].

4.1.1 Potential Solutions to Enhance Chemical Stability

Commercializing SSBs requires enhancing SSE and interfacial stability through multiple strategies. Additives like LiF suppress Li_2CO_3 formation on LLZO, reducing interfacial resistance [69]. Oxides (e.g., Li_2O, Fe_2O_3, ZnO, Bi_2O_3) absorb H_2S in sulfide SSEs [70, 71], though some may increase activation energy, reducing ionic conductivity. Composition optimization, for e.g., using $Li_7La_3Sn_2O_{12}$ for humidity and resistance, and perovskite-type $Li_{0.38}Sr_{0.44}Ta_{0.7}Hf_{0.3}O_{2.95}F_{0.05}$ for pH stability [72], shows promise. Substituted sulfides like As-Li_4SnS_4 enhance stability without compromising conductivity. Novel sulfides (e.g., $Li_4Cu_8Ge_3S_{12}$ and $Li_{4-x}Sb_xSn_{1-x}S_4$), exhibit high air stability based on HSAB theory [73].

Interface engineering solutions, such as ALD and buffer layers, stabilize ion-conductive interfaces, suppressing decomposition [74]. Post-synthesis treatments, including polishing and heat treatments to reduce surface impurities and enhance lithium wettability [55, 75]. Li_2CO_3-free garnet surfaces can be achieved via 900 °C sintering, though Li loss remains a challenge. Acid treatment has facilitated stable Li plating/stripping performance exceeding 700 h [76]. Recent studies emphasize designing electronic-insulating but ion-conductive interphases to prevent interfacial degradation and enhance cycling stability [77].

4.2 Mechanical Instability

Mechanical challenges in batteries stem from volume changes at the electrode/electrolyte interfaces during cycling, causing stress, cracks, and voids that lead to electrode degradation. Dendrite growth in LIBs can penetrate the separator, causing short circuits [78]. Repeated expansion/contraction of the electrode material can further deteriorate the mechanical stability, resulting in capacity fading over prolonged cycles. Addressing these challenges is essential for commercializing SSBs and requires flexible and robust electrode materials, advanced binders, and designed electrolytes that can withstand mechanical deformations without impairing performance.

4.2.1 Lithium Dendrite Growth

The formation of lithium dendrites in Li-metal anodes poses severe safety hazards in SSBs, including internal short circuits, thermal runaway, fire, and explosion [79, 80]. Dendrites form due to non-uniform Li-ion deposition, driven by volume fluctuations and SEI instability [81, 82]. High-shear-modulus SSEs still experience dendrite penetration, raising concerns over grain boundary resistance and interfacial instability [83]. Studies on Al, Nb, and Ta-doped LLZO electrolytes show increased boundary resistance extends lifespan, but poor interfacial contact and high current densities still promote dendrites [84–86]. Interfacial modifications, including Au layers, improve lithium deposition uniformity [83]. SSE design is further complicated by electronic conductivity and microstructural defects, acting as pathways for lithium penetration [87]. Beyond safety risks, dendrite growth degrades battery performance, reducing coulombic efficiency, increasing impedance, and limiting cycling stability, hindering SSBs commercialization (Fig. 4) [88].

4.2.2 Cracking and Volume Expansion

Cracking and volume expansion hinder commercialization by affecting performance, safety, and lifespan. Mechanical stress from cycling causes strain accumulation, cracks, and electrode/electrolyte contact loss, while thermal expansion weakens interfaces [89–91]. Oxide SSEs, with $B/G < 1.7$ and $KIC \sim 1\ MPa \cdot m^{1/2}$, exhibit high brittleness, whereas sulfide electrolytes offer better ductility but remain mechanically unreliable [91, 92]. Interfacial reactions forming secondary phases like Li_2CO_3

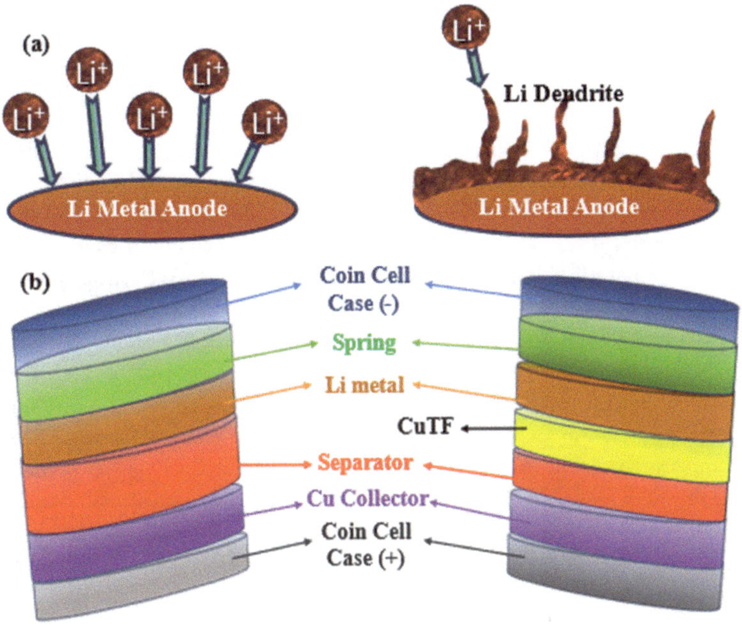

Fig. 4 Schematic illustration of a metal-coated separator designed to inhibit lithium dendrite growth in solid-state batteries (SSBs)

and $La_2Zr_2O_7$ in LLZO induce volume shrinkage and cracking [93–95]. Dendrite penetration through cracks further degrades ionic conductivity, raising the risk of short-circuiting. Theoretical models suggest that cracks propagate when expansion exceeds 7% and fracture energy drops below 4 $J{\cdot}m^{-2}$, requiring external pressure for stability [96]. X-ray tomography and modeling by Tippens et al. [97] confirms SSEs with fracture toughness > 4 $J{\cdot}m^{-2}$ endure up to 7.5% volume expansion before cracking [96, 97], highlighting the need for robust SSE designs.

4.2.3 Stress

Stress development in SSBs hinders commercialization due to rigid electrode–electrolyte interfaces. Unlike conventional LIBs with liquid electrolytes that provide stress relaxation, SSBs experience higher mechanical stresses during cycling. These arise from the brittle nature of ceramic-based SSEs, non-uniform electrode–electrolyte contact, interfacial decomposition, and the AM volume change during lithiation/delithiation [98]. Stress-induced degradation accelerates fatigue, increases interfacial resistance, and induces microcracks, compromising battery longevity. Additionally, diffusion-induced stress from lithium concentration gradients raises overpotentials, affecting electrochemical performance [99].

4.2.4 Potential Solutions for Mechanical Instability

To commercialize SSBs, integrated strategies addressing dendrite growth, cracking, and stress management are needed. High-strength solid electrolytes like garnet-type $Li_7La_3Zr_2O_{12}$ (LLZO) and sulfide-based Li_3PS_4 prevent dendrite penetration, though issues persist at high current densities [98]. Surface modifications, such as carbon-functionalized layers (CFLs) and thin buffer layers (TBLs), enhance interfacial stability and lithium-ion conductivity [100]. Metallic coatings (e.g., Cu films on polymeric separators) regulate lithium deposition, improve cycling stability, and enhance coulombic efficiency (95.6% over 300 cycles). Protective separators with ultra-thin carbon layers or buffer layers composed of LiF and PVDF-HFP improve adhesion, ionic conductivity, and dendrite suppression [101, 102].

To mitigate cracking and stress, optimizing mechanical properties is crucial. PEG-coated $LiFePO_4$ absorbs stress and minimizes crack formation, while rigid SSEs help sustain stress and prevent excessive strain [96]. Heat-treated $LiNi_{0.33}Co_{0.33}Mn_{0.33}O_2$ (NCM) and $LiCoO_2$ (LCO) cathodes reduce void formation and stress concentration [103]. Doping strategies, including zirconium-doped silicon anodes and polyethylene oxide (PEO)-doped LCO, enhance structural stability and crack resistance. Controlled grain size in $Li_{1.3}Al_{0.3}Ti_{1.7}(PO_4)_3$ (LATP) minimizes residual stress and cracking [104, 105]. Polymer-based or hybrid electrolytes offer flexibility, accommodating volume changes in active materials [106], while interface engineering with buffer layers reduces stress concentrations and interfacial degradation [107]. Computational modeling, such as finite element methods (FEM), aids in designing stress-resilient materials by simulating electro-chemo-mechanical interactions [96]. In addition, advanced manufacturing techniques, including precision sintering and additive manufacturing, enhance long-term stability (Fig. 5).

4.3 *Thermal Instability*

Thermal stability is crucial for the commercialization of SSBs, affecting safety, longevity, and performance. It refers to the ability of battery components, including electrolytes and electrodes, to withstand high temperatures without degradation or exothermic reactions. Unlike LIBs with flammable organic electrolytes, SSBs use solid electrolytes, such as oxide-based (LLZO, LATP) and sulfide-based materials ($Li_{10}SnP_2S_{12}$), which offer superior stability due to higher decomposition temperatures, reducing thermal risks. However, balancing thermal stability with high ionic conductivity remains a key challenge. Developing thermally robust yet high-performance materials is essential for commercialization, requiring extensive research into optimized electrolyte composition and structural modifications [108, 109].

Heat generation is one of the primary challenges in commercializing SSBs arising from Joule heating due to ionic resistance in solid electrolytes, interfacial resistance at electrode–electrolyte contacts, and exothermic side reactions. Unlike liquid

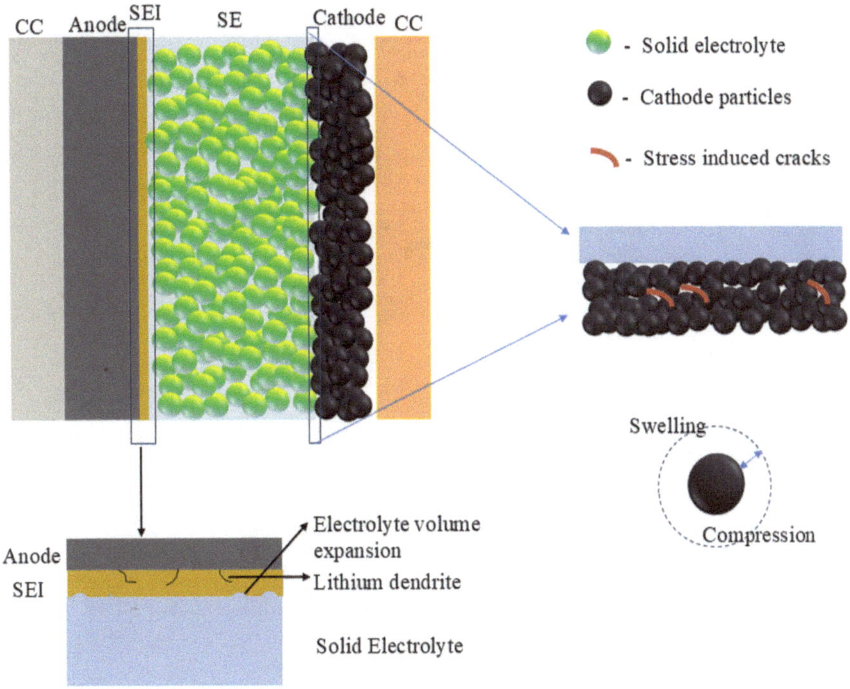

Fig. 5 Schematic representation of stress distribution in SSBs

electrolytes that can dissipate heat more effectively, solid electrolytes struggle with heat dissipation, causing localized hotspots and uneven temperature distribution. Oxide electrolytes offer superior thermal stability, with decomposition temperatures exceeding 1000 °C, while sulfide electrolytes decompose at lower temperatures (~220 °C for $75Li_2S{\cdot}25P_2S_5$) [109, 110]. Polymer electrolytes further impact heat generation, as lithium salts-polymer interactions reduce thermal stability. However, advancements in material engineering and interface design have improved thermal management, enhancing SSB performance and safety [93]. Thermal runaway is another major obstacle to the large-scale commercialization of SSBs. Despite improved safety over LIBs, exothermic reactions can still occur at high temperatures, especially at solid electrolyte-lithium metal interfaces. Studies show that oxide electrolytes trigger thermal runaway at 200 °C, forming resistive byproducts like Li_2S and Li_3P, leading to localized heating and failure. The LATP-oxide cathodes mixture decomposes at 500 °C, lower than their individual thresholds. Additionally, the lack of standardized SSB architectures complicates thermal behavior, which complicates the prediction of thermal behavior of electrolytes. Optimizing electrode–electrolyte interfaces and ensuring uniform heat dissipation remain critical challenges, requiring further research to enhance commercial viability [108, 110] (Fig. 6).

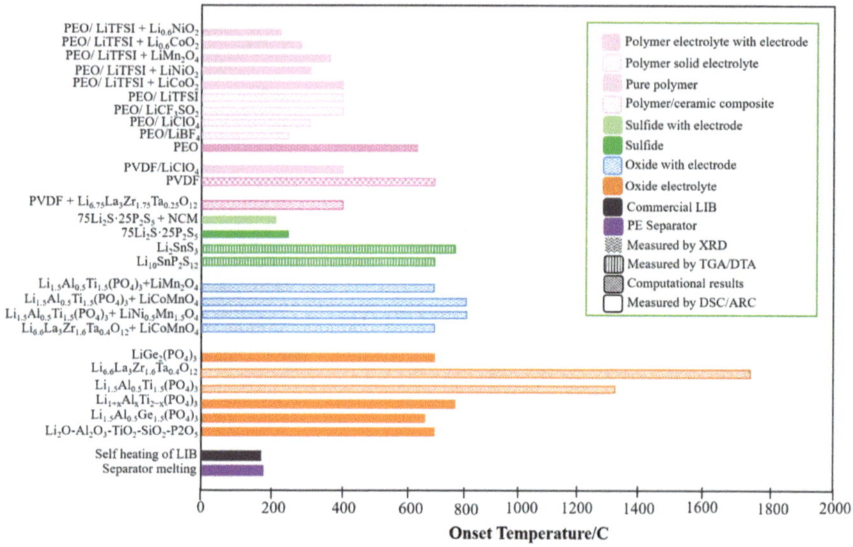

Fig. 6 Comparison of onset temperatures for exothermic reactions in various solid-state electrolytes (SSEs). Different colors indicate distinct sample compositions, while column patterns represent values obtained through different measurement methods

4.3.1 Potential Solution for Thermal Stability

Enhancing thermal stability is crucial for commercialization, with several strategies under exploration. Developing advanced solid electrolytes with improved heat resistance and low interfacial resistance, like Nb-LLZO, that remain stable up to 480 °C. Another strategy involves optimizing the battery architecture to ensure better electrode/electrolyte contact, thereby minimizing interfacial resistance and mitigating localized heat generation. Additionally, integrating thermal management strategies, such as heat-resistant coatings, phase-change materials, and integrated cooling systems, can further mitigate thermal issues. Finally, safety evaluations, such as accelerating rate calorimetry (ARC) and differential scanning calorimetry (DSC), should be incorporated into the design and testing phases to ensure robust thermal performance across different operating conditions. These advancements collectively contribute to improving the thermal safety and commercial viability of SSBs [108].

4.4 Electrochemical Instability

Electrochemical stability is a critical element for bringing SSBs to market, given its impact on durability, safety, and compatibility of electrolytes with electrodes over time. Unlike conventional liquid electrolytes, SSEs are expected to exhibit wide electrochemical stability windows, reducing side reactions and enhancing battery lifespan. However, real-world applications show discrepancies between theoretical and experimental electrochemical stability, leading to performance issues. For example, although some SSEs exhibit electrochemical windows up to 5 V, incompatibility with next-generation cathodes, operating at higher voltages restricts their practicality in commercial batteries.

A key hurdle to commercialization lies in balancing ionic conductivity with interfacial stability. Several SSEs offer either significant ionic conductivity with limited stability or wide electrochemical stability windows with low conductivity. Among sulfide-based electrolytes, $Li_{10}GeP_2S_{12}$ is particularly noteworthy, which can demonstrate high ionic conductivity (1.2×10^{-2} S cm^{-1}) but suffers from instability at high voltages due to interfacial decomposition. Oxide-based electrolytes like LLZO ($Li_7La_3Zr_2O_{12}$) are more electrochemically stable but have lower ionic conductivity due to grain boundary resistance, affecting lithium-ion transport [43, 111, 112]. At the cathode interface, SSEs tend to form resistive interphases, increasing polarization and reducing battery efficiency. Sulfide-based electrolytes, for instance, react with transition metal oxides in cathodes, leading to interphase formation and increased resistance. In the anode region, the intimate contact of Li-metal with SSEs can result in lithium dendrite growth, causing short circuits. Even oxide electrolytes, though more stable, face challenges due to their rigid structure, leading to poor interfacial contact and higher resistance. These interfacial reactions limit battery cycle life and performance, making it essential to develop compatible interfaces for commercial viability [43].

4.4.1 Potential Solution for Electrochemical Stability

To tackle these challenges, efforts focus on developing novel electrolyte formulations and interface modification techniques. Protective coatings, such as Li_3PO_4 and $LiNbO_3$, are reported to enhance interface robustness by preventing direct reactions between electrodes and SSEs. Composite electrolytes, which combine oxides, sulfides, and polymers, ensure a synergistic balance of ionic conductivity and stability. Additionally, doping strategies in oxide SSEs can reduce grain boundary resistance, improving lithium-ion transport. Another approach is to develop flexible solid-state electrolytes that improve mechanical contact with electrodes, minimizing interfacial resistance and enhancing battery performance for commercial applications [98].

5 Recent Advances in Solid-State Batteries

SSBs represent an evolution of traditional liquid-based batteries, offering improved safety features. Nevertheless, numerous obstacles persist in solid-state battery technology, as previously outlined, which need to be addressed to enable scalability and commercial adoption. Scientists are continually striving to address these challenges through the exploration of novel materials. Solid-state electrolytes, for example, demand high levels of ionic conductivity to facilitate enhanced Li-ion movement and resilient mechanical properties to resist dendrite propagation. Zhao et al. formulated a composite electrolyte using in-situ polymerization of cross-linked polyvinyl carbonate (PVC) on a $Li_{1.4}Al_{0.2}Ti_{1.8}Si_{0.2}P_{2.8}O_{12}$ (LATSP)-coated polypropylene (PP) separator. This composite delivers enhanced structural strength with a fast ion-conducting framework. Through the combination of a $LiFePO_4$ (LFP) cathode and lithium-metal anode, they achieved a capacity of 141.11 mAh g^{-1} at 2C rate, maintaining 81.56% capacity after 800 cycles, indicating significant potential for solid-state battery commercialization [113]. To tackle issues of poor conductivity and interface stability, researchers have incorporated two-dimensional materials as fillers in polyethylene oxide (PEO)-based electrolytes. By embedding $NiFe_2O_4$ (NFO) nanosheets within a polymer framework, Shi et al. designed vertically aligned NFO-PEO composite solid-state electrolytes (CSEs), leveraging a magnetic field-directed alignment technique. With this approach, an ionic conductivity (9.16×10^{-4} S cm^{-1} at 60 °C) was achieved and, when coupled with a lithium-metal anode and LFP cathode, produced a capacity of 132.90 mAh g^{-1} with 95.82% retention over 60 cycles [114]. The electronic insulation and slow kinetics of sulfide materials like Li_2S present performance challenges in sulfur-based cathodes for solid-state batteries. Hu et al. introduced a highly conductive Li_2S-$NbSe_2$ material that enhances electrochemical performance through bidirectional self-activation. This material accelerates electrochemical processes by modulating the conductive behavior of sulfur species and transformation pathways. Employing this cathode material, an energy density of 394 Wh kg^{-1} and a power density of 524 W kg^{-1} were achieved at 1C (4.35 mA cm^{-2}) and 25 °C, achieving 99.3% of its capacity after 100 cycles at 0.5 C with negligible degradation [115]. While oxide cathode materials are commonly employed in solid-state batteries, their low capacity and ionic conductivity restrict energy density. Zhou and colleagues proposed Li_xFeX_{x+2} (X = Cl, Br) families as high-capacity, high-ionic-conductivity materials derived from intercalation-conversion reactions. Their findings revealed that $LiFeCl_3$, containing 95 wt.% active material, achieves an impressive capacity of 446 mAh g^{-1} and an energy density of 912 Wh kg^{-1}. These findings represent a notable step forward in the study of intercalation-conversion halide cathode materials for high-energy–density SSBs [116]. The integration of 3D printing methods in the fabrication of polymer electrolytes has garnered significant attention in recent scientific investigations. An illustrative case is provided by the work of He et al., who employed stereolithography (SLA) three-dimensional (3D) printing to produce a solid polymer electrolyte. A 3D-Archimedean spiral architecture was utilized in their design, offering distinct advantages: it facilitated faster Li-ion transfer pathway from electrolyte to electrode, promoted stronger interfacial adhesion, and allowed greater incorporation of active materials. The study reported impressive outcomes, with ionic conductivity enhanced to 3.7×10^{-4} S cm^{-1} at 25 °C. Furthermore, in the presence of a Li-metal anode combined with an LFP cathode, the cells resulted in reduced interfacial impedance and a sustained specific capacity of 128 mAh g^{-1} over 250 cycles. These results reaffirm the promising potential of 3D printing in advancing battery technology [117].

With researchers persistently addressing the wide-ranging challenges of solid-state batteries, the primary obstacle from a commercialization standpoint remains the development of cost-effective solutions to enhance scalability. Various battery manufacturers are actively pursuing this objective. A case in point is the US-based company Sakuu, which has developed the Kavian printing platform. This innovative system utilizes 3D printers to produce SSBs that are both lighter and more economical than those manufactured using conventional methods. The Kavian platform offers significant advantages, including a 33% reduction in manufacturing costs and a 44% reduction in factory footprint. In addition, it aims for complete recyclability through a powder-to-powder recycling system. Through AI-assisted quality control, Sakuu's process serves to minimize waste generation and toxic byproducts in both fabrication and recycling. This technology enables high-volume, rapid battery production with reduced environmental footprint [118].

6 Future Perspective of Solid-State Battery

The commercialization of SSBs requires significant advancements in multiple domains, including chemical stability, mechanical durability, thermal resilience, and electrochemical performance. Addressing these challenges will enable the extensive utilization of SSBs in EVs, energy grid storage, and consumer electronics by enhancing safety, cycle life, and energy efficiency. The development of intrinsically stable SSEs is critical for commercial viability. Advanced computational modeling, such as density functional theory (DFT), can guide material selection and design by predicting thermodynamic stability and resistance to environmental factors like moisture and air exposure [119]. High-entropy SSEs, which incorporate multiple cations, present a promising pathway to reducing phase decomposition and interfacial reactivity. Additionally, dopant engineering in garnet and NASICON electrolytes can mitigate Li^+/H^+ exchange and the formation of resistive layers, thereby enhancing long-term performance [120]. Scalable surface modification techniques, such as atomic layer deposition (ALD) and protective coatings, should be optimized to create passivated interfaces without compromising ionic conductivity. Future research should focus on evaluating these materials under operational conditions, including temperature and humidity variations, aimed at connecting laboratory progress with real-world implementation. One of the major obstacles to the commercialization of SSBs is the suppression of lithium dendrite growth, as their presence can result in short circuits and battery failure. Future research should explore in situ electron microscopy and synchrotron X-ray tomography to visualize dendrite evolution in real time [121]. Computational modeling employing DFT and molecular-scale simulations can shed light on dendrite emergence and evolution mechanisms. The development of hybrid solid electrolytes, which combine organic and inorganic components, is another promising strategy for improving both ionic conductivity and mechanical strength [122]. Additionally, separator engineering with ceramic or polymer-functionalized coatings should be investigated for effective dendrite mitigation. Scalable manufacturing techniques, including molecular layer deposition (MLD) and atomic layer deposition (ALD), are also necessary to create thin, uniform protective layers that can be integrated into commercial battery production.

Ensuring thermal resilience is crucial for the large-scale deployment of SSBs, especially important for EVs operating under wide temperature ranges. Future research should leverage operando X-ray diffraction and neutron scattering to probe structural evolution under thermal stress. Engineering next-generation SSEs characterized by improved thermal resilience and electrochemical stability should be prioritized. Furthermore, scalable manufacturing techniques that ensure uniform material properties and interface compatibility must be explored to enhance commercial feasibility. Establishing standardized safety protocols and performance benchmarks will be essential in ensuring that future SSB technologies meet industrial and consumer demands. To achieve commercialization, SSBs must exhibit high ionic conductivity and wide electrochemical stability windows. Electrochemical processes depend on mixed ion–electron conductive frameworks. Unlike liquid electrolytes, solid electrolytes require intimate contact to enable continuous ion transport in ASSBs. Nanosizing SEs and matching particle sizes with electrodes are the future research direction that enhances efficiency. To support eco-friendly production, dry electrode technology (powder to film) enables large-scale fabrication while eliminating solvents, simplifying preparation, and reducing energy consumption. Future research should focus on hybrid electrolytes that incorporate both sulfide and oxide components to leverage their respective advantages. In situ characterization techniques, such as atomic-scale imaging and real-time interfacial analysis, will be critical in understanding and mitigating interfacial degradation. Additionally, scalable manufacturing techniques must be developed to ensure uniform electrolyte–electrode interfaces, reducing variability and improving reliability. Overcoming these challenges will be fundamental for positioning SSBs as a commercially viable alternative to conventional LIBs in the marketplace. Research on solid electrolyte-anode interactions highlights that capacity degradation and cell failure often result from interface reactions and remains a key future research area. Studies on sodium-anode and sulfide-based electrolytes emphasize the need for chemically stable electrolytes compatible with both anodes and cathodes. Applying fabrication and stack pressure also remain a key research area to increase ion diffusion and critical current density, improve the loose structure, release the stress, and inhibit volume expansion.

The progression from laboratory experimentation to industrial-scale production is one of the most significant hurdles in commercializing SSBs. Future research should focus on developing cost-effective and scalable synthesis methods, such as rapid sintering and non-equilibrium processing, to enhance phase purity and grain boundary chemistry. Additive manufacturing techniques, including 3D printing and roll-to-roll processing, could be employed to minimize internal stress and improve the mechanical integrity of SSB components [123]. Furthermore, automated quality control and real-time monitoring systems should be implemented in production lines to ensure consistency and reduce manufacturing defects. Addressing the multidisciplinary challenges in SSB research through computational, experimental, and engineering approaches will be key to realizing next-generation batteries with better safety, durability, and energy efficiency. By integrating advanced materials, innovative manufacturing techniques, and comprehensive safety assessments, future research can narrow the gap between lab-scale innovation and commercial deployment. The successful commercialization of SSBs will not only revolutionize energy storage technologies but also contribute significantly to sustainable energy solutions for the future.

7 Conclusion

In summary, solid-state batteries have significant advantages and promising applications in the fields of electric vehicles (EVs), end-user electronic systems, medical devices, grid energy storage, aerospace, and defense. Compared to conventional Li-ion batteries, SSB has many advantages, such as no leakage problem, occupies less volume that is needed for electric vehicles, less hazardous, low self-discharge rate, and remains operational across a broader temperature range. Electrolytes for SSBs are discussed in detail, offering a comprehensive discussion of the challenges and corresponding solutions. PEO-based solid electrolytes provide notable advantages and hold considerable promise for ASSBs applications. Despite their advantages, further application is constrained due to low ionic conductivity, a restricted electrochemical window, and problems at the electrode interface. With ongoing research, technological advancements, and material development, the challenges hindering PEO's widespread application are expected to be overcome. Composite solid electrolytes leverage the advantages of different components, making them a promising development. Synergistic reactions between inorganic and polymer phases promote amorphization, suppress crystallinity, and improve electrochemical properties. Composite electrolytes show moderate performance with ionic conductivity, interfacial, and mechanical properties. Sulfide electrolytes are characterized by greater ionic conductivity than their polymer and oxide-based counterparts. Garnets and NASICON electrolytes show better thermal stability than others.

Furthermore, four types of stability issues with SSBs are classified and discussed from a commercialization perspective, including chemical instability, mechanical instability, thermal instability, and electrochemical instability. Thermal runaway is a critical obstacle in the large-scale commercialization of SSBs. Optimizing battery architecture and thermal management are key strategies for this. Safety issues are mainly related to mechanical instability, including dendrite growth, stress, cracking, and volume expansion. Metallic coatings help regulate dendrite propagation, and polymer coatings can help reduce stress and minimize cracks. NASICON-type SSEs are more chemically stable than other polymer, oxide, and sulfide-based electrolytes. Sulfide electrolytes are known for their good ionic conductivity, whereas oxide-based electrolytes exhibit a high electrochemical stability window. Composite solid electrolytes can solve this problem, which has both high ionic conductivity and an enhanced electrochemical stability window. This chapter briefly outlines recent research trends in solid-state batteries aimed at accelerating their commercialization. Advancements in anode technology are crucial, with lithium metal as a key candidate for ASSB anodes due to its high theoretical capacity. Research is also exploring lithium-free anodes [124, 125] and low-reactivity silicon anodes [126], showing promising performance by employing additive-free, void-free silicon wafer electrodes without electrolytes.

The global battery market is set for rapid growth, driven by EV demand, with SSBs emerging as a key sector. Although SSBs are expected to hold just 1% of the market share by 2035, they remain strong competitors to LIBs, offering advantages in EV range and charging time. Overall, this study provides valuable insight into recent advances, challenges, and the future of SSBs.

Acknowledgements The authors acknowledge the Scheme for Young Scientists & Technologists (SYST project {Ref. No. SP-YO-2021-2489 (G)} funded by the Department of Science and Technology (DST), Government of India, and the Research Initiation Grant project of IIT Bhilai for carrying out the research work.

References

1. Yao X, Huang B, Yin J, Peng G, Huang Z, Gao C, Liu D, Xu X (2016) All-solid-state lithium batteries with inorganic solid electrolytes: review of fundamental science. Chin Phys B 25:018802
2. Sung J, Heo J, Kim D-H, Jo S, Ha Y-C, Kim D, Ahn S, Park J-W (2024) Recent advances in all-solid-state batteries for commercialization. Mater Chem Front 8:1861–1887
3. Vincent C (2000) Lithium batteries: a 50-year perspective, 1959–2009. Solid State Ion 134:159–167
4. Chen R, Li Q, Yu X, Chen L, Li H (2020) Approaching practically accessible solid-state batteries: stability issues related to solid electrolytes and interfaces. Chem Rev 120:6820–6877
5. Li Z, Fu J, Guo X (2023) How to commercialize solid-state batteries: a perspective from solid electrolytes. Natl Sci Open 2:20220036
6. Kim JG, Son B, Mukherjee S, Schuppert N, Bates A, Kwon O, Choi MJ, Chung HY, Park S (2015) A review of lithium and non-lithium based solid state batteries. J Power Sources 282:299–322
7. Chen-Yang Y (2002) Polyacrylonitrile electrolytes 1. A novel high-conductivity composite polymer electrolyte based on PAN, $LiClO_4$ and α-Al_2O_3. Solid State Ion 150:327–335
8. Shi P, Ma J, Liu M, Guo S, Huang Y, Wang S, Zhang L, Chen L, Yang K, Liu X, Li Y, An X, Zhang D, Cheng X, Li Q, Lv W, Zhong G, He Y-B, Kang F (2023) A dielectric electrolyte composite with high lithium-ion conductivity for high-voltage solid-state lithium metal batteries. Nat Nanotechnol 18:602–610
9. Wang S, Sun Q, Zhang Q, Li C, Xu C, Ma Y, Shi X, Zhang H, Song D, Zhang L (2023) Li-Ion transfer mechanism of ambient-temperature solid polymer electrolyte toward lithium metal battery. Adv Energy Mater 13:2204036
10. Yang H, Wu N (2022) Ionic conductivity and ion transport mechanisms of solid-state lithium-ion battery electrolytes: a review. Energy Sci Eng 10:1643–1671
11. Xu K (2004) Nonaqueous liquid electrolytes for lithium-based rechargeable batteries. Chem Rev 104:4303–4418

12. Zhang L, Fu L, Qin W, He Y, Liu H, Hu H (2023) Enhancing the ionic conductivity and mechanical properties of PEO-based solid electrolytes through thermal pre-stretching treatment. Phys Chem Chem Phys 25:18297–18309
13. Li Z, Fu J, Zhou X, Gui S, Wei L, Yang H, Li H, Guo X (2023) Ionic conduction in polymer-based solid electrolytes. Adv Sci 10:2201718
14. Wang C, Yang Y, Liu X, Zhong H, Xu H, Xu Z, Shao H, Ding F (2017) Suppression of lithium dendrite formation by using LAGP-PEO (LiTFSI) composite solid electrolyte and lithium metal anode modified by PEO (LiTFSI) in all-solid-state lithium batteries. ACS Appl Mater Interfaces 9:13694–13702
15. Wang H, Hou T, Cheng H, Jiang B, Xu H, Huang Y (2022) Bifunctional LiI additive for poly(ethylene oxide) electrolyte with high ionic conductivity and stable interfacial chemistry. J Energy Chem 71:218–224
16. Yoon H-K, Chung W-S, Jo N-J (2004) Study on ionic transport mechanism and interactions between salt and polymer chain in PAN based solid polymer electrolytes containing $LiCF_3SO_3$. Electrochim Acta 50:289–293
17. Yuan F, Chen H-Z, Yang H-Y, Li H-Y, Wang M (2005) PAN–PEO solid polymer electrolytes with high ionic conductivity. Mater Chem Phys 89:390–394
18. Zhang Z, Wang X, Li X, Zhao J, Liu G, Yu W, Dong X, Wang J (2023) Review on composite solid electrolytes for solid-state lithium-ion batteries. Mater Today Sustain 21:100316
19. Croce F, Appetecchi GB, Persi L, Scrosati B (1998) Nanocomposite polymer electrolytes for lithium batteries. Nature 394:456–458
20. Zhang X, Liu T, Zhang S, Huang X, Xu B, Lin Y, Xu B, Li L, Nan C-W, Shen Y (2017) Synergistic coupling between $Li_{6.75}La_3Zr_{1.75}Ta_{0.25}O_{12}$ and poly(vinylidene fluoride) induces high ionic conductivity, mechanical strength, and thermal stability of solid composite electrolytes. J Am Chem Soc 139:13779–13785
21. Zheng X, Yang T, Wei J, Wang C, Chen M (2021) Co-contribution of quenching and nanocrystallization on ionic-conductivity improvement of a composite electrolyte of polyethylene Oxide/$Li_7La_3Zr_2O_{12}$ nanofibers at 45 °C for all-solid-state Li metal batteries. J Power Sources 496:229843
22. Zheng F, Kotobuki M, Song S, Lai MO, Lu L (2018) Review on solid electrolytes for all-solid-state lithium-ion batteries. J Power Sources 389:198–213
23. Goodenough JB, Hong HY-P, Kafalas JA (1976) Fast Na^+-ion transport in skeleton structures. Mater Res Bull 11:203–220
24. Aono H, Sugimoto E, Sadaoka Y, Imanaka N, Adachi G (1991) Electrical property and sinterability of $LiTi_2(PO_4)_3$ mixed with lithium salt (Li_3PO_4 or Li_3BO_3). Solid State Ion 47:257–264
25. Kasper HM (1969) Series of rare earth garnets $Ln^{3+}{}_3M_2Li^+{}_3O_{12}$ (M=Te, W). Inorg Chem 8:1000–1002
26. Thangadurai V, Kaack H, Weppner WJF (2003) Novel fast lithium ion conduction in garnet-type $Li_5La_3M_2O_{12}$ (M = Nb, Ta). J Am Ceram Soc 86:437–440
27. Thangadurai V, Weppner W (2005) $Li_6ALa_2Ta_2O_{12}$ (A = Sr, Ba): novel garnet-like oxides for fast lithium ion conduction. Adv Funct Mater 15:107–112
28. Murugan R, Thangadurai V, Weppner W (2007) Fast lithium ion conduction in garnet-type $Li_7La_3Zr_2O_{12}$. Angew Chem Int Ed 46:7778–7781
29. Buannic L, Orayech B, López Del Amo J-M, Carrasco J, Katcho NA, Aguesse F, Manalastas W, Zhang W, Kilner J, Llordés A (2017) Dual substitution strategy to enhance Li^+ ionic conductivity in $Li_7La_3Zr_2O_{12}$ solid electrolyte. Chem Mater 29:1769–1778
30. Wu J-F, Pang WK, Peterson VK, Wei L, Guo X (2017) Garnet-type fast Li-ion conductors with high ionic conductivities for all-solid-state batteries. ACS Appl Mater Interfaces 9:12461–12468
31. Inaguma Y, Liquan C, Itoh M, Nakamura T, Uchida T, Ikuta H, Wakihara M (1993) High ionic conductivity in lithium lanthanum titanate. Solid State Commun 86:689–693
32. Chung H (1998) Dependence of the lithium ionic conductivity on the B-site ion substitution in $(Li_{0.5}La_{0.5})Ti_{1-x}M_xO_3$ (M=Sn, Zr, Mn, Ge). Solid State Ion 107:153–160

33. Chen C (2004) Stable lithium-ion conducting perovskite lithium–strontium–tantalum–zirconium–oxide system. Solid State Ion 167:263–272
34. Hong HY-P (1978) Crystal structure and ionic conductivity of $Li_{14}Zn(GeO_4)_4$ and other new Li^+ superionic conductors. Mater Res Bull 13:117–124
35. Kuwano J, West AR (1980) New Li^+ ion conductors in the system, Li_4GeO_4-Li_3VO_4. Mater Res Bull 15:1661–1667
36. Hu Y-W, Raistrick ID, Huggins RA (1977) Ionic conductivity of lithium orthosilicate—lithium phosphate solid solutions. J Electrochem Soc 124:1240–1242
37. Kamaya N, Homma K, Yamakawa Y, Hirayama M, Kanno R, Yonemura M, Kamiyama T, Kato Y, Hama S, Kawamoto K, Mitsui A (2011) A lithium superionic conductor. Nat Mater 10:682–686
38. Suzuki N, Inaba T, Shiga T (2012) Electrochemical properties of LiPON films made from a mixed powder target of Li_3PO_4 and Li_2O. Thin Solid Films 520:1821–1825
39. Huggins RA (1977) Recent results on lithium ion conductors. Electrochim Acta 22:773–781
40. Kanno R, Hata T, Kawamoto Y, Irie M (2000) Synthesis of a new lithium ionic conductor, thio-LISICON–lithium germanium sulfide system. Solid State Ion 130:97–104
41. Sun Y, Suzuki K, Hara K, Hori S, Yano T, Hara M, Hirayama M, Kanno R (2016) Oxygen substitution effects in $Li_{10}GeP_2S_{12}$ solid electrolyte. J Power Sources 324:798–803
42. Kato Y, Hori S, Saito T, Suzuki K, Hirayama M, Mitsui A, Yonemura M, Iba H, Kanno R (2016) High-power all-solid-state batteries using sulfide superionic conductors. Nat Energy 1:16030
43. Mizuno F, Hayashi A, Tadanaga K, Tatsumisago M (2005) New, highly ion-conductive crystals precipitated from Li_2S-P_2S_5 Glasses. Adv Mater 17:918–921
44. Deiseroth H, Kong S, Eckert H, Vannahme J, Reiner C, Zaiß T, Schlosser M (2008) Li_6PS_5X: a class of crystalline Li-rich solids with an unusually high Li^+ mobility. Angew Chem Int Ed 47:755–758
45. Schneider H, Du H, Kelley T, Leitner K, ter Maat J, Scordilis-Kelley C, Sanchez-Carrera R, Kovalev I, Mudalige A, Kulisch J, Safont-Sempere MM, Hartmann P, Weiβ T, Schneider L, Hinrichsen B (2017) A novel class of halogen-free, super-conductive lithium argyrodites: synthesis and characterization. J Power Sources 366:151–160
46. Chen M, Rao RP, Adams S (2014) High capacity all-solid-state Cu–Li_2S/Li_6PS_5Br/In batteries. Solid State Ion 262:183–187
47. Chen M, Adams S (2015) High performance all-solid-state lithium/sulfur batteries using lithium argyrodite electrolyte. J Solid State Electrochem 19:697–702
48. Zhao Y, Daemen LL (2012) Superionic conductivity in lithium-rich anti-perovskites. J Am Chem Soc 134:15042–15047
49. Zhang Y, Zhao Y, Chen C (2013) Ab initio study of the stabilities of and mechanism of superionic transport in lithium-rich antiperovskites. Phys Rev B 87:134303
50. Fang H, Jena P (2017) Li-rich antiperovskite superionic conductors based on cluster ions. Proc Natl Acad Sci 114:11046–11051
51. Lü X, Howard JW, Chen A, Zhu J, Li S, Wu G, Dowden P, Xu H, Zhao Y, Jia Q (2016) Antiperovskite Li_3OCl superionic conductor films for solid-state Li-ion batteries. Adv Sci 3:1500359
52. Muramatsu H, Hayashi A, Ohtomo T, Hama S, Tatsumisago M (2011) Structural change of Li_2S–P_2S_5 sulfide solid electrolytes in the atmosphere. Solid State Ion 182:116–119
53. Sahu G, Lin Z, Li J, Liu Z, Dudney N, Liang C (2014) Air-stable, high-conduction solid electrolytes of arsenic-substituted Li_4SnS_4. Energy Environ Sci 7:1053–1058
54. Cheng L, Crumlin EJ, Chen W, Qiao R, Hou H, Franz Lux S, Zorba V, Russo R, Kostecki R, Liu Z, Persson K, Yang W, Cabana J, Richardson T, Chen G, Doeff M (2014) The origin of high electrolyte–electrode interfacial resistances in lithium cells containing garnet type solid electrolytes. Phys Chem Chem Phys 16:18294–18300
55. Sharafi A, Kazyak E, Davis AL, Yu S, Thompson T, Siegel DJ, Dasgupta NP, Sakamoto J (2017) Surface chemistry mechanism of ultra-low interfacial resistance in the solid-state electrolyte $Li_7La_3Zr_2O_{12}$. Chem Mater 29:7961–7968

56. Xia W, Xu B, Duan H, Tang X, Guo Y, Kang H, Li H, Liu H (2017) Reaction mechanisms of lithium garnet pellets in ambient air: the effect of humidity and CO_2. J Am Ceram Soc 100:2832–2839
57. Durán T, Climent-Pascual E, Pérez-Prior MT, Levenfeld B, Varez A, Sobrados I, Sanz J (2017) Aqueous and non-aqueous Li^+/H^+ ion exchange in $Li_{0.44}La_{0.52}TiO_3$ perovskite. Adv Powder Technol 28:514–520
58. Harding JR, Amanchukwu CV, Hammond PT, Shao-Horn Y (2015) Instability of poly(ethylene oxide) upon oxidation in lithium-air batteries. J Phys Chem C 119:6947–6955
59. Wen-Yuan L, Chi-Lin L, Zheng-Wen F (2006) Stability of lithium phosphorous oxynitride thin films in humid air. Acta Phys Chim Sin 22:1413–1418
60. Hasegawa S, Imanishi N, Zhang T, Xie J, Hirano A, Takeda Y, Yamamoto O (2009) Study on lithium/air secondary batteries—stability of NASICON-type lithium ion conducting glass–ceramics with water. J Power Sources 189:371–377
61. Dashjav E, Ma Q, Xu Q, Tsai C-L, Giarola M, Mariotto G, Tietz F (2018) The influence of water on the electrical conductivity of aluminum-substituted lithium titanium phosphates. Solid State Ion 321:83–90
62. Zhang T, Imanishi N, Shimonishi Y, Hirano A, Takeda Y, Yamamoto O, Sammes N (2010) A novel high energy density rechargeable lithium/air battery. Chem Commun 46:1661
63. Kim KH, Iriyama Y, Yamamoto K, Kumazaki S, Asaka T, Tanabe K, Fisher CAJ, Hirayama T, Murugan R, Ogumi Z (2011) Characterization of the interface between $LiCoO_2$ and $Li_7La_3Zr_2O_{12}$ in an all-solid-state rechargeable lithium battery. J Power Sources 196:764–767
64. Sakuda A, Hayashi A, Tatsumisago M (2010) Interfacial observation between $LiCoO_2$ electrode and $Li_2S-P_2S_5$ solid electrolytes of all-solid-state lithium secondary batteries using transmission electron microscopy. Chem Mater 22:949–956
65. Haruyama J, Sodeyama K, Han L, Takada K, Tateyama Y (2014) Space-charge layer effect at interface between oxide cathode and sulfide electrolyte in all-solid-state lithium-ion battery. Chem Mater 26:4248–4255
66. Yamamoto K, Iriyama Y, Asaka T, Hirayama T, Fujita H, Fisher CAJ, Nonaka K, Sugita Y, Ogumi Z (2010) Dynamic visualization of the electric potential in an all-solid-state rechargeable lithium battery. Angew Chem 122:4516–4519
67. Hartmann P, Leichtweiss T, Busche MR, Schneider M, Reich M, Sann J, Adelhelm P, Janek J (2013) Degradation of NASICON-type materials in contact with lithium metal: formation of mixed conducting interphases (MCI) on solid electrolytes. J Phys Chem C 117:21064–21074
68. Rettenwander D, Wagner R, Reyer A, Bonta M, Cheng L, Doeff MM, Limbeck A, Wilkening M, Amthauer G (2018) Interface instability of Fe-stabilized $Li_7La_3Zr_2O_{12}$ versus Li metal. J Phys Chem C 122:3780–3785
69. Li Y, Xu B, Xu H, Duan H, Lü X, Xin S, Zhou W, Xue L, Fu G, Manthiram A, Goodenough JB (2017) Hybrid polymer/garnet electrolyte with a small interfacial resistance for lithium-ion batteries. Angew Chem Int Ed 56:753–756
70. Ohtomo T, Hayashi A, Tatsumisago M, Kawamoto K (2013) Glass electrolytes with high ion conductivity and high chemical stability in the system $LiI-Li_2O-Li_2S-P_2S_5$. Electrochemistry 81:428–431
71. Chen S, Xie D, Liu G, Mwizerwa JP, Zhang Q, Zhao Y, Xu X, Yao X (2018) Sulfide solid electrolytes for all-solid-state lithium batteries: Structure, conductivity, stability and application. Energy Storage Mater 14:58–74
72. Li Y, Xu H, Chien P, Wu N, Xin S, Xue L, Park K, Hu Y, Goodenough JB (2018) A perovskite electrolyte that is stable in moist air for lithium-ion batteries. Angew Chem Int Ed 57:8587–8591
73. Zhang Z, Zhang J, Sun Y, Jia H, Peng L, Zhang Y, Xie J (2020) $Li_{4-x}Sb_xSn_{1-x}S_4$ solid solutions for air-stable solid electrolytes. J Energy Chem 41:171–176
74. Peled E, Golodnitsky D, Ardel G (1997) Advanced model for solid electrolyte interphase electrodes in liquid and polymer electrolytes. J Electrochem Soc 144:L208–L210
75. Wu J-F, Pu B-W, Wang D, Shi S-Q, Zhao N, Guo X, Guo X (2019) In situ formed shields enabling Li_2CO_3-free solid electrolytes: a new route to uncover the intrinsic lithiophilicity of

garnet electrolytes for dendrite-free Li-metal batteries. ACS Appl Mater Interfaces 11:898–905
76. Huo H, Chen Y, Zhao N, Lin X, Luo J, Yang X, Liu Y, Guo X, Sun X (2019) In-situ formed Li_2CO_3-free garnet/Li interface by rapid acid treatment for dendrite-free solid-state batteries. Nano Energy 61:119–125
77. Han F, Westover AS, Yue J, Fan X, Wang F, Chi M, Leonard DN, Dudney NJ, Wang H, Wang C (2019) High electronic conductivity as the origin of lithium dendrite formation within solid electrolytes. Nat Energy 4:187–196
78. Koerver R, Aygün I, Leichtweiß T, Dietrich C, Zhang W, Binder JO, Hartmann P, Zeier WG, Janek J (2017) Capacity fade in solid-state batteries: interphase formation and chemomechanical processes in Nickel-Rich layered oxide cathodes and lithium thiophosphate solid electrolytes. Chem Mater 29:5574–5582
79. Wu B, Wang S, Lochala J, Desrochers D, Liu B, Zhang W, Yang J, Xiao J (2018) The role of the solid electrolyte interphase layer in preventing Li dendrite growth in solid-state batteries. Energy Environ Sci 11:1803–1810
80. Tian H-K, Liu Z, Ji Y, Chen L-Q, Qi Y (2019) Interfacial electronic properties dictate Li dendrite growth in solid electrolytes. Chem Mater 31:7351–7359
81. Nishikawa K, Mori T, Nishida T, Fukunaka Y, Rosso M, Homma T (2010) In situ observation of dendrite growth of electrodeposited Li metal. J Electrochem Soc 157:A1212
82. Tantratian K, Cao D, Abdelaziz A, Sun X, Sheng J, Natan A, Chen L, Zhu H (2020) Stable Li metal anode enabled by space confinement and uniform curvature through lithiophilic nanotube arrays. Adv Energy Mater 10:1902819
83. Tsai C-L, Roddatis V, Chandran CV, Ma Q, Uhlenbruck S, Bram M, Heitjans P, Guillon O (2016) $Li_7La_3Zr_2O_{12}$ interface modification for Li dendrite prevention. ACS Appl Mater Interfaces 8:10617–10626
84. Suzuki Y, Kami K, Watanabe K, Watanabe A, Saito N, Ohnishi T, Takada K, Sudo R, Imanishi N (2015) Transparent cubic garnet-type solid electrolyte of Al_2O_3-doped $Li_7La_3Zr_2O_{12}$. Solid State Ion 278:172–176
85. Ishiguro K, Nemori H, Sunahiro S, Nakata Y, Sudo R, Matsui M, Takeda Y, Yamamoto O, Imanishi N (2014) Ta-doped $Li_7La_3Zr_2O_{12}$ for water-stable lithium electrode of lithium-air batteries. J Electrochem Soc 161:A668–A674
86. Cheng L, Chen W, Kunz M, Persson K, Tamura N, Chen G, Doeff M (2015) Effect of surface microstructure on electrochemical performance of garnet solid electrolytes. ACS Appl Mater Interfaces 7:2073–2081
87. Tian H-K, Xu B, Qi Y (2018) Computational study of lithium nucleation tendency in $Li_7La_3Zr_2O_{12}$ (LLZO) and rational design of interlayer materials to prevent lithium dendrites. J Power Sources 392:79–86
88. Rice RW, Wu CC, Boichelt F (1994) Hardness–grain-size relations in ceramics. J Am Ceram Soc 77:2539–2553
89. Li Q, Yang D, Chen H, Lv X, Jiang Y, Feng Y, Rui X, Yu Y (2021) Advances in metal phosphides for sodium-ion batteries. SusMat 1:359–392
90. Chen Y, Wang Z, Li X, Yao X, Wang C, Li Y, Xue W, Yu D, Kim SY, Yang F, Kushima A, Zhang G, Huang H, Wu N, Mai Y-W, Goodenough JB, Li J (2020) Li metal deposition and stripping in a solid-state battery via Coble creep. Nature 578:251–255
91. Jackman SD, Cutler RA (2012) Effect of microcracking on ionic conductivity in LATP. J Power Sources 218:65–72
92. McGrogan FP, Swamy T, Bishop SR, Eggleton E, Porz L, Chen X, Chiang Y, Van Vliet KJ (2017) Compliant yet brittle mechanical behavior of Li_2S-P_2S_5 lithium-ion-conducting solid electrolyte. Adv Energy Mater 7:12
93. Chung H, Kang B (2017) Mechanical and thermal failure induced by contact between a $Li_{1.5}Al_{0.5}Ge_{1.5}(PO_4)_3$ solid electrolyte and Li metal in an all solid-state Li cell. Chem Mater 29:8611–8619
94. Leite MS, Ruzmetov D, Li Z, Bendersky LA, Bartelt NC, Kolmakov A, Talin AA (2014) Insights into capacity loss mechanisms of all-solid-state Li-ion batteries with Al anodes. J Mater Chem A 2:20552–20559

95. Kobi S, Mukhopadhyay A (2018) Structural (in)stability and spontaneous cracking of Li-La-zirconate cubic garnet upon exposure to ambient atmosphere. J Eur Ceram Soc 38:4707–4718
96. Bucci G, Swamy T, Chiang Y-M, Carter WC (2017) Modeling of internal mechanical failure of all-solid-state batteries during electrochemical cycling, and implications for battery design. J Mater Chem A Mater 5:19422–19430
97. Tippens J, Miers JC, Afshar A, Lewis JA, Cortes FJQ, Qiao H, Marchese TS, Di Leo CV, Saldana C, McDowell MT (2019) Visualizing chemomechanical degradation of a solid-state battery electrolyte. ACS Energy Lett 4:1475–1483
98. Moradi Z, Lanjan A, Tyagi R, Srinivasan S (2023) Review on current state, challenges, and potential solutions in solid-state batteries research. J Energy Storage 73:109048
99. Piper DM, Yersak TA, Lee S-H (2013) Effect of compressive stress on electrochemical performance of silicon anodes. J Electrochem Soc 160:A77–A81
100. Li J, Jia H, Li H, Zhao X, Sun G, Chang Z, Li L, Jin M, Wang Z, Feng M (2021) Thin buffer layer assist carbon-modifying separator for long-life lithium metal anodes. J Energy Chem 57:61–68
101. Lee H, Ren X, Niu C, Yu L, Engelhard MH, Cho I, Ryou M, Jin HS, Kim H, Liu J, Xu W, Zhang J (2017) Suppressing lithium dendrite growth by metallic coating on a separator. Adv Funct Mater 27:1704391
102. Lanjan A, Moradi Z, Srinivasan S (2021) Multiscale investigation of the diffusion mechanism within the solid-electrolyte interface layer: coupling quantum mechanics, molecular dynamics, and macroscale mathematical modeling. ACS Appl Mater Interfaces 13:42220–42229
103. Ito Y, Otoyama M, Hayashi A, Ohtomo T, Tatsumisago M (2017) Electrochemical and structural evaluation for bulk-type all-solid-state batteries using Li_4GeS_4-Li_3PS_4 electrolyte coating on $LiCoO_2$ particles. J Power Sources 360:328–335
104. Jeon B-S, Lee S-M (2009) Fabrication and electrochemical characteristics of crack-resistant Si-based anode materials for all-solid-state thin-film batteries. Electron Mater Lett 5:13–17
105. Hupfer T, Bucharsky EC, Schell KG, Senyshyn A, Monchak M, Hoffmann MJ, Ehrenberg H (2016) Evolution of microstructure and its relation to ionic conductivity in $Li_{1+x}Al_xTi_{2-x}(PO_4)_3$. Solid State Ion 288:235–239
106. Hao F, Mukherjee PP (2018) Mesoscale analysis of the electrolyte-electrode interface in all-solid-state Li-ion batteries. J Electrochem Soc 165:A1857–A1864
107. Tian H-K, Qi Y (2017) Simulation of the effect of contact area loss in all-solid-state Li-ion batteries. J Electrochem Soc 164:E3512–E3521
108. Perea A, Dontigny M, Zaghib K (2017) Safety of solid-state Li metal battery: solid polymer versus liquid electrolyte. J Power Sources 359:182–185
109. Miara L, Windmüller A, Tsai C-L, Richards WD, Ma Q, Uhlenbruck S, Guillon O, Ceder G (2016) About the compatibility between high voltage spinel cathode materials and solid oxide electrolytes as a function of temperature. ACS Appl Mater Interfaces 8:26842–26850
110. Inoue T, Mukai K (2017) Are all-solid-state lithium-ion batteries really safe?–verification by differential scanning calorimetry with an all-inclusive microcell. ACS Appl Mater Interfaces 9:1507–1515
111. Rangasamy E, Liu Z, Gobet M, Pilar K, Sahu G, Zhou W, Wu H, Greenbaum S, Liang C (2015) An iodide-based $Li_7P_2S_8I$ superionic conductor. J Am Chem Soc 137:1384–1387
112. Liu Z, Fu W, Payzant EA, Yu X, Wu Z, Dudney NJ, Kiggans J, Hong K, Rondinone AJ, Liang C (2013) Anomalous high ionic conductivity of nanoporous β-Li_3PS_4. J Am Chem Soc 135:975–978
113. Zhao W, Wang H, Dong Q, Shao H, Zhang Y, Tang Y, Shen Y, Chen L (2025) Mechanical stable composite electrolyte for solid-state lithium metal batteries. Chem Eng J 505:159662
114. Shi J, Gui F, Huang K, Zhou X, Li X, Yang L, Huang J, Wang G, Xu G (2025) Magnetic field-assisted vertically aligned $NiFe_2O_4$ nanosheets in composite solid polymer electrolytes for advanced all solid-state lithium metal batteries. J Colloid Interface Sci 678:583–592
115. Hu Y, Liu Y, Lu Y, Zhang Z, Liu S, He F, Liu Y, Chen Y, Liu F (2025) Conductive Li_2S-$NbSe_2$ cathode material capable of bidirectional self-activation for high-performance all-solid-state lithium metal batteries. Adv Funct Mater 35:2412070

116. Zhou X, Jiang M, Duan Y, Jia Z, Yuan C, Feng K, Fu Q, Zhang L, Yang X, Li X (2025) Multi-electron transfer halide cathode materials based on intercalation-conversion reaction towards all-solid-state lithium batteries. Angew Chem Int Ed 64:e202416635
117. He Y, Chen S, Nie L, Sun Z, Wu X, Liu W (2020) Stereolithography three-dimensional printing solid polymer electrolytes for all-solid-state lithium metal batteries. Nano Lett 20:7136–7143
118. Sakuu to become publicly traded company Via Business Com bination with plum acquisition corp. I. Business Wire. (2023, March 2). Retrieved April 12, 2023, from https://www.businesswire.com/news/home/20230302005883/en/
119. Zhu Y, He X, Mo Y (2015) Origin of outstanding stability in the lithium solid electrolyte materials: insights from thermodynamic analyses based on first-principles calculations. ACS Appl Mater Interfaces 7:23685–23693
120. Zhu Y, Connell JG, Tepavcevic S, Zapol P, Garcia-Mendez R, Taylor NJ, Sakamoto J, Ingram BJ, Curtiss LA, Freeland JW, Fong DD, Markovic NM (2019) Dopant-dependent stability of garnet solid electrolyte interfaces with lithium metal. Adv Energy Mater 9:1803440
121. Xiao Y, Xu R, Yan C, Liang Y, Ding J-F, Huang J-Q (2020) Waterproof lithium metal anode enabled by cross-linking encapsulation. Sci Bull (Beijing) 65:909–916
122. Sun C, Lin A, Li W, Jin J, Sun Y, Yang J, Wen Z (2020) In situ conversion of Cu_3P nanowires to mixed ion/electron-conducting skeleton for homogeneous lithium deposition. Adv Energy Mater 10:1902989
123. Zhu Y, He X, Mo Y (2016) First principles study on electrochemical and chemical stability of solid electrolyte–electrode interfaces in all-solid-state Li-ion batteries. J Mater Chem A Mater 4:3253–3266
124. Lee Y-G, Fujiki S, Jung C, Suzuki N, Yashiro N, Omoda R, Ko D-S, Shiratsuchi T, Sugimoto T, Ryu S, Ku JH, Watanabe T, Park Y, Aihara Y, Im D, Han IT (2020) High-energy long-cycling all-solid-state lithium metal batteries enabled by silver–carbon composite anodes. Nat Energy 5:299–308
125. Park SW, Choi HJ, Yoo Y, Lim H, Park J, Lee Y, Ha Y, Lee S, Kim BG (2022) Stable cycling of all-solid-state batteries with sacrificial cathode and lithium-free indium layer. Adv Funct Mater 32:2108203
126. Na I, Kim H, Kunze S, Nam C, Jo S, Choi H, Oh S, Choi E, Song YB, Jung YS, Lee YS, Lim J (2023) Monolithic 100% silicon wafer anode for all-solid-state batteries achieving high areal capacity at room temperature. ACS Energy Lett 8:1936–1943

Safety Aspects of Solid-State Batteries

Vivek Mishra and Apoorva Singh

Abstract This chapter delves into the safety aspects of solid-state batteries (SSBs), highlighting their potential to revolutionize energy storage while addressing critical safety concerns. As the demand for higher energy density and improved safety in battery technologies escalates, SSBs present a promising alternative to traditional lithium-ion batteries. The chapter begins by examining the fundamental structure and materials of solid-state batteries, focusing on how these elements contribute to enhanced safety profiles, including reduced flammability and thermal stability. We explore common failure modes, such as dendrite formation and electrolyte degradation, and discuss the innovative strategies employed to mitigate these risks. Additionally, the chapter explores recent advancements in solid-state electrolyte formulations and protective coatings that enhance mechanical integrity and ionic conductivity. By assessing both laboratory studies and real-world applications, we provide a comprehensive overview of the safety implications of SSB technology, ultimately concluding with insights into future research directions aimed at ensuring the safe adoption of solid-state batteries in consumer electronics, electric vehicles, and large-scale energy storage systems.

Keywords Li-ion batteries · Solid-state batteries · Electrolyte · Anode · Cathode · Dendrite formation

1 Introduction

Solid-state batteries (SSBs) represent a revolutionary advancement in energy storage technology, offering significant improvements in **safety**, **energy density**, and **longevity** over traditional lithium-ion battery (LIB) systems. Unlike conventional LIBs, which utilize flammable liquid electrolytes that can leak, ignite, or degrade over

V. Mishra (✉) · A. Singh
Amity Institute of Click Chemistry Research and Studies (AICCRS), Amity University Uttar Pradesh, Noida 201313, Uttar Pradesh, India
e-mail: vmishra@amity.edu

A. K. Rai et al. (eds.), *Electrochemistry of Next-Generation Solid-State Batteries*, Green Energy and Technology, https://doi.org/10.1007/978-981-95-7603-6_10

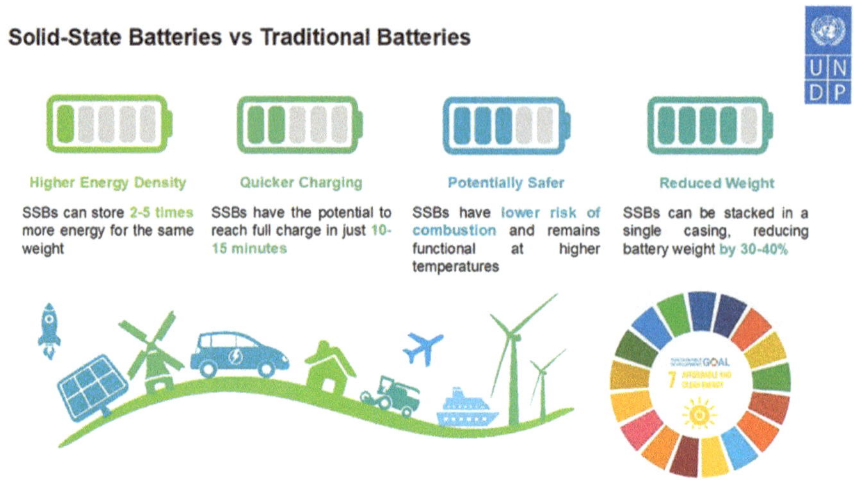

Fig. 1 Representative image of solid-state batteries versus traditional batteries

time, SSBs employ solid electrolytes that mitigate many of these risks. This fundamental shift in battery architecture contributes to enhanced thermal stability, reduced flammability, and increased durability, making SSBs a compelling solution across a broad spectrum of applications, including consumer electronics, electric vehicles (EVs), and grid-scale energy storage systems [1]. The global push for decarbonization and greener energy systems underscores the importance of advancing safer and more efficient battery technologies. LIBs, despite their commercial dominance, suffer from critical drawbacks such as **thermal runaway**, **dendritic lithium formation**, **limited cycle life**, and **restricted energy capacity** [2]. These challenges hinder their adoption in high-performance or mission-critical applications. In contrast, SSBs overcome many of these limitations by delivering superior safety profiles, greater energy densities, and longer lifespans, positioning them as a pivotal technology in the global transition to sustainable energy infrastructure (Fig. 1).

1.1 Background and Importance of Solid-State Batteries

The concept of solid-state batteries has existed for decades, with early research emerging in the **mid-twentieth century**. Initial investigations focused on incorporating solid electrolytes into electrochemical cells, but these efforts were constrained by low ionic conductivity and poor electrode-electrolyte interface stability. Due to these challenges and high manufacturing costs, early SSB prototypes remained limited to niche applications. However, breakthroughs in materials science, interface engineering, and nanotechnology have since catalyzed the maturation of SSBs [3]. Notable materials discoveries, such as **lithium phosphorus oxynitride**

(LiPON), **garnet-type oxides** like **$Li_7La_3Zr_2O_{12}$ (LLZO)**, and **sulfide-based electrolytes** like **$Li_{10}GeP_2S_{12}$ (LGPS)**, have significantly improved the ionic conductivity and mechanical robustness of solid electrolytes. Garnet-type oxides offer high conductivity and air-stability, while sulfide-based electrolytes provide ultra-high conductivities but are sensitive to moisture and oxygen.

Composite electrolyte systems and hybrid interfaces have further enhanced the functionality of SSBs by addressing mechanical rigidity and ion transport limitations. From a safety standpoint, the use of non-volatile solid electrolytes drastically reduces the risk of electrolyte leakage and fire. The elimination of flammable solvents allows SSBs to operate safely over a broader temperature range, and the solid interface acts as a physical barrier to suppress lithium dendrite propagation. This enables the safe use of lithium metal anodes, which offer much higher theoretical capacities than the graphite anodes used in LIBs. Consequently, SSBs can provide significantly higher energy densities and longer operational lifetimes. Beyond energy and safety performance, SSBs show exceptional promise in real-world applications [4]. In automotive sectors, SSBs could lead to lighter, safer electric vehicles with extended driving ranges and rapid charging capabilities. In consumer electronics, they enable compact, high-capacity batteries with minimal overheating risks. Moreover, stationary storage systems using SSBs can play a critical role in balancing energy loads and stabilizing power grids in the face of variable renewable generation. Environmental sustainability is another cornerstone benefit of SSB technology. By minimizing fire risks and chemical leakage, SSBs improve environmental safety during both usage and disposal phases. Their longer service life reduces the frequency of replacements, thus cutting down on material consumption and waste generation. Additionally, researchers are working to incorporate non-toxic, abundant materials, and develop sustainable supply chains and recycling frameworks to ensure the ecological integrity of SSB production and end-of-life management. In conclusion, SSBs are more than an incremental improvement over lithium-ion systems—they represent a paradigm shift in how we think about energy storage. Through their enhanced safety, superior energy capacity, and environmental advantages, SSBs are poised to become a foundational technology in the clean energy transition, transforming everything from electric mobility to renewable energy integration.

1.2 Comparison with Conventional Lithium-Ion Batteries

When comparing solid-state batteries with traditional lithium-ion batteries (LIBs), several stark contrasts emerge, particularly in terms of safety, performance, and structural stability. Conventional LIBs utilize liquid electrolytes, typically composed of organic solvents like ethylene carbonate, which are highly volatile and flammable. These electrolytes, while effective in conducting ions, contribute to several critical safety risks. Under conditions such as overcharging, overheating, or mechanical damage, the liquid electrolyte can decompose exothermically, releasing gases that may lead to swelling, rupture, or even explosion. This scenario, known as thermal runaway, is one of the most severe hazards associated with LIBs. In addition, liquid electrolytes are prone to leakage and evaporation over time, which can cause performance degradation and pose environmental hazards. On the other hand, SSBs use solid electrolytes that are chemically stable, non-volatile, and non-flammable [5] (Fig. 2).

This significantly reduces the risk of fire even under abusive conditions such as high temperature or electrical stress. Another fundamental difference lies in the electrochemical performance: solid-state designs allow the use of lithium metal anodes, which are highly desirable for their energy density but incompatible with liquid electrolytes due to dendrite formation. Solid electrolytes act as physical barriers that inhibit dendrite penetration, enabling the safe use of lithium metal. Furthermore, SSBs tend to have higher thermal and electrochemical stability windows, allowing them to operate safely across a broader range of temperatures and voltages. However, these advantages do come with engineering challenges. Solid electrolytes often exhibit lower ionic conductivity compared to their liquid counterparts and require precise material integration and interface engineering. Additionally, the mechanical brittleness of certain ceramic electrolytes poses design challenges in maintaining structural integrity under stress. Despite these obstacles, the net benefit in terms of safety, energy density, and lifecycle performance positions SSBs as a significantly superior alternative to conventional LIBs, particularly for high-demand applications such as electric mobility and grid storage [6].

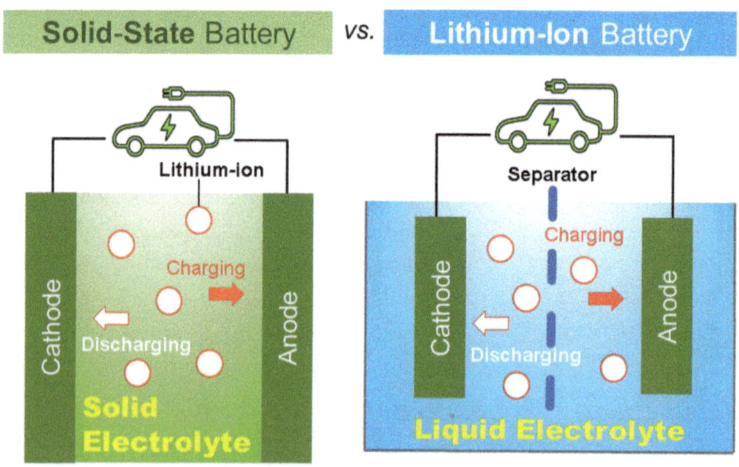

Fig. 2 Comparison of solid-state battery versus lithium-ion battery

1.3 Scope and Objectives of the Chapter

This chapter aims to provide an exhaustive overview of the safety aspects of solid-state batteries, offering both a technical foundation and a practical outlook on their implementation. The primary objective is to elucidate how the unique structural and material characteristics of SSBs contribute to their enhanced safety profile, distinguishing them from conventional lithium-ion batteries. To achieve this, the chapter begins with a comprehensive examination of the fundamental architecture of SSBs, focusing on the role and selection of solid electrolytes, electrodes, and interface materials. Understanding these components and their interactions is crucial for appreciating the inherent safety benefits and the challenges that remain. Next, the discussion moves to a detailed analysis of specific safety advantages, including reduced flammability, resistance to thermal runaway, and mechanical robustness [7]. These qualities are essential for applications where reliability and failure tolerance are non-negotiable. However, SSBs are not without their own vulnerabilities. The chapter will also explore common failure modes such as dendrite formation, interface degradation, and mechanical fracture, all of which can compromise battery safety and performance if not properly managed. To address these issues, the chapter outlines a range of mitigation strategies and technological innovations, including advanced electrolyte formulations, protective coatings, interface engineering, and intelligent battery management systems. The inclusion of real-world case studies will further illustrate how SSB safety features perform under practical conditions in various sectors including consumer electronics, electric vehicles, and large-scale energy storage. Finally, the chapter concludes with a forward-looking perspective on future research directions, commercial scalability, and regulatory frameworks necessary for the widespread adoption of SSBs. Through this comprehensive approach, the chapter seeks to equip researchers, engineers, and stakeholders with a nuanced understanding of both the potential and the pitfalls of this promising battery technology [8].

2 Fundamental Structure and Materials of Solid-State Batteries

2.1 Key Components: Solid Electrolytes, Electrodes, and Interfaces

The fundamental architecture of a solid-state battery comprises three main components: the solid electrolyte, the electrodes (anode and cathode), and the interfaces between them. Each of these components plays a vital role in determining the battery's overall performance, safety, and longevity [9]. The solid electrolyte is perhaps the most defining feature of SSBs. It serves as the ion-conducting medium that enables lithium ions to shuttle between the anode and cathode during charge and discharge cycles [3].

Unlike liquid electrolytes, solid electrolytes can be fabricated from ceramics, glassy materials, sulfides, or polymers, each with distinct physical and electrochemical characteristics. The selection of the electrolyte material significantly influences the ionic conductivity, electrochemical stability, and mechanical strength of the battery. On either side of the electrolyte are the electrodes. The anode in SSBs often comprises lithium metal, owing to its high theoretical capacity and low electrochemical potential, while the cathode typically consists of layered oxides, phosphates, or sulfides. However, the interface between the solid electrolyte and electrodes remains a critical area of concern. Inadequate contact between surfaces, mismatched thermal expansion, and chemical incompatibility can lead to increased resistance, localized heating, and safety issues such as dendrite penetration or delamination [10]. Effective engineering of these interfaces is therefore essential to ensure stable and safe battery operation.

2.2 Material Selection and Its Impact on Safety

In recent years, polymer-based solid electrolytes have gained attention due to their flexibility and processability, offering a balance between mechanical robustness and ionic transport. Moreover, the incorporation of nanomaterials and catalytic additives has been explored to enhance ion conductivity and interfacial stability, aligning with the broader goals of green and sustainable materials development [11–20]. Material selection is a cornerstone in the design of safe and efficient solid-state batteries. The safety profile of a solid-state battery is intrinsically tied to the physical, chemical, and electrochemical properties of its constituent materials. For instance, ceramic electrolytes such as garnet-type lithium lanthanum zirconate (LLZO) and NASICON-type structures [21] are highly stable and non-flammable, offering excellent thermal and electrochemical stability.

However, their brittleness and high interfacial resistance pose challenges for mechanical integrity and ionic transport. On the other hand, sulfide-based electrolytes such as $Li_{10}GeP_2S_{12}$ (LGPS) boast high ionic conductivity and softness, which facilitate better interfacial contact, but they are highly sensitive to moisture and can release toxic hydrogen sulfide gas upon exposure to air. Polymer electrolytes like polyethylene oxide (PEO) offer flexibility and ease of processing, but they suffer from low room-temperature ionic conductivity and limited electrochemical stability. The electrodes must also be carefully chosen; lithium metal anodes, while highly desirable for their energy density, can lead to dendrite growth if not properly managed. Cathode materials must be compatible with the electrolyte to avoid side reactions and phase changes that could compromise safety [22]. Consequently, the interplay between these materials—solid electrolyte, anode, cathode, and interface layers—must be meticulously optimized to balance performance and safety. Material innovations, such as composite electrolytes and protective coatings, are actively being explored to mitigate these challenges and improve the robustness of solid-state batteries.

2.3 *Thermal and Mechanical Properties*

The thermal and mechanical properties of the materials used in solid-state batteries are pivotal to their safe and reliable operation. From a thermal standpoint, one of the primary advantages of SSBs is their enhanced resistance to thermal runaway, which is largely attributed to the use of non-volatile and non-flammable solid electrolytes. Unlike organic solvents used in liquid electrolytes, solid materials do not easily vaporize or ignite, thus significantly reducing the risk of fire or explosion under thermal abuse conditions.

The ability to operate across a wide temperature range without significant degradation also adds to the thermal resilience of SSBs. However, these benefits can be compromised if the mechanical properties of the solid-state components are not adequately addressed.

Solid electrolytes, particularly ceramics, are inherently brittle and susceptible to cracking under mechanical stress or due to volumetric changes during cycling. Such cracks can propagate over time, creating pathways for lithium dendrites or causing internal short circuits. Mechanical mismatch at the interfaces, thermal expansion discrepancies, and poor adhesion between layers can further exacerbate structural integrity issues. Therefore, the mechanical design of solid-state batteries must incorporate considerations for stress distribution, flexibility, and fracture toughness. Figure 3 shows progression of thermal runaway leading to catastrophic failure in a conventional lithium-ion battery deployed in an electric vehicle. Advanced materials such as flexible composite electrolytes, as well as architectural innovations like gradient interlayers or compliant interface layers, are being investigated to enhance the mechanical robustness of SSBs. Ultimately, a deep understanding of the thermal and mechanical behaviors of all battery components is essential for designing safe, durable, and high-performance solid-state batteries [24].

3 Safety Advantages of Solid-State Batteries

3.1 *Reduced Flammability and Thermal Runaway Risks*

One of the most compelling safety benefits of solid-state batteries lies in their significantly reduced flammability and lower risk of thermal runaway. Traditional lithium-ion batteries use volatile organic liquid electrolytes that can easily ignite when exposed to heat, overcharging, or physical damage. These electrolytes serve as fuel for fire and contribute to rapid escalation once thermal runaway begins. In

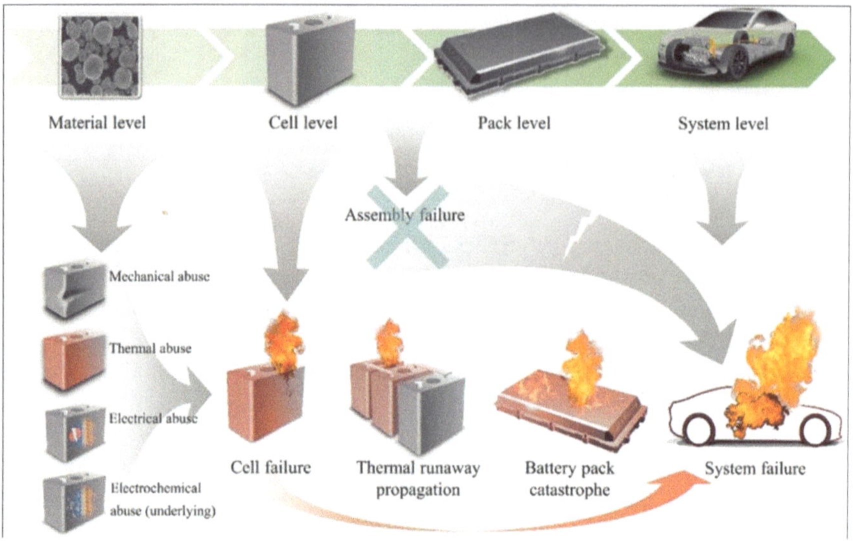

Fig. 3 Progression of thermal runaway leading to catastrophic failure in a conventional lithium-ion battery deployed in an electric vehicle. Reproduced from Ref. [23] with permission

stark contrast, solid-state batteries replace these combustible liquids with solid electrolytes, which are generally non-volatile, non-flammable, and stable over a broader range of temperatures. This critical shift in material chemistry drastically reduces the chances of internal combustion or fire even under severe abuse conditions. Furthermore, solid electrolytes do not undergo the same types of decomposition reactions as their liquid counterparts, thereby eliminating many of the exothermic reactions that can exacerbate thermal events [25]. The non-flammability of these solid components is especially vital in applications like electric vehicles, aerospace systems, and grid storage where battery safety is paramount. By removing one of the major triggers for battery fires, solid-state technology offers an inherently safer alternative, making it a preferred choice for next-generation high-energy storage systems where both performance and safety cannot be compromised.

3.2 *Enhanced Thermal Stability*

Solid-state batteries exhibit superior thermal stability compared to their liquid electrolyte-based counterparts, primarily due to the robust nature of their solid electrolytes and electrode interfaces. Unlike liquid electrolytes that have narrow thermal operating windows and tend to degrade or evaporate at elevated temperatures, solid electrolytes—particularly those made from ceramics and sulfides—can maintain their structural and electrochemical integrity over a much wider temperature range.

This enhanced thermal stability translates into safer operation under extreme environmental conditions or high-current charging scenarios [3]. In electric vehicles and industrial applications, where batteries are subjected to fluctuating thermal loads, this resilience can prevent temperature-induced failures or performance degradation. Moreover, the stability of solid-state components at elevated temperatures minimizes the risk of gas formation, swelling, or pressure buildup within the cell, which are common precursors to catastrophic failure in liquid-based systems. The thermally stable interfaces in SSBs also reduce the likelihood of heat-induced side reactions that could compromise battery performance or safety. By maintaining their operational characteristics across diverse thermal conditions, solid-state batteries provide a much-needed level of predictability and safety in applications demanding both high-energy density and reliability.

3.3 Structural Integrity and Mechanical Robustness

The architectural configuration of solid-state batteries lends itself to superior structural integrity and mechanical robustness, which are essential characteristics for ensuring long-term operational safety. Unlike conventional batteries that rely on soft and pliable liquid components, SSBs utilize rigid or semi-rigid solid materials that are less susceptible to deformation, leakage, or rupture. This enhanced mechanical structure offers multiple safety advantages, particularly in scenarios involving mechanical shock, vibration, or physical impact. Solid electrolytes, especially those based on ceramics or reinforced polymers, provide a more stable matrix that can resist compression, puncturing, or expansion-related failures. This is especially critical in electric vehicles and aerospace systems where mechanical stresses are frequent and unpredictable. Furthermore, the elimination of flammable liquids removes the risk of electrolyte leakage in the event of casing damage or rupture [4, 26]. Mechanical integrity also contributes to the prevention of internal short circuits by minimizing the movement and misalignment of battery components. Advanced solid-state designs may include flexible composite layers, graded interfaces, and shock-absorbing materials to further bolster mechanical resilience. In essence, the solid-state architecture not only enhances battery safety under normal operational conditions but also provides robust defense mechanisms against accidental or external physical stresses that could otherwise lead to hazardous failures [27].

4 Common Failure Modes and Safety

Solid-state batteries (SSBs) are often hailed as the future of safer, higher-performing energy storage devices. However, while their advantages over traditional lithium-ion batteries are clear, the failure modes they encounter are unique and complex, often linked directly to the rigid nature of their materials and their interfacial behaviors.

Understanding these failure modes is not only essential for improving battery safety but also for enabling their large-scale implementation in demanding applications such as electric vehicles and grid energy storage [28]. This section delves deep into the most critical safety concerns in SSBs, explaining how these issues arise, why they are significant, and what their broader implications are for real-world applications.

4.1 Introduction to Safety Hazards

Despite the well-documented safety advantages of solid-state batteries (SSBs) over traditional lithium-ion batteries (LIBs), several safety hazards continue to challenge their practical application and commercial scalability. These hazards are deeply rooted in the unique physicochemical interactions and mechanical behavior associated with the solid materials that constitute SSBs. The very characteristics that make these batteries advantageous—such as the use of lithium metal anodes and rigid ceramic electrolytes—also introduce complex safety risks if not properly managed. The transition from liquid to solid electrolytes brings forth a shift in failure modes, demanding a new understanding of how materials behave under electrochemical and mechanical stresses [26]. Key issues such as lithium dendrite growth, interfacial instability between electrodes and electrolytes, mechanical degradation during electrochemical cycling, and the risk of overcharging or operational failure modes must be meticulously studied and engineered to prevent catastrophic failures. Understanding these challenges is crucial not only for the optimization of battery performance but also for ensuring their safe deployment across consumer electronics, electric vehicles, and grid-scale energy storage systems [29]. This section delves into these predominant safety concerns, exploring their causes, manifestations, and the current research strategies aimed at their mitigation, providing a comprehensive overview of the multifaceted risks associated with SSBs.

4.2 Dendrite Formation

One of the most widely discussed and potentially catastrophic failure modes in solid-state batteries is lithium dendrite formation. In simple terms, dendrites are thin, tree-like structures made of metallic lithium that can grow from the anode toward the cathode during charging. Ideally, lithium ions should deposit uniformly on the anode surface. However, under high-current densities, uneven surface conditions, or interfacial voids, the deposition becomes localized, giving rise to these needle-like dendritic structures. These dendrites are particularly dangerous because they can pierce the solid electrolyte layer—something that contradicts the traditional belief that solid electrolytes are immune to dendritic penetration due to their rigidity. As a matter of fact, numerous solid electrolytes, especially sulfide and even some oxide-based ceramics, are susceptible to microcracks, grain boundary defects, or voids that facilitate dendrite growth [30].

Once a dendrite grows long enough to span the gap between the anode and cathode, it can cause an internal short circuit. This leads to a sudden surge in current flow, localized heating, and in extreme cases, thermal runaway—a condition where the battery becomes uncontrollably hot, potentially catching fire or exploding. The problem is further worsened by the fact that solid electrolytes cannot flow or self-heal like liquid electrolytes. This means that once a crack or defect forms, it becomes a permanent weak spot, allowing future dendrites to exploit the same path. Preventing dendrite formation requires an intricate balancing act involving material design, current density control, stack pressure optimization, and innovative interfacial engineering—all of which are areas of intense research and innovation [31, 32].

Unlike liquid electrolyte systems, where dendrites can be mitigated using separator layers or electrolyte additives, SSBs with solid electrolytes present a paradox: although theoretically resistant to dendrite penetration due to their mechanical strength, in practice, many solid electrolytes—especially those with relatively low shear modulus or microstructural defects—still permit dendritic growth. This is particularly true in the case of lithium metal anodes, which are favored for their high theoretical capacity and low electrochemical potential but also promote aggressive dendritic deposition under unfavorable operating conditions. Once these dendrites pierce the solid electrolyte and bridge the anode-cathode gap, they can create internal short circuits, leading to localized heating, thermal runaway, and potentially hazardous battery failure, including fire or explosion.

Multiple factors influence dendrite formation, including inhomogeneous current distribution, interfacial voids, uneven surface morphology of the lithium metal, and the inherent brittleness or porosity of certain ceramic or sulfide-based electrolytes. The challenge is exacerbated by the inability of solid electrolytes to self-heal or flow around defects, as liquid electrolytes can. Efforts to counteract dendrite growth include engineering buffer layers between the anode and electrolyte, utilizing solid electrolytes with higher shear modulus such as garnet-type oxides, applying stack pressure to maintain contact, and optimizing charging protocols to reduce overpotentials. Moreover, researchers are developing hybrid electrolyte systems that combine the mechanical stability of ceramics with the flexibility of polymers to suppress dendrite formation. Advanced in situ imaging techniques, such as X-ray tomography and neutron diffraction, alongside computational modeling, are also being employed to monitor and understand dendrite initiation and growth in real time, enabling the design of more dendrite-resistant architectures [33, 34].

4.3 *Interfacial Instability*

Interfacial instability is another major challenge that threatens the safety, efficiency, and longevity of solid-state batteries. The interfaces between the solid electrolyte and electrodes are critical regions that must permit seamless ionic transport while remaining chemically and mechanically stable under repeated cycling and a range of thermal conditions. However, in real-world scenarios, several adverse phenomena occur at these interfaces due to mismatched chemical potentials, lattice incompatibilities, differing thermal expansion coefficients, and electrochemical reactions. These issues can result in the formation of high-resistance interphases, interfacial voids, or even delamination and cracking, all of which severely hinder ion flow, promote local heating, and serve as initiation points for dendrite growth or electrical breakdown. Chemical degradation at the electrode-electrolyte interface, especially at the lithium metal-solid electrolyte boundary, can lead to the formation of passivation layers such as lithium oxide, lithium carbonate, or lithium sulfide (Fig. 4), depending on the composition of the solid electrolyte [2–4].

While some interphases may be beneficial by providing a stable solid electrolyte interphase (SEI), many are electronically conductive or mechanically weak, thereby leading to ongoing degradation, impedance rise, and safety risks. Moreover, physical contact between solid layers can deteriorate over time due to volume changes, particularly in the case of lithium metal or conversion-type cathodes, resulting in the propagation of cracks and loss of electrochemical contact. Solutions to mitigate interfacial instability involve various strategies such as applying chemically and mechanically stable interfacial coatings (e.g., $LiNbO_3$, Li_3PO_4), doping solid electrolytes to improve interfacial adhesion and reduce interfacial reactivity, and designing composite or gradient interlayers that facilitate smooth ionic transport while accommodating mechanical stress. Additionally, tailoring the microstructure of electrode and electrolyte surfaces to enhance physical contact and reduce void formation is proving to be effective. New fabrication techniques, such as atomic layer deposition (ALD), spark plasma sintering, and cold sintering, are being explored to create seamless interfaces and minimize interfacial resistance [36].

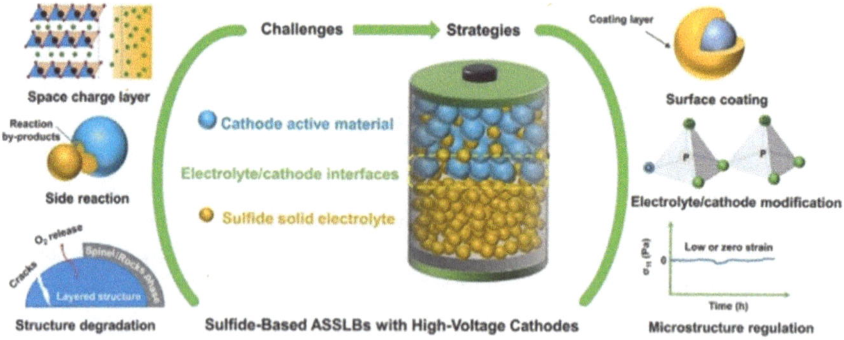

Fig. 4 Schematic diagram detailing various challenges faced by sulfide-based SSEs and solution strategies. Reproduced from Ref. [35] with permission. Copyright, Elsevier 2024

4.4 Mechanical Stress and Fracture Risks

Solid-state batteries, by their very nature, involve the use of brittle materials—particularly ceramic-based solid electrolytes—that are not as forgiving as the liquid systems they aim to replace. This introduces a new and deeply structural failure mode: mechanical stress and fracture. As the battery charges and discharges, the active materials in the electrodes expand and contract due to lithium insertion and removal. While this volume change might be modest in traditional systems, it becomes a major issue in SSBs because the solid electrolyte cannot accommodate the stress without cracking or fracturing. Even minute, repeated stresses over hundreds of cycles can gradually lead to the formation of microcracks that compromise the structural integrity of the battery [4, 37, 38]. These mechanical fractures are more than just physical defects—they can have serious electrochemical consequences. Cracks disrupt the continuity of the electrolyte, creating areas of high resistance and local current concentration, which may, in turn, lead to dendrite growth. They can also allow lithium to become trapped or to deposit unevenly, reducing battery efficiency and capacity over time. What's more, during battery manufacturing and operation, high stack pressures are often applied to ensure good contact between layers. If not properly controlled, these pressures can accelerate mechanical wear or even trigger catastrophic cracking in thin or poorly supported cells [39]. To mitigate these risks, researchers are exploring a variety of strategies, such as using flexible polymer-ceramic composites, designing compliant interlayers, or even engineering cell architectures that distribute stress more evenly. Additionally, real-time monitoring tools like acoustic emission sensors and strain gauges are being tested to detect fracture events early, allowing for predictive maintenance and improved reliability. The interplay between mechanical and electrochemical factors in SSBs is one of the most challenging and exciting areas of current battery research [40]. Here, Table 1 gives an overview of the major challenges faced by different types of solid-state electrolytes, including issues like poor ionic movement, interface instability, and sensitivity to air or moisture. It also shows how each electrolyte category comes with its own strengths and weaknesses, making it difficult to achieve the ideal balance needed for practical solid-state battery applications.

Table 1 Challenges of various solid-state electrolyte categories [40]

Polymer-based SSE	Oxide-based SSE	Sulfide-based SSE
Low mechanical strength	Mismatch interface	Chemical/electrochemical instabilities
Low ionic conductivity at R.T.	Volume change	Environmental sensitivity
Shuttle effect	Poor contact	Lithium dendrites

4.5 Overcharging, Short Circuits, and Other Operational Hazards

Even though SSBs are designed to be safer than traditional batteries, operational errors such as overcharging or high-rate cycling can still introduce serious hazards. Overcharging can push the electrolyte and electrode materials beyond their electrochemical stability limits, leading to irreversible decomposition reactions, gas formation, and thermal degradation. These reactions not only damage the battery but can also trigger runaway reactions that result in fire or explosion, especially in the presence of flammable decomposition products. Short circuits—caused by dendrites, manufacturing defects, or mechanical damage—can similarly result in sudden heating and catastrophic failure.

Moreover, SSBs are sensitive to temperature fluctuations and pressure changes. Excessive heat, either from the environment or from internal resistance, can degrade the electrolyte or cause the growth of interfacial layers. On the other hand, cold temperatures can slow ion transport dramatically, increasing the risk of lithium plating or uneven current distribution. Poor thermal management, therefore, becomes a hidden but critical risk factor. Additionally, operational hazards may also arise from the use of incompatible battery management systems (BMS), which are traditionally optimized for liquid-based lithium-ion cells. Without real-time sensing of the specific failure modes unique to SSBs, such as interfacial delamination or internal cracking, these systems might fail to detect early warning signs of failure [3, 41–43].

To combat these operational hazards, manufacturers are working to integrate next-generation battery management systems equipped with predictive analytics, temperature and pressure sensors, and self-regulating protocols. Safety features such as shutdown layers, thermal cutoffs, and current interrupters are being adapted for use in solid-state platforms. Furthermore, rigorous quality control in the production of solid-state components—ensuring consistent layer thicknesses, uniform surface morphology, and minimal defects—is essential to reducing the incidence of latent manufacturing flaws that can evolve into major safety concerns. In essence, while SSBs offer a path to safer energy storage, their operational safety hinges on an ecosystem of intelligent design, careful manufacturing, and precise control.

5 Mitigation Strategies and Innovations

5.1 Advanced Solid-State Electrolyte Formulations

Recent efforts in polymer chemistry [44–57] and nanocatalyst design have opened new avenues for tailoring solid electrolytes with higher conductivity and reduced environmental impact, reflecting the convergence of materials science and green chemistry [58–61]. Solid-state electrolytes (SSEs) are at the core of the technological shift toward safer, higher-performing batteries. They replace the liquid electrolytes used in traditional lithium-ion batteries, significantly reducing the risk of leakage, flammability, and other hazards associated with conventional batteries. However, despite the immense potential of solid-state batteries (SSBs), achieving a balance between performance and safety requires continuous innovations in solid-state electrolyte formulations. This has become a critical area of research in the battery industry.

One of the key areas of focus in advanced solid-state electrolyte formulations is the development of materials with higher ionic conductivity. The ultimate goal is to create solid electrolytes that can match or surpass the performance of liquid electrolytes, which are currently the standard in terms of energy efficiency. Researchers are exploring a wide range of materials for solid electrolytes, including ceramics, sulfides, and polymers. Ceramic-based electrolytes, like lithium lanthanum zirconate (LLZO), are highly stable and offer excellent ionic conductivity, but they face challenges in terms of processability and mechanical brittleness. On the other hand, sulfide-based electrolytes, such as lithium phosphorus oxynitride (LiPON), show great promise due to their flexibility, making them more resilient to mechanical stresses [3, 41, 62]. Researchers are also exploring composite electrolytes, which combine different materials to optimize both ionic conductivity and mechanical stability, while also addressing issues like dendrite formation that can lead to short circuits.

Another critical aspect of developing advanced SSEs is ensuring their compatibility with high-performance electrodes. Solid-state batteries often involve complex interactions between the electrolyte and electrodes, which can cause interfacial resistance and performance degradation. By developing electrolyte formulations that can form stable interfaces with electrodes, researchers can enhance the overall efficiency and lifespan of solid-state batteries [63]. Additionally, efforts are underway to improve the scalability of these advanced electrolytes, ensuring that they can be manufactured efficiently and cost-effectively to meet the demands of commercial production.

5.2 *Protective Coatings and Interface Engineering*

Protective coatings and interface engineering are pivotal in enhancing the safety and longevity of solid-state batteries. The interfaces between the solid electrolyte and the electrodes are particularly critical, as they are often the site of degradation, dendrite formation, and other failure mechanisms. Improving these interfaces through coatings and innovative engineering techniques can substantially mitigate these risks. One of the main strategies in interface engineering is the use of protective coatings on the solid electrolyte or electrodes to prevent direct contact with reactive components. For example, the use of thin, stable layers of lithium phosphate or other inorganic materials on the surface of the electrolyte can help mitigate the formation of dendrites, a common issue that can lead to short-circuiting [6, 64]. These protective coatings not only prevent dendrite growth but also provide a barrier against the diffusion of lithium or other active ions, which could lead to degradation at the interface.

Moreover, engineers are also exploring the use of interface modifiers, which can enhance the mechanical and chemical stability of the solid-state interface. By modifying the surface properties of the electrolyte, these modifiers can reduce interfacial resistance and improve the overall efficiency of ion transport. These innovations help to mitigate one of the most significant challenges in solid-state battery technology, the instability of the electrolyte-electrode interface, and contribute to improving the overall safety and reliability of solid-state batteries.

Another promising area of development in interface engineering is the use of advanced bonding techniques. Ensuring a solid, durable bond between the electrolyte and the electrodes is essential to prevent delamination or the formation of voids that could lead to performance loss or safety hazards [6, 65, 66]. Techniques such as atomic layer deposition (ALD) and sputtering are being used to deposit ultra-thin, uniform coatings that provide superior adhesion and stability at the interface. These bonding techniques also allow for precise control over the thickness and composition of the coatings, optimizing the material properties for enhanced safety and performance.

5.3 Design Innovations for Mechanical Integrity

Mechanical integrity is a crucial factor for ensuring the safety and durability of solid-state batteries. Unlike liquid-based batteries, solid-state batteries rely on rigid materials that must maintain their structural integrity over long cycles and under various external stresses, such as thermal expansion, vibrations, and physical impacts. Any failure in mechanical integrity could lead to performance degradation, short internal circuits, or even catastrophic failure, making it essential to design solid-state batteries that can withstand such conditions. One of the primary concerns regarding mechanical integrity in solid-state batteries is the risk of cracking or fracture of the solid electrolyte during the battery's operation. This can occur due to stress induced by thermal cycling or volume changes during charge and discharge cycles. To mitigate these risks, researchers are exploring novel electrolyte materials that are more resilient to mechanical stress. For instance, materials like glassy lithium sulfide or nanocomposite electrolytes are being developed to improve flexibility while maintaining high ionic conductivity [6, 26]. Additionally, advanced fabrication techniques such as 3D printing or ceramic tape casting are being investigated to enable the creation of electrolytes with tailored mechanical properties, allowing for more uniform stress distribution.

Another significant challenge in solid-state battery design is the prevention of electrode and electrolyte delamination. This can occur due to the mismatch in thermal expansion coefficients between the materials or from the accumulation of stress over time. To address this, researchers are developing new strategies such as optimizing the microstructure of the electrode and electrolyte materials to improve their compatibility. Moreover, the use of buffer layers and hybrid materials that combine the advantages of both rigid and flexible materials is being explored to enhance mechanical stability and prevent delamination [67]. In addition to addressing material challenges, mechanical integrity can also be improved through better packaging and battery design. Innovations in the battery's outer casing and internal components can help protect the solid electrolyte and electrodes from external shocks, pressure, and other environmental factors. Enhanced packaging designs that allow for better heat dissipation and pressure management will be key to improving the safety and longevity of solid-state batteries.

5.4 Smart Battery Management Systems

A critical aspect of ensuring the safety of solid-state batteries in real-world applications is the integration of smart battery management systems (BMS). These systems monitor the performance and health of batteries in real-time, making them crucial for preventing safety issues such as overcharging, overheating, or short circuits. For solid-state batteries, which are often used in high-stakes environments like electric vehicles (EVs) and renewable energy storage, a sophisticated BMS is essential for maximizing both performance and safety. Smart BMS rely on sensors to collect data on various parameters such as voltage, temperature, current, and the state of charge (SOC) of the battery. By continuously monitoring these parameters, the BMS can make real-time decisions about the operation of the battery, such as adjusting charging rates, activating cooling systems, or shutting down the system in case of an anomaly. For example, if a solid-state battery starts to overheat, the BMS can trigger cooling mechanisms to prevent thermal runaway, or if the battery shows signs of overcharging, it can automatically cut off power to protect the battery from damage. Advanced BMS also play a significant role in preventing mechanical degradation in solid-state batteries by monitoring stress and strain on the battery materials. Sensors embedded within the battery can detect the formation of cracks, voids, or delamination within the electrolyte or electrodes, providing early warning signs of failure. This data is then used by the BMS to adjust operating conditions, such as reducing the charging or discharging rate, to prevent further damage [68, 69].

Another important feature of a smart BMS is its ability to manage the charging and discharging cycles efficiently, ensuring that the battery is never subjected to excessive stresses that could compromise its safety or longevity. For instance, adaptive charging protocols can be employed, where the BMS adjusts the charging speed based on the current state of the battery, optimizing the balance between fast charging and maintaining the integrity of the solid electrolyte.

6 Case Studies and Real-World Applications

6.1 Laboratory Studies on Safety Performance

Laboratory studies play a crucial role in advancing the safety performance of solid-state batteries (SSBs). These controlled experiments are designed to simulate a range of operational and extreme conditions that batteries might face in real-world applications. Researchers examine various failure modes that could compromise the safety and efficiency of solid-state batteries, such as thermal runaway, dendrite formation, short-circuiting, or electrolyte degradation. These studies often focus on testing batteries under abusive conditions, such as high charge/discharge rates, overcharging, deep discharge, or exposure to extreme temperatures. Through these experiments, valuable data is collected that can inform the development of mitigation strategies, battery management systems, and other safety features that will ensure the integrity of solid-state batteries in commercial settings [6, 70].

Moreover, laboratory studies often involve the analysis of different types of solid-state electrolytes and their behavior under stress. Researchers are exploring a variety of solid electrolytes, such as ceramics, sulfides, and polymers, to determine their reliability in preventing common safety hazards, like dendrite growth, which is one of the leading causes of battery failure. These experiments are vital in providing insight into the longevity of solid-state batteries, as they track performance degradation over thousands of charge and discharge cycles. The data derived from these studies helps to pinpoint areas where solid-state batteries can be improved in terms of stability, mechanical integrity, and thermal management. Figure 5 shows an overview of the evolution of battery technology. Ultimately, laboratory studies are instrumental in bridging the gap between theoretical research and practical, commercial applications by providing evidence-based solutions for the challenges faced by solid-state batteries.

6.2 Solid-State Batteries in Consumer Electronics

The application of solid-state batteries in consumer electronics represents one of the most promising prospects for this emerging technology. Consumer electronics, including smartphones, laptops, tablets, and wearable devices, are constantly pushing for batteries that offer greater energy density, longer lifespan, and superior safety. Traditional lithium-ion batteries, while efficient, come with inherent risks such as leakage, flammability, and degradation over time. These risks are especially problematic in consumer devices, where battery safety and reliability are critical. Solid-state batteries offer a safer alternative because they replace the liquid electrolyte with a solid, non-flammable material that reduces the likelihood of thermal runaway, leaks, and other safety hazards [2, 3]. In consumer electronics, the application of solid-state batteries could significantly enhance the user experience by offering batteries that last

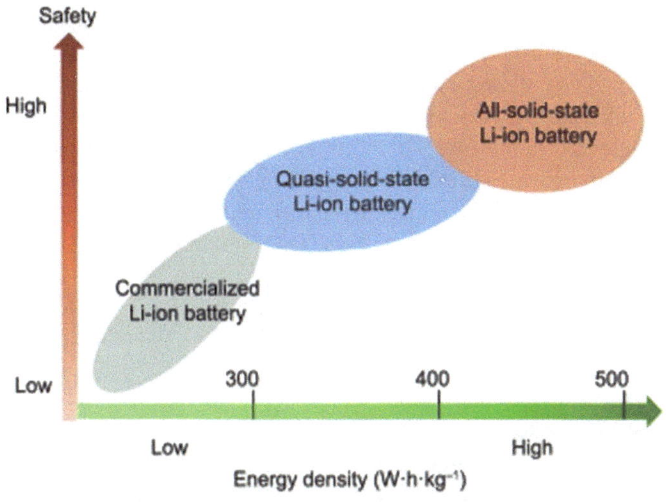

Fig. 5 Overview of the evolution of battery technology

longer, charge faster, and are more durable over time. Additionally, because solid-state batteries can be made smaller and lighter than traditional lithium-ion batteries while maintaining or even increasing energy density, they are ideal for compact devices that require both portability and high performance. The non-flammable nature of solid-state batteries also makes them particularly suited for devices that are frequently near users, such as wearables. However, while solid-state batteries hold promise, challenges remain in terms of cost and scalability. Manufacturing solid-state batteries at a competitive price point for consumer electronics is a significant hurdle that still requires innovative solutions in both material science and production methods.

Despite these challenges, the push towards incorporating solid-state batteries into consumer electronics is driven by the growing demand for better battery technology. Several companies are already conducting feasibility studies and pilot projects to integrate solid-state batteries into commercial devices, with some prototypes already showcasing the benefits of solid-state technology. As research continues and economies of scale are achieved, the widespread adoption of solid-state batteries in consumer electronics may soon become a reality, revolutionizing the industry with more powerful, efficient, and safer devices.

6.3 *Applications in Electric Vehicles*

Electric vehicles (EVs) stand to benefit significantly from the adoption of solid-state battery technology. One of the primary concerns with current lithium-ion batteries in EVs is their relatively limited energy density, which restricts driving range, and their susceptibility to degradation and safety risks over time. Solid-state batteries, with their higher energy densities and enhanced safety profiles, present a clear advantage (Fig. 6). Representing the advantageous applications of SSBs over LIBs. They can provide EVs with the potential for longer ranges on a single charge, faster charging times, and improved durability, which are key factors in making electric vehicles more practical and attractive to consumers [71]. The enhanced energy density of solid-state batteries could extend the driving range of EVs, addressing one of the most significant challenges in EV adoption: range anxiety.

Fig. 6 Advantageous applications of SSBs over LIBs

Furthermore, the solid-state nature of these batteries eliminates the risks associated with liquid electrolytes, such as leakage and flammability, offering a safer alternative for EVs. Solid-state batteries are also less prone to thermal runaway, a major safety concern in conventional lithium-ion batteries, which can be prone to overheating during rapid charging or discharging. This makes solid-state batteries particularly well-suited for use in EVs, where high-performance demands are coupled with the need for stringent safety standards.

However, while the prospects are exciting, there are still challenges in terms of mass production, cost, and integration into existing EV platforms. The manufacturing processes for solid-state batteries are more complex and expensive than those for conventional lithium-ion batteries, which makes it difficult to scale the technology for widespread use in the automotive industry. Additionally, issues such as the interface between the solid-state electrolyte and the electrodes, as well as the risk of dendrite formation, still need to be addressed before solid-state batteries can be deployed on a large scale in EVs [72]. Despite these challenges, major automotive manufacturers and battery companies are already investing heavily in solid-state battery research, and some are targeting the commercialization of solid-state batteries for EVs within the next decade.

6.4 Large-Scale Energy Storage Systems

Solid-state batteries also show tremendous promise in large-scale energy storage systems (ESS), which are essential for storing energy generated from renewable sources such as wind and solar power. These systems allow for the storage of excess energy during periods of high generation, which can then be released to the grid during periods of low generation, helping to stabilize the grid and support the transition to a renewable energy future. Solid-state batteries are particularly well-suited for ESS applications because they can provide high-energy density, long cycle life, and superior safety compared to conventional lithium-ion batteries. This makes them ideal for long-duration energy storage, where high efficiency, safety, and reliability are paramount [73].

Unlike lithium-ion batteries, which can degrade over time with repeated cycling, solid-state batteries are more resistant to wear and tear, providing a longer lifespan and reducing the need for frequent replacement. Furthermore, the solid-state nature of these batteries eliminates many of the safety concerns associated with liquid electrolytes, such as leakage and flammability, making them a safer choice for large-scale energy storage applications, where a battery failure could have catastrophic consequences.

Additionally, solid-state batteries tend to have lower self-discharge rates, meaning they can hold their charge for extended periods without significant loss of energy, which is critical for large-scale storage systems that may need to store energy for weeks or months. Despite these advantages, the widespread deployment of solid-state batteries in ESS is still in the early stages. Challenges remain in terms of scaling up manufacturing processes and reducing costs to make these batteries competitive with existing technologies [74]. However, with growing demand for efficient, sustainable energy storage solutions, the development of solid-state batteries for ESS applications holds significant potential for transforming the energy storage landscape.

7 Future Research Directions

7.1 Emerging Materials and Technologies

Emerging research emphasizes the synergy of catalysis [75–81], nanomaterials, and polymer science in designing multifunctional solid-state systems that offer both safety and environmental compatibility. The future of solid-state batteries hinges significantly on the exploration of emerging materials and technologies that can enhance their performance, safety, and scalability. As the demand for higher energy densities, improved safety, and longer lifespans increases, researchers are looking beyond traditional materials to discover new compounds and structures that could unlock these advancements. For instance, materials such as sulfur-based electrolytes and lithium metal anodes offer great promise in increasing the energy density of solid-state batteries while also improving their thermal stability. Additionally, advances in nanotechnology, such as the use of nanostructured electrodes or nanoscale solid electrolytes, could provide improved ionic conductivity and mechanical flexibility, addressing some of the key challenges currently faced by solid-state batteries [82].

One area of intense research is the development of more stable and conductive solid-state electrolytes. Researchers are exploring materials like lithium sulfide and lithium superoxide, which could help enhance battery performance. The integration of 2D materials, such as graphene, into solid-state electrolytes is another promising direction, as it may increase conductivity while maintaining the required structural integrity. Moreover, advances in electrode materials, including high-capacity cathodes and dendrite-resistant anodes, will be crucial in increasing the overall efficiency of solid-state batteries. To complement these material innovations, researchers are also focusing on new manufacturing techniques. Methods such as 3D printing, thin-film deposition, and atomic layer deposition (ALD) are being explored to create complex battery structures with enhanced performance characteristics. These advanced techniques could help to optimize the microstructure of solid-state batteries, allowing for better ion transport, mechanical stability, and interface performance.

7.2 Challenges in Scaling and Commercialization

The commercialization of solid-state batteries faces several challenges, with scaling up production being one of the most significant hurdles. The manufacturing processes for solid-state batteries are currently more complex and costly than those for conventional lithium-ion batteries, making it difficult to produce them at a competitive price point for mass-market applications. This challenge is exacerbated by the need for high-purity materials and precise manufacturing processes, which can increase both the time and cost of production. Furthermore, ensuring consistent quality in large-scale production is critical, as even slight variations in the battery's components or construction can lead to significant variations in performance and safety.

Another major challenge is the integration of solid-state batteries into existing products and infrastructure. For example, in electric vehicles, solid-state batteries must be designed to be compatible with current battery management systems (BMS), charging technologies, and vehicle architectures. Additionally, solid-state batteries must meet strict regulatory and safety standards, which can vary from region to region, adding another layer of complexity to their commercialization process [4, 7, 83].

Despite these challenges, researchers and companies are actively working to overcome these barriers by investing in new manufacturing technologies, improving the scalability of solid-state battery production, and streamlining supply chains. As the industry moves towards mass production, it is expected that economies of scale will help reduce the costs of solid-state batteries, making them more accessible for widespread use.

7.3 Regulatory and Standardization Efforts

As solid-state batteries move closer to commercialization, the development of regulatory frameworks and industry standards will be crucial. These standards will ensure that solid-state batteries meet safety, performance, and environmental requirements before they are released into the market. Currently, the regulatory landscape for solid-state batteries is still evolving, as existing standards are primarily focused on liquid-based batteries. Regulatory bodies will need to adapt and create new guidelines that specifically address the unique characteristics of solid-state batteries, such as their thermal stability, mechanical properties, and performance under extreme conditions.

The creation of standardized testing protocols will also be essential for ensuring that solid-state batteries are safe and reliable. These standards will provide manufacturers with clear guidelines for designing, testing, and certifying solid-state batteries, which will help accelerate their adoption in various industries [63, 84]. Furthermore, regulatory frameworks will play a key role in ensuring the environmental sustainability of solid-state batteries, particularly regarding material sourcing, recycling, and disposal [41, 62, 84].

7.4 Vision for Safe, Widespread Adoption

The vision for the future of solid-state batteries is one of safe, widespread adoption across multiple sectors, including consumer electronics, electric vehicles, and large-scale energy storage. To achieve this vision, significant progress must be made in overcoming the current challenges of cost, scalability, and integration. However, with ongoing research, technological advancements, and increasing investments from both the private and public sectors, the future of solid-state batteries looks bright [85].

Researchers are optimistic that within the next decade, solid-state batteries will become a mainstream energy storage technology, offering superior performance and safety compared to traditional lithium-ion batteries. As new materials, manufacturing processes, and regulatory frameworks emerge, the path to widespread adoption will become clearer, paving the way for a future where solid-state batteries power everything from our smartphones to electric vehicles and renewable energy systems.

8 Conclusion

8.1 Summary of Key Insights

This chapter has provided an in-depth look at the current state of solid-state battery technology, focusing on its safety, performance, and real-world applications. Solid-state batteries offer significant advantages over traditional lithium-ion batteries, particularly in terms of safety, energy density, and durability. Key advancements in solid-state electrolyte formulations, protective coatings, and interface engineering are addressing the challenges of dendrite formation, mechanical stress, and thermal instability, which have historically limited the performance and safety of solid-state batteries.

The real-world applications of solid-state batteries, including consumer electronics, electric vehicles, and large-scale energy storage systems, highlight the vast potential of this technology. Despite the promise, there are still hurdles to overcome, such as manufacturing complexity, cost, and scalability. However, ongoing research and innovation are bringing us closer to realizing the commercial potential of solid-state batteries, and the continued development of regulatory standards will help ensure that these batteries are safe, reliable, and environmentally sustainable [2–8, 10, 24–27, 29, 30, 32–34, 36, 38, 39, 41, 42, 62, 63, 65, 66, 70, 74, 82–91].

8.2 Final Thoughts on Safety in Solid-State Batteries

Safety remains at the forefront of solid-state battery development. Unlike conventional batteries that rely on flammable liquid electrolytes, solid-state batteries use solid materials that are inherently more stable and less prone to thermal runaway, leakage, or explosion. Nevertheless, safety cannot be taken for granted. Each component of the battery—from the choice of electrolyte to the design of interfaces and cell architecture—plays a crucial role in ensuring long-term performance and preventing catastrophic failures. Future innovations must continue to prioritize safety while pushing the boundaries of energy density and performance.

8.3 The Road Ahead

The road ahead for solid-state batteries is filled with promise and potential. As scientific understanding deepens and technological advancements continue, the vision of a safer, more efficient, and widely adopted battery technology is becoming more achievable. Collaboration among researchers, manufacturers, policymakers, and regulatory bodies will be essential in overcoming the remaining barriers to commercial adoption. With sustained investment and innovation, solid-state batteries have the potential to transform the global energy landscape, supporting everything from clean transportation to resilient power grids, all while ensuring the highest standards of safety and reliability.

References

1. Mishra V, Mishra A (2025) Metal-organic frameworks as forensic detectors. Springer, pp 21–33
2. Rawat S, Thakur J, Phogat P, Shreya, Jha R, Singh S (2025) J Alloys Compd 1020:179387
3. Nzereogu PU, Oyesanya A, Ogba SN, Ayanwunmi SO, Sobajo MS, Chimsunum VC, Ayanwunmi VO, Amoo MO, Adefemi OT, Chukwudi CC (2025) Hybrid Adv 8:100339
4. Machín A, Morant C, Márquez F (2024) Batteries 10:29
5. Liu M, Song A, Zhang X, Wang J, Fan Y, Wang G, Tian H, Ma Z, Shao G (2025) Nano Energy 136:110749
6. Joshi A, Mishra DK, Singh R, Zhang J, Ding Y (2025) Appl Energy 386:125546
7. Sung J, Heo J, Kim D-H, Jo S, Ha Y-C, Kim D, Ahn S, Park J-W (2024) Mater Chem Front 8:1861–1887
8. De Carne G, Maroufi SM, Beiranvand H, De Angelis V, D'Arco S, Gevorgian V, Waczowicz S, Mather B, Liserre M, Hagenmeyer V (2024) Electr Power Syst Res 236:110963
9. Xu L, Tang S, Cheng Y, Wang K, Liang J, Liu C, Cao YC, Wei F, Mai L (2018) Joule 2:1991–2015
10. Zhou D, Shanmukaraj D, Tkacheva A, Armand M, Wang G (2019) Chem 5:2326–2352
11. Kumari A, Yadav N, Tyagi D, Mishra V (2025) Microchem J 217:114921
12. Bharti T, Rathi AK, Mishra V (2025) Anal Chim Acta 345036
13. Khandelwal D, Bhattacharya A, Kumari V, Gupta S, Ranjan KR, Mishra V (2025) J Mater Chem B 13:802–820
14. Kalra C, Ranjan KR, Mishra V (2026) Curr Nanomater 11:268–286
15. Yadav N, Gaikwad RP, Mishra V, Gawande MB (2022) Bull Chem Soc Jpn 95:1638–1679
16. Yadav N, Mudgal D, Mishra M, Mishra V (2024) Chem Biodivers e202400891
17. Yadav N, Sharma A, Yano T, Anthopoulos TD, Mishra V (2026) Coord Chem Rev 548:217152
18. Yadav N, Mudgal D, Mishra V (2023) Anal Chim Acta 1272:341502

19. Yadav N, Mudgal D, Mishra V (2024) Indian J Microbiol 65:424–438
20. Yadav N, Mudgal D, Mishra A, Shukla S, Malik T, Mishra V (2024) PLoS One 19:e0296270
21. Öksüzoğlu F, Ateş Ş, Özkendir OM, Çelik G, Eker YR, Baveghar H (2024) Ceram Int 50:31435–31441
22. Sazvar A, Hajibandeh M, Vafaei P, Hosseinzadeh E, Jabbari M (2024) J Energy Storage 101:113863
23. Zhao J, Feng X, Pang Q, Fowler M, Lian Y, Ouyang M, Burke AF (2024) Prog Energy Combust Sci 102:101142
24. Klinsmann M, Hildebrand FE, Ganser M, McMeeking RM (2019) J Power Sources 442:227226
25. Wu M, Han S, Liu S, Zhao J, Xie W (2024) Energy Storage Mater 66:103174
26. Boaretto N, Garbayo I, Valiyaveettil-SobhanRaj S, Quintela A, Li C, Casas-Cabanas M, Aguesse F (2021) J Power Sources 502:229919
27. Zhang X, Sahraei E, Wang K (2016) Sci Rep 6:32578
28. Bates AM, Preger Y, Torres-Castro L, Harrison KL, Harris SJ, Hewson J (2022) Joule 6:742–755
29. Jeon DH, Kim S, Hempelmann R (2025) J Power Sources 646:237242
30. Pant BR, Ren Y, Cao Y (2024) ACS Appl Mater Interfaces 16:56947–56956
31. Deng J, Yang X, Zhang G (2022) Mater Today Commun 31:103570
32. Zhu L, Li Y, Zhao J, Liu J, Wang L, Lei J (2023) Green Energy Environ 8:1279–1307
33. Cao D, Sun X, Li Q, Natan A, Xiang P, Zhu H (2020) Matter 3:57–94
34. Porz L, Swamy T, Sheldon BW, Rettenwander D, Frömling T, Thaman HL, Berendts S, Uecker R, Carter WC, Chiang Y (2017) Adv Energy Mater 7:1701003
35. Hu J, Yang S, Pei Y, Wang X, Liao Y, Li S, Yue A, Huang JQ, Yuan H (2024) Particuology 86:55–66
36. Khezraqa H, Golshan M, Salami-Kalajahi M (2025) Appl Energy 384:125509
37. Chang H, Zhang X, Li W, Liu H, Hu H, Liu Z, Liu W, Jin Y (2025) Next Mater 6:100307
38. Huang J, Li C, Jiang D, Gao J, Cheng L, Li G, Luo H, Xu Z, Shin D, Wang Y, Lu Y, Kim Y (2025) Adv Funct Mater 35:2411171
39. Tu Q, Barroso-Luque L, Shi T, Ceder G (2020) Cell Rep Phys Sci 1:100106
40. Liang X, Wang L, Wu X, Feng X, Wu Q, Sun Y, Xiang H, Wang J (2022) J Energy Chem 73:370–386
41. Antony Jose S, Gallant A, Gomez PL, Jaggers Z, Johansson E, LaPierre Z, Menezes PL (2025) Batteries 11:90
42. Dai Y, Panahi A (2025) Next Energy 6:100186
43. Yang S, Hu J, Jiang F, Yuan H, Park HS, Huang J (2024) InfoMat 6:e12512
44. Mishra V, Jung S-H, Jeong HM, Lee H (2014) Polym Chem 5:2411–2416
45. Ghosal K, Bhattacharyya SK, Mishra V, Zuilhof H (2024) Chem Rev 124:13216–13300
46. Singh M, Gulamabbas T, Ahumuza B, Singh NP, Mishra V. Applications of polymers in aerospace. Wiley-Scrivener
47. Srivastava A, Mishra V, Singh P, Kumar R (2012) J Appl Polym Sci 126:395–407
48. Mishra V, Kumar R (2012) J Sci Res 56:141–176
49. Mishra V, Pandey S, Aggarwal S (2022) Electroactive polymeric materials. CRC Press, pp 271–298
50. Mishra V, Kumar R (2011) J Appl Polym Sci 124:4475–4485
51. Aggrawal S, Pandey S, Mishra V (2022) Porous polymers science and applications. In: Inamuddin MI et al (eds) CRC Press U.K., 296 pp. ISBN 9780367770587
52. Mishra V, Kumar R (2011) Carbohydr Polym 86:296–303
53. Mishra V, Kumar R (2013) J Appl Polym Sci 128:3295–3307
54. Mishra V, Kumar R (2019) Polym Sci Ser B 61:753–761
55. Mishra V, Kumar R (2011) Carbohydr Polym 83:1534–1540
56. Srivastava A, Mishra V, Singh SK, Kumar R (2009) e-Polymers 6:1–14
57. Srivastava A, Mishra V, Singh SK, Kumar R (2010) J Appl Polym Sci 115:2375–2385
58. Kumar V, Saha R, Chatterjee S, Mishra V, Kumar A, Yadav AK, Mishra V, Kumar D, Kaushal A, Bhukya J, Yadav AK, Mishra V, Kumar D (2023) React Chem Eng 7:154742

59. Chatterjee S, Kumar N, Sehrawat H, Yadav N, Mishra V (2021) Curr Res Green Sustain Chem 4:100064
60. Chatterjee S, Mishra V (2020) Curr Res Green Sustain Chem 3:100025
61. Truong CC, Mishra DK, Mishra V (2021) Green sustainable process for chemical and environmental engineering and science. Elsevier, pp 253–275
62. Zhu Q, Ye C, Mao D (2022) Nano 12:3612
63. Dolla TH, Ajayi SO, Sikeyi LL, Mathe MK, Palaniyandy N (2025) J Energy Storage 106:114737
64. Richards WD, Miara LJ, Wang Y, Kim JC, Ceder G (2016) Chem Mater 28:266–273
65. Bouguern MD, Anil Kumar MR, Zaghib K (2024) J Power Sources 623:235457
66. Yang H, Wu N (2022) Energy Sci Eng 10:1643–1671
67. Dong S, Sheng L, Wang L, Liang J, Zhang H, Chen Z, Xu H, He X (2023) Adv Funct Mater 33:2304371
68. Gabbar H, Othman A, Abdussami M (2021) Technologies (Basel) 9:28
69. Naseem H, Hadi A, Kang M, Seok J (2025) Energy Technol 13:2500100
70. Machín A, Morant C, Márquez F (2024) Batteries 10:29
71. Şen M, Özcan M, Eker YR (2024) Next Sustain 3:100036
72. Koech AK, Mwandila G, Mulolani F (2024) Heliyon 10:e34806
73. Njema GG, Ouma RBO, Kibet JK (2024) J Renew Energy 2024:1–35
74. Mulpuri SK, Sah B, Kumar P (2023) iScience 26:107770
75. Mudgal D, Yadav N, Singh J, Srivastava GK, Mishra V (2023) Int J Biol Macromol 127491
76. Mudgal D, Singh RP, Bharti T, Dwivedi P, Yadav N, Mishra V (2025) J Mol Struct 1327:141102
77. Mudgal D, Singh RP, Yadav N, Mishra V (2023) SynOpen 7:570–579
78. Mudgal D, Yadav N, Mishra V (2024) Environ Sci Pollut Res 31:36748–36760
79. Oh SG, Mishra V, Cho JK, Kim B-J, Kim YJ (2014) Catal Commun 43:79–83
80. Son GM, Truong CC, Mishra DK, Mishra V, Kim YJ (2018) Bull Korean Chem Soc 39:174–183
81. Mishra V, Cho JK, Shin S-H, Suh Y-W, Kim HS, Kim YJ (2014) Appl Catal A Gen 487:82–90
82. Sarfraz N, Kanwal N, Ali M, Ali K, Hasnain A, Ashraf M, Ayaz M, Ifthikar J, Ali S, Hendi A, Baig N, Ehsan MF, Shah SS, Khan R, Khan I (2024) Energy Storage Mater 71:103619
83. Minkiewicz J, Jones GM, Ghanizadeh S, Bostanchi S, Wasely TJ, Yamini SA, Nekouie V (2023) Open Ceram 16:100497
84. Machín A, Cotto MC, Díaz F, Duconge J, Morant C, Márquez F (2024) Batteries 10:255
85. Bridgelall R (2024) Sustainability 16:10994
86. Tuli NT, Khatun S, Bin Rashid A (2024) Heliyon 10:e27328
87. Osman A, Lu J (2023) Mater Sci Eng R Rep 154:100734
88. Famprikis T, Canepa P, Dawson JA, Islam MS, Masquelier C (2019) Nat Mater 18:1278–1291
89. Janek J, Zeier WG (2016) Nat Energy 1:16141
90. Daems K, Yadav P, Dermenci KB, Van Mierlo J, Berecibar M (2024) Renew Sustain Energy Rev 191:114136
91. Kansal N, Tripathi G (2025) Future Batteries 6:100057

MIX
Papier | Fördert gute Waldnutzung
FSC® C083411

Kolibri 360 GmbH
Ferdinand-Jühlke-Straße 7
99095 Erfurt, Deutschland
produktsicherheit@kolibri360.de